An Introduction to the Philosophy of Science

SECOND EDITION

This thoroughly updated second edition guides readers through the central concepts and debates in the philosophy of science. Using concrete examples from the history of science, Kent W. Staley addresses questions about what science is, why it is important, and the basis for trust in scientific results. Part I of the book introduces the central concepts of philosophy of science, with updated discussions of the problem of induction, underdetermination, rationality, scientific progress, and important movements such as falsificationism, logical empiricism, and postpositivism, together with a new chapter on social constructionism. Part II offers updated chapters on probability, scientific realism, explanation, and values in science, along with new discussions of the role of models in science, science in policymaking, and feminist philosophy of science. This broad yet detailed overview will give readers a strong grounding in the philosophy of science while also providing opportunities for further exploration.

KENT W. STALEY is Professor of Philosophy at Saint Louis University. He is the author of *The Evidence for the Top Quark: Objectivity and Bias in Collaborative Experimentation* (Cambridge University Press, 2004).

An Introduction to the Philosophy of Science

SECOND EDITION

KENT W. STALEY

Saint Louis University, Missouri

Shaftesbury Road, Cambridge CB2 8EA, United Kingdom

One Liberty Plaza, 20th Floor, New York, NY 10006, USA

477 Williamstown Road, Port Melbourne, VIC 3207, Australia

314–321, 3rd Floor, Plot 3, Splendor Forum, Jasola District Centre,
New Delhi – 110025, India

103 Penang Road, #05-06/07, Visioncrest Commercial, Singapore 238467

Cambridge University Press is part of Cambridge University Press & Assessment,
a department of the University of Cambridge.

We share the University's mission to contribute to society through the pursuit of
education, learning and research at the highest international levels of excellence.

www.cambridge.org
Information on this title: www.cambridge.org/highereducation/isbn/9781009098250

DOI: 10.1017/9781009099387

When citing this work, please include a reference to the DOI 10.1017/9781009099387

First published 2014

Second edition 2025

Cover image: KTSDESIGN/SCIENCE PHOTO LIBRARY/Getty Images

A catalogue record for this publication is available from the British Library

A Cataloging-in-Publication data record for this book is available from the Library of Congress

ISBN 978-1-009-09825-0 Hardback
ISBN 978-1-009-09796-3 Paperback

To the many students I have learned from.

Contents

Figures

Tables

Preface: Philosophy of Science for Philosophers, Scientists, and Everyone Else

For better and for worse, in ways both obvious and subtle, the work of scientists has helped to shape the world around us. The obvious impacts of science on our lives include the technologies that depend on scientific research. Our approaches to communicating, eating, becoming or staying healthy, moving from one place to another, reproducing, entertaining ourselves – all of these have been changed by the findings of scientific research. Science also changes how we think about ourselves and our world. The concepts and ways of thinking that scientists employ as they engage in research and seek to explain phenomena they uncover tend to migrate out of the specialist settings where they originate. Though they tend to be transformed in the process, scientific concepts and methods become part of how we describe and explain the world.

Clearly, science is a big deal. This book is not just about science, though; it is about the philosophy of science. Even if you agree that one should care about science, you might wonder whether you should care about the philosophy of science. The point of this preface is to begin to persuade you that you should care about it. I will divide my arguments based on different types of readers. My first argument will be directed at those who either are or will become working scientists. My second argument will be directed at those people *and* at those who are interested in philosophy but unsure about the philosophy of science. My last argument will be directed at anyone – whether or not a scientist, whether or not philosophically inclined – who lives in a world as strongly influenced by scientific work as ours.

First, a caveat: These arguments are not meant to be fully persuasive as such. Full arguments regarding important practical matters – such as whether to devote one's valuable time to reading a particular book – can rarely be stated succinctly (and I do not wish to linger overlong here in the preface but to get on to the subject at hand). Here I articulate some promissory notes that make more or less explicit the value of philosophy of science for assorted concerns that I take to be rather widely shared. It is my hope that reading through the rest of the book will leave you fully persuaded that this is a worthwhile subject matter. I grant that some measure of trust is required to get started.

For those who are or will become working scientists, the argument is quite simple: Whether performing experiments, analyzing data, developing theories, building instruments, or writing papers, scientists at some point in their work implicitly *rely on* philosophical assumptions about their discipline and their subject matter. This is perhaps most easily shown in the case of those scientists who would most vocally *oppose* the importance of philosophy in science, who espouse a 'shut up and calculate' approach. To endorse such a view is to take a certain position regarding the value of attempting to interpret the meaning of a scientific theory; it is, in short, a philosophical claim to the effect that the value of theory resides only in its usefulness for calculating outcomes in application, whether experimental or technological. This position is a form of what philosophers call *instrumentalism*, and the instrumentalists constitute a long-standing and important tradition within the philosophy of science, discussed in Chapter 10. The anti-philosophy scientist is embracing a philosophy, and a defense of their position will be a philosophical defense. There is no escaping philosophy. The question for the scientist is not whether to engage in philosophy, but only whether to do it well or badly.

One particularly important context in which philosophical considerations are important to the practice of science concerns the use of statistics. Many experimental results are couched in terms that result from using statistical tools. Theoretical explanations of phenomena involving, for example, quantum mechanics, population biology, or meteorology invoke statistical concepts. These statistical notions are developed within frameworks that make assumptions about the meaning of concepts like probability, randomness, and others that are a matter of significant philosophical dispute. The use of statistics without an awareness of these philosophical issues is likely to result in confused or misleading statements about the meaning of one's research results or the character of natural phenomena. This book devotes significant attention to philosophical issues regarding the use of statistics.

Finally, and perhaps most fundamentally, I would like to advance the idea that the pursuit of scientific knowledge is already, at least in part, a kind of philosophical inquiry in the broad sense. This might look like a trick of definition-mongering in which I have simply chosen to give 'philosophy' such a broad meaning that this claim becomes trivially true. But there is a good historical reason for taking empirical science to be continuous with philosophy in this way: Much of what we would today call 'science' was once called 'natural philosophy.' It makes sense, too: Among the aims of scientific theorizing are (arguably) the understanding of the processes of the natural world, the kinds of things or structures that might be found there, and how these would appear to us under various circumstances, react to our intervention in them, and develop over time. It makes sense to call something that does these things a 'philosophy of nature.'

Perhaps, however, you are not a scientist and are not planning to become one. Let us suppose that you are instead someone with an interest in philosophy more generally. Any well-developed philosophical view of the world will include a philosophy of science at least implicitly. Suppose for example, your philosophical interests center on whether human actions are truly chosen freely. Any position regarding this issue of free will must respond *somehow* to the results of scientific research in such domains as neurophysiology and cognitive science. To be sure, one possible response would be simply to ignore these scientific results and develop one's position on purely a priori grounds derived from introspection. The point is that whether one does this or tries to be sensitive to the scientific findings in some more or less substantive manner, one must have *reasons* for treating the scientific results in the way one does, and these reasons must involve some views about the relationship between empirical science and philosophical theorizing – a philosophy of science, in short.

This brings us to 'everyone else.' Perhaps you have no particular scientific training and do not plan on receiving any. Or perhaps you do not plan on engaging in an extended philosophical reflection or argumentation as a significant part of your life's undertakings. For you the philosophy of science will remain important at least for practical reasons – as a kind of instrument of self-defense if nothing else. Politicians, marketers, even scientists themselves will attempt to persuade you to act in certain ways based on what they present as 'sound science.' Without expertise in the relevant scientific field, you might find yourself wondering about the soundness of their claims. Although we often rely on the testimony of experts to form our own opinions, and it is often right to do so, the very question of when to defer to authorities and when to doubt them relies on the ability to critically evaluate such testimony. What is the nature of the authority of empirical science? What are the limits of that authority? What questions are important for scrutinizing a claimed scientific result? Training in the philosophy of science should equip you with some basis for thinking soundly about these issues as they arise in the public discourse of and around science.

At its best, the scientific enterprise exemplifies the bold yet responsible pursuit of knowledge. Appropriately weighted and clearly understood, the best of our scientific heritage should stand as a check on our too-often-indulged propensity to let demagoguery and obfuscation lead us into unreasoned, even dangerous ways of thinking. An understanding of what makes for good scientific thinking may help us to hold ourselves to a higher standard in how we understand our world and our place in it.

Finally, studying science from a philosophical perspective should highlight and explicate a fact about scientific inquiry that, though in some sense obvious, might be easily forgotten when confronted with the difficulty of understanding the technical language in which unfamiliar scientific claims are presented: Science is a human undertaking, an extension of our common search for understanding. Scientists extend our ordinary ways of inquiring and knowing into realms that are remote in distance, or size, or degree of abstraction, or complexity. Knowing just how they do so is important to evaluating the results of their inquiry. Moreover, perhaps understanding how inquiry is conducted in these remote realms will shed light back onto our everyday knowledge-seeking practices. In any case, to understand science philosophically is to understand something important about what it is to be human.

Plan of the Book

This book divides roughly into two parts. The first six chapters ('Background and Basic Concepts') introduce both some important concepts that will be used throughout the discussion and some important figures and movements in twentieth-century philosophy of science that have helped shape the discipline and which remain relevant to current debates. We will consider the arguments of writers such as Pierre Duhem, Karl Popper, Rudolf Carnap, Thomas Kuhn, Imre Lakatos, and Paul Feyerabend. This will allow us to gain some familiarity with the *falsificationist* (Popper) and *logical positivist* or *logical empiricist* (Carnap) schools of thought that loomed large in the first half of the twentieth century, while also considering later developments involving *post-positivist* thinkers (Kuhn, Lakatos, Feyerabend) that offered alternatives to the falsificationist and logical empiricist points of view, as well as the *social constructionist* approaches that developed from the social-scientific study of science.

Chapters 7 through 13 ('Ongoing Investigations') will survey some of the areas in which lively debates are ongoing in current philosophy of science. These surveys will seek to connect these current areas of debate with the historical precedents that are the focus of the first part of the book, but will emphasize foundational concepts and more or less recent developments. Chapter 7 deals with the wide variety of roles that models play in science and the implications these have for how we think about scientific knowledge. Chapters 8 and 9 will concern competing approaches to the use of probability ideas in science. Probability is crucial to the analysis of data and to the understanding of scientific reasoning. These chapters will, as you might expect, have the most mathematical content, but are meant to be self-contained. Chapters 10 and 11 will survey current debates concerning scientific realism vs. antirealism, and explanation, respectively. Chapter 12 concerns the role of values and value judgments in science and the role of science and scientists in policymaking. Chapter 13 presents a survey of feminist approaches to philosophizing about science, subject of a lively discourse with important connections to many of the other issues in this book.

Acknowledgments

This second edition has benefited from conversations with and feedback from many people over a rather long time. I apologize to anyone I neglect to mention. The first edition was reviewed in helpful and insightful ways by Viorel Paslaru (in *Philosophy of Science*) and Evelyn Brister (in *Metascience*). Along with the many people who helped me in writing the first edition, the second edition has benefited, directly or indirectly, from exchanges with Robert Cousins, Michela Massimi, Johannes Mierau, Carl Craver, Scott Berman, my collaborators Bill Rehg, Sophie Ritson, and Hugo Beauchemin, and many students over the years in my undergraduate courses on philosophy of science, and gender and science.

I am grateful to my editor Hilary Gaskin at Cambridge University Press, first for inviting me to propose a second edition of this text, and especially for her support and patience as I labored to bring it about. Senior editorial assistant Abi Sears provided important assistance in the publication process.

Finally I wish to express love and gratitude to my wife Dianne and son Charlie for making my life good beyond measure. Charlie also lent his superior artistic judgment to help come up with suggestions for the cover image.

Acknowledgments for the First Edition

I am very fortunate to have had the opportunity to write this book. I thought that I was simply creating a potentially useful teaching resource for myself and (hopefully) for others. The project turned into an opportunity to revisit long-standing interests and discover some new literature as well. I am grateful to Hilary Gaskin at Cambridge University Press for her sage guidance and immense patience. Her assistant Anna Lowe, and subsequently Rosemary Crawley, provided wonderful support. The final text benefited from the skillful copy-editing of Jon Billam.

As I have written, unwritten, and rewritten, I have found many friends and colleagues willing to share their comments on various drafts. I would like to thank Carl Craver, Helen Longino, Deborah Mayo, Bill Rehg, Richard Richards, and Tom Pashby for their comments on assorted chapters of the book. Carl Craver invited me to guest lecture on Chapters 8 and 9 to a group of graduate students and postdocs at Washington University, and the ensuing discussion was very helpful. I am also thankful to the undergraduate students who have been enrolled in my Philosophy of Science survey class over the past couple of years; they have helped me field-test earlier drafts of the chapters in this book. The members of the History and Philosophy of Science Reading Group at Saint Louis University graciously read the entire text during the fall of 2013, and our discussions were tremendously helpful. I am especially grateful to reading group stalwarts Scott Berman, Ben Conover, Mike Mazza, Jonathan Reibsamen, and Chong Yuan. In spite of having access to such collective wisdom, I probably have left in some things that are dubious or just plain dumb, and I have no one to blame but myself.

Dianne Brain has, as ever, advised, sustained, and encouraged me. I rely daily on her good judgment and patience. Our son Charlie has provided many delightful distractions from my work on this text.

Abbreviations

Abbreviation	Meaning	Chapters
ACh	acetylcholine	11
CDC	Centers for Disease Control	12
CERN	Conseil Européen pour la Recherche Nucléaire	9, 12
CL	Covering Law	11
CM	Causal-Mechanical	11
D-N	Deductive-Nomological	11
DHHS	Department of Health and Human Services	9
DNE1	Empirical requirement of Deductive-Nomological Account	11
DNL1–3	Logical requirements of Deductive-Nomological Account	11
DR	Differential Refrangibility	2
DRNH	Differential Refrangibility restricted to the Northern Hemisphere	2
EPA	Environmental Protection Agency	9
GTR	General Theory of Relativity	9
HH	Hodgkin–Huxley	11
I-S	Inductive-Statistical	11
IBE	Inference to the Best Explanation	10
IID	Independent and Identically Distributed	9
LDD	Law of Doughnut Delectability	1, 2
LDD′	alternate formulation of Law of Doughnut Delectability	2, 4
$LDD'_{a, b, c, d}$	the same, applied to individuals named *a*, *b*, *c*, *d*	2
LHC	Large Hadron Collider	9
MDC	Machamer, Darden, and Craver	11
MSRP	Methodology of Scientific Research Programs	5

(*cont.*)

Abbreviation	Meaning	Chapters
NCM	Newtonian Classical Mechanics	5
NIH	National Institutes of Health	12
NMF	Naïve Methodological Falsificationism	5
NOA	Natural Ontological Attitude	10
NP	Neyman–Pearson	9
OD	Origin and Development derivation	11
PUDP	Principle of Uniformity of Doughnut Properties	1
PUN	Principle of Uniformity of Nature	1
RF	Resonant Frequency derivation	11
SCC	squamous cell carcinoma	9
SMF	Sophisticated Methodological Falsificationism	5
SSR	*The Structure of Scientific Revolutions*	5
UD	Underdetermination	3

Part I

Background and Basic Concepts

1 Some Problems of Induction

1.1 Introduction

To the lay reader, much of what is written by scientists can seem barely comprehensible. Even to someone who has had some science courses in school, a sentence like "The M2 proton channel ... has a 40-residue region that interacts with membranes consisting of a transmembrane helix (which mediates tetramerization, drug-binding, and channel activity), followed by a basic amphiphilic helix important for budding of the virus from cellular membranes and release (scission)" will seem as though it has been written in a language not quite identical to English (Fiorin, Carnevale, & DeGrado, 2010). As a result, the nonspecialist may find much of what scientists accomplish mysterious and esoteric.

Philosophers of science have sometimes attempted to 'humanize' scientific work by portraying scientific methods as extensions of our ordinary ways of knowing things. It is true that scientists use technologies both material and mathematical to make observations and draw conclusions that we could never achieve otherwise. Nonetheless, they observe, conjecture, infer, and decide just as we all do, if perhaps in a more systematic and sophisticated way. Although a certain kind of training is needed to understand some of the language scientists use to report their findings, those findings are not the result of magic or trickery, but of an especially refined and disciplined application of widely shared human cognitive resources.

But if scientists use and extend the cognitive abilities of ordinary knowers, they also inherit the philosophical problems of ordinary knowers. One of the thorniest and most discussed of these – the problem of induction – is the topic of this chapter. We will see how this general problem about knowledge arises in both nonscientific and scientific contexts. An example from the history of optics will show that scientific experimentation and careful use of scientific methods can improve our ability to give well-reasoned answers to questions, but do not suffice to solve the problem of induction itself.

1.2 The Problem of Induction about Doughnuts

The problem of induction is the problem of how we learn from experience. Consider how it arises in the life of a somewhat ordinary person, whom we shall call Zig. Zig likes doughnuts, let's suppose, and one day decides to try a new doughnut shop. This doughnut shop has introduced a new horseradish-flavored doughnut, something that Zig has never tried before. Zig expects, however, to like the horseradish-flavored doughnut. After all, he has enjoyed nearly all of the doughnuts he has eaten in his life (he has eaten many!). In fact, Zig believes there to be a general Law of Doughnut Delectability (LDD):

Law 1 (LDD) *All doughnuts are delicious.*

Since horseradish doughnuts are doughnuts, it follows that horseradish doughnuts must also be delicious.

Zig's reasoning here exhibits a logical trait of considerable significance. He is using an argument that is *deductively valid*, meaning that *it is impossible that the premises are true and the conclusion false.* This can be seen clearly enough in the case of Zig's argument. Suppose the LDD is true, and that it is true (as it seems it must be) that all horseradish doughnuts are doughnuts. Could there possibly be a nondelicious horseradish doughnut? Suppose there were. If we insist on the law that all doughnuts are delicious (first premise), then this nondelicious horseradish doughnut must not really be a doughnut after all. On the other hand, if we insist that all horseradish doughnuts really are doughnuts (second premise), then our nondelicious horseradish doughnut stands in contradiction to the law that all doughnuts are delicious. We simply cannot hang on to the truth of our premises while denying the conclusion of this argument.

Deductive arguments are sometimes said to be 'truth-preserving' because a deductively valid argument is guaranteed not to lead you from true premises to a false conclusion. This makes them powerful as logical tools of persuasion; as long as the person you are trying to convince agrees with you that your premises are true, she cannot resist the conclusion of a valid argument from those premises without contradicting herself. A deductively valid argument with true premises is said to be *sound*.

To his very great disappointment, Zig finds the horseradish doughnut revolting. He has just encountered a limitation of deductively valid arguments. Although the truth of their premises guarantees the truth of their conclusions, there is nothing about validity itself that guarantees the truth of the premises (soundness). Zig proceeds to make a new deductively valid inference: All horseradish doughnuts are doughnuts. Some horseradish doughnuts are not delicious (to say the least!). Therefore, some doughnuts are not delicious. With this argument, Zig has refuted the LDD.[1]

Now Zig begins to wonder: Was that one horrendous-tasting horseradish doughnut a fluke? The next day he returns to the doughnut shop and orders another one. The second horseradish doughnut turns out to be equally vile. After ten more visits to the same shop, each time ordering a horseradish doughnut that he finds quite disgusting, Zig begins to visit other doughnut shops selling horseradish doughnuts. Zig proceeds to sample 12 horseradish doughnuts from each of 20 different doughnut shops and finds all 240 of them to be repellent.

Surely Zig is overdoing it! He has, you would think, more than enough evidence to justify an expectation that *any* horseradish doughnut he tries will not taste good to him. But suppose he remains skeptical of this claim. He freely admits the unpleasant taste of all horseradish doughnuts he has eaten in the past, but does not think that this has any bearing on what he should expect of those he has not yet tasted. What logical argument could we offer to persuade him?

The problem of induction amounts to just this skeptical problem: It seems that no amount of unpleasant experience with horrid horseradish doughnuts would suffice to justify any belief regarding the deliciousness or putridity of one that is yet to be tasted.

Consider the premises we have at our disposal for reasoning with Zig. "Look, Zig," we might say. "You have eaten 240 horseradish doughnuts from a variety of doughnut shops in a variety of locations. You agree that each of those doughnuts was deeply unpleasant. Does it not follow that the next horseradish doughnut will be similarly awful?" Zig would be correct to reply

[1] You may have noticed that Zig could have saved the LDD if he had been willing to consider giving up the claim that all horseradish doughnuts are doughnuts. Maybe the terrible-tasting thing that he ate was not really a doughnut after all? We will return to the possibility of such evasions in Chapter 3.

"No, it does not follow. The premises you offer are consistent both with the next one being tasty and with it being terrible." In other words, the suggested inference to the conclusion that all horseradish doughnuts are not delicious is not deductively valid. Our inference regarding the next horseradish doughnut is an inductive not a deductive argument, and as presented it does not exhibit validity.

One might try to make it into a deductive argument by filling in some additional premises. Consider, for example, the following argument: All horseradish doughnuts in a large and varied sample are not delicious. All horseradish doughnuts that have not been tasted are similar to those in the sample with respect to their taste. Therefore, all horseradish doughnuts are not delicious.

Now we have a valid deductive argument, but we also have a *new premise*, and it is one that cannot be considered to report any observation that Zig has made. "All horseradish doughnuts that have not been tasted are similar to those in the sample with respect to their taste" asserts a generalization about horseradish doughnuts, and Zig would rightly ask us to tell him what reasoning would justify his accepting this new premise. Like our original conclusion, it is not a deductive consequence of Zig's observations of the bad taste of the horseradish doughnuts he has already eaten.

Perhaps this new premise can be supported by some separate argument that we think Zig should find compelling. One could, for example, regard it as a special case of a general Principle of Uniformity of Doughnut Properties (PUDP): All doughnuts of a given kind resemble one another with regard to their taste. And this general principle, we could argue, is supported by past doughnut-eating experiences, not just of Zig but of the general doughnut-eating public, which has found in the past that all raspberry-filled doughnuts tasted similar, all chocolate cake doughnuts tasted similar, all sour cream doughnuts tasted similar, and so on.

Clearly, however, this justification for the PUDP is itself an inductive argument. Having declined to accept the conclusion of our original induction, Zig is under no logical compulsion to accept the conclusion of this argument either. We could try to patch up the logical gap with some other general principle such as the Uniformity of Food Properties, but then *this* principle will need to be supported by an inductive argument from uniformity in past food experiences.

At each stage in our effort to close the logical gap between premises and conclusion in an inductive argument, we will find ourselves invoking a principle that in turn has to be justified inductively. Now either this series of justifications ends or it does not. If it ends, then it must end with us appealing to some general principle for which no further inductive justification is offered. This most general such principle is sometimes simply called the Principle of the Uniformity of Nature (PUN). Just how to formulate the PUN so that it is both strong enough to perform its logical task and plausibly true is difficult to say.[2] But we do not need to formulate it to see the problem: It is a general principle that cannot simply be true by definition (like 'carnivores eat meat') or be derived deductively from other premises that are true by definition. If it does not lend itself to such a deductive proof, and we have no inductive argument to support it, then it must simply be unjustified by any argument whatsoever, and so then are any of its consequences.

If, on the other hand, the series of justifications does not end, then we have an infinite regress of inductive arguments, so that the justification for our original conclusion can never be completed. Either way, the conclusion of any inductive argument (for there is nothing special about our doughnut example) turns out to lack any justification.

Note that in this illustration of the problem of induction we did not equip Zig with any specialized methods or concepts of the sort that scientists bring to bear on their research problems. The problem of induction appears not to be a problem of science so much as it is a problem of knowledge in general.

1.3 Reasoning about Light: Isaac Newton's Induction

Here, however, let us set aside the problems posed by our imaginary doughnut investigator's skepticism and consider an example of historically significant scientific reasoning. What we will see is that although the investigator – Isaac Newton – gave no explicit attention to the problem of induction in this general form that we have just considered, he devoted considerable attention to the question of just what he could justifiably infer from his observations and the assumptions that would be required to justify those inferences.

[2] The difficulties can be seen already emerging in the PUDP: How strongly do the doughnuts in a given kind resemble one another? If the principle asserts that they are exactly alike, then the PUDP seems to be false; if they are only somewhat similar in a very broad sense, then the principle will be too weak to support the conclusions that are meant to be drawn from it.

We might say that, although he did not concern himself with *the* problem of induction, he did concern himself with several specific problems of induction, that is, problems connected with particular ways in which he might go wrong in drawing his conclusions.

> In a very dark Chamber, at a round Hole, about one third Part of an Inch broad, made in the Shut of a Window, I placed a Glass Prism, whereby the Beam of the Sun's Light, which came in at that Hole, might be refracted upwards toward the opposite Wall of the Chamber, and there form a colour'd Image of the Sun. (Newton, 1979, 26)

Thus begins Isaac Newton's 'Proof by Experiments' of his second proposition in his 1704 work *Opticks*. The proposition in question states, "The Light of the Sun consists of Rays differently Refrangible" (Newton, 1979, 26). In other words, sunlight can be decomposed into rays of light that differ from one another in the angle at which they are refracted through a given medium (they have different *refrangibility*).[3]

Newton proceeds to describe how he arranged his prism so that its axis was perpendicular to the ray of light incident upon it, and so that the angles of refraction at the incident face of the prism and at the face from which the rays exited the prism were equal (see Figure 1.1). He describes the image made by the refracted light on a sheet of white paper as "Oblong and not Oval, but terminated with two Rectilinear and Parallel Sides, and two Semicircular Ends" (Newton, 1979, 28) and proceeds to report in detail on its dimensions. He reports that when the refracting angle of the prism was reduced from its initial value of 64 degrees, the length of the image was reduced, but its breadth was unchanged. If the prism was turned so as to make the rays exiting the prism emerge at a more oblique angle, the image was lengthened. He notes some irregularities in the glass of his first prism and thus tries the same experiment with different prisms "with the same

[3] Newton had first reported on his experiments with prisms in his inaugural lectures as the newly appointed Lucasian Professor of Mathematics at Cambridge in 1669 (Newton, 1984). His first published paper on the topic appeared in the *Philosophical Transactions of the Royal Society* in February 1671. There he describes the experiments as having been performed five years earlier "in the beginning of the year 1666," many years before writing the *Opticks* (Newton, 1978). The correct historical sequence of Newton's optical experiments and the trustworthiness of his narrative are matters of historical dispute (Guerlac, 1983; Newton, 1984).

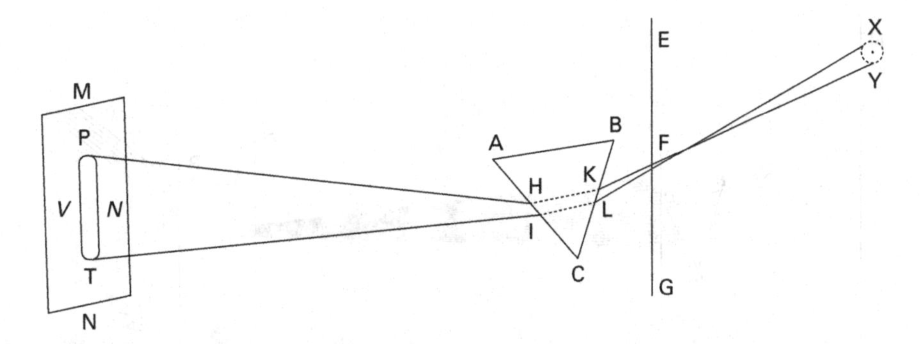

Figure 1.1 Newton's first experimental arrangement. F is the hole through which light from the sun (XY) passes, on its way to the prism (ABC). On a sheet of paper (MN) the light casts an image (PT) of the sun. From Newton (1979, 27). By permission of Dover Publications.

Success." "And because it is easy to commit a Mistake in placing the Prism in its due Posture, I repeated the Experiment four or five Times, and always found the Length of the Image" to correspond to that which he first reported (Newton, 1979, 30).

Newton then draws an intermediate conclusion from his measurements: The rays exiting the prism diverged from one another in such a manner as to constitute the length of the image, forming an angle "of more than two degrees and an half" (Newton, 1979, 31). That result Newton shows to be incompatible with "the Laws of Opticks vulgarly received," according to which the angles at which the rays exit the prism should diverge at an angle corresponding to the angle formed by the rays arriving at the hole from the extreme points of the sun's diameter – about half a degree. In other words, the 'vulgar' prediction is that the round shape of the sun should be reproduced in a round shape of the image, with a length equal to its breadth. What has gone wrong?

Newton points us to the answer by noting that if the rays that constitute the image were "alike refrangible," we should expect a round image rather than an elongated one. "And therefore seeing by Experience it is found that the Image is not round, but about five times longer than broad, the Rays which going to the upper end . . . of the Image suffer the greatest Refraction, must be more refrangible than those which to go the lower end" (Newton, 1979, 32).

There follows – in the *Opticks* – Newton's description of a series of experiments involving various arrangements of prisms through which Newton

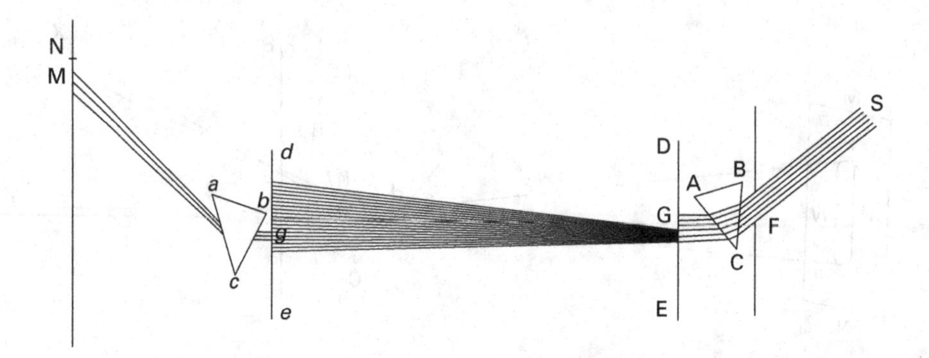

Figure 1.2 Newton's two-prism experimental arrangement. *S* is the Sun, light from which passes through an aperture *F* to prism 1 (ABC). A second aperture (G) can be used to select a portion of the refracted image of *S* to be sent through a third aperture (g) to prism 2 (*abc*). The refracted light from *abc* forms an image on the wall MN. From Newton (1979, 47). By permission of Dover Publications.

explores the phenomenon just described. Here I will just describe one such permutation that seems particularly relevant for understanding Newton's defense of his proposition that the sun's light consists of rays that differ in their refrangibility. In this arrangement (see Figure 1.2), Newton allowed sunlight to enter through "a much broader hole" before passing into a first prism, the orientation of which could be adjusted about its axis. The refracted light is then directed at a board with a hole about one-third of an inch in diameter. The beam of light thus formed is then projected onto a second board "and there paint[s] such an oblong coloured Image of the Sun" as had been found in the previously described experiment. This second board also has a small hole in it, and light that passes through that hole is then sent into a second prism, with its axis parallel to that of the first prism, which is held at a fixed position.

By varying the orientation of the first prism, Newton effectively selects different parts of the spectrum created by the first prism to send into the second prism. In this way he is able to more stringently focus on the question of whether these rays that emerge at different angles from the prism really do – *considered separately from one another* – refract at different angles. To test whether this is so, he observes the location on the opposite wall of the room at which the resultant ray of light arrives, having passed through both prisms. He found that indeed light from the higher part of the spectrum projected onto the second board refracted through the second prism to a higher location than light from the lower part of the spectrum, while light from intermediate parts of the spectrum arrived at intermediate locations on the wall. Newton thus shows that, although each ray arrives at the second prism at the same angle, "yet in that common Incidence some of the Rays were more refracted, and others less. And those were more refracted in this Prism, which by a greater Refraction in the first Prism were more turned out of the way, and therefore for their Constancy of being more refracted are deservedly called more refrangible" (Newton, 1979, 48).

Newton defends a general proposition here. He does not claim merely that the particular rays of solar light that he observed are composed of rays that vary in their refrangibility. Neither is his claim limited to the sun's light as observed in Cambridge, or as observed in the late seventeenth century. Indeed, although his explicit claim makes reference to the light of our Sun, his motivation for this research and subsequent application of his findings has to do with limitations on the resolution of telescopes, which he attributes at least in part to the differential refraction of the component rays of light from celestial objects, including stars. Clearly, then, he thought that the same proposition applied to light coming from at least some stars other than our own sun. His proposition states much more than can be summarized by simply noting what he observed in the particular instances he describes.

1.3.1 Newton's Problems of Induction

Yet Newton shows no concern about the problem of induction as we articulated it above. There is no evidence that Newton was at all worried about whether light might behave differently on Mondays than on Thursdays, or that his results might come out differently if he tried his experiment in the countryside instead of in town, or in Bristol rather than Cambridge. If anyone else worried about these matters, Newton's argument includes no overt reasoning that would allay those concerns. Given just the description of the experiments that Newton did perform, we would have to say that it remains *logically possible* that the Sun's light observed at different times or under different conditions does *not* consist of rays that differ in their refrangibility.

But Newton clearly is concerned about other possible ways in which his claim might go wrong and designs his experiments explicitly to rule out these sources of error. For example, Newton considers the oblong (rather than circular) shape of the image created by the prism to be significant, as it differs from the prediction from "the Laws of Opticks vulgarly received" (Newton, 1979, 31). What if that shape were somehow an artifact of the particular prism that Newton used, perhaps because of a defect in the glass? Newton must have been concerned about *this* possible error, because he describes how he repeated the experiment using different prisms and got the same result. What if Newton had made a mistake in lining up his prisms in the correct configuration? Again, he repeated his experiments to rule out small but unnoticed errors in his alignment of the prisms.

These two possible sources of error – flaws in the prisms, or errors of prism placement – might be thought of as *local* in the sense that they concern details of the particular instances from which Newton seeks to draw conclusions. But sources of error that are more *global* in nature are also relevant for him.

For example, why does Newton bother with the second experimental arrangement just described, in which he passes light refracted from the first prism through two more apertures before it enters the second prism? Although Newton does not state explicitly his motivation for this experiment, one might consider it as addressing a concern about a potentially systematic misunderstanding of the phenomenon thus far studied. Although the image resulting from refraction by the prism clearly indicates that light exits the prism at a range of angles, one might worry that somehow the phenomenon cannot really be analyzed in terms of the differing properties of refrangibility of rays in the incident light. Perhaps the spread somehow involves all of the light together, such that if some part of the spectrum is taken away, the whole phenomenon is destroyed. A worry of this sort would be addressed by Newton's second experimental arrangement, which allows him to demonstrate that individual segments of the spectrum consist of light that, just by itself, refracts at an angle different from that of the light in other parts of the spectrum.

Another sort of worry might arise from concerns about just what is implied by Newton's conclusion. Just what does Newton mean by saying that the Sun's light "consists of Rays differently Refrangible"? Here one might take him to be making a claim about the nature of light itself. Maybe Newton is committing himself to a view of light as a kind of particle, the trajectory of which is described by a ray. But if Newton means to imply such a view of light, then he would seem to be going well beyond what his experiments can show, since there does not seem to be anything in these experiments as described that would rule out other theories about the nature of light, such as that it consists of some kind of wave. (Theories of this sort had been advocated by Newton's contemporaries Christian Huygens and Robert Hooke.)

It was in fact Hooke himself who challenged Newton on this score, charging that his conclusions about color were part of a particle theory of light. In response to Hooke, Newton drew a distinction between hypothetical theories about the nature of light that would causally explain what he had observed experimentally and a description of those properties of light that he considered to be revealed in his experiments. Recognizing that his observations could be explained by 'Mechanicall' theories other than the particle account he himself favored, he wrote: "Therefore I chose to decline them all, & speake of light in generall termes, considering it abstractedly as something or other propogated every way in streight line from luminous bodies, without determining what that thing is" (quoted in Shapiro, 2002, 232).

Thus Newton denies that we should draw any inferences from his use of the term 'Rays' to conclusions about the nature of light itself. He thus secures his argument against objections against any such theories about light by making his conclusion independent of such assumptions.

To solve *the* problem of induction, Newton would have had to either show how, through an inductive argument, he could eliminate all possibilities of error in his conclusion, or else show – without falling into a regress or a vicious circle – how his conclusion could be justified *without* eliminating all possibilities of error. Newton did not attempt either of these things and thus did not attempt to solve the problem of induction. He did, however, attempt to solve problems relating to *particular* ways he might go wrong in drawing his conclusion. The example of Newton's optical experimentation illustrates how, even if one does not worry overtly about the problem of induction in general, a convincing experimental argument for a general claim does require that one be able to rule out sources of error that are judged to be relevant to one's conclusion.

1.4 Conclusion

In philosophical discussions of the problem of induction, attention typically focuses on cases like our doughnut example, in which the inductive inference consists of a simple generalization from observations made in some finite sample to some broader population of instances (sometimes called *enumerative induction).* We have seen how Newton, although he does generalize in his conclusion, undertook to justify his conclusion by using experimental *probing* to explore the various ways in which his conclusion might be in error. This *active* error-probing[4] reflects more accurately the reasoning of experimental science than do philosophy textbook stories of passive extrapolation like Zig and the doughnuts.

This still leaves us with some tough questions: Which possible sources of error should one consider relevant? Perhaps Newton justifiably disregards worries that the light from the Sun might behave differently at different times or in different places. Why, however, disregard those possibilities but explicitly address those about the possibilities of error in the alignment of prisms or of flaws in the prism glass? If one argues that Newton justifiably assumes that the Sun's light behaves uniformly in different places and at different times, then one has to give a justification for this that does not thrust us back into a regress or vicious circle of inductive arguments. On the other hand, if the difference between these possible errors is simply that no one – neither Newton nor his audience – was worried about the former but that they were – or plausibly would be – worried about the latter, then Newton's justification seems to depend on nothing other than the opinions of his contemporaries. That would have the rather unpalatable consequence that one's inductive arguments would be stronger, all else being equal, simply in virtue of one's audience being less critical!

Worse, it seems that although Newton does not worry about the problem of induction, he *should!* Simply ignoring a problem does not make it go away, and we are still left with the unsatisfactory situation of relying on an inductive form of argument that we ultimately seem unable to defend or justify. In Chapter 2 we will consider an approach to science that purports to do without induction entirely, but allows science to proceed without using a method that we cannot justify.

[4] For more on this, see Chapter 9, Section 9.4.

2 Falsificationism

Science without Induction?

2.1 Growth through Falsification

The philosopher of science Karl Popper considered the problem of induction so grave as to require the *elimination* of induction from the methods employed in science. He went so far as to assert that "in my view there is no such thing as induction" (Popper, 1992, 40).

In light of our discussion of inductive arguments in Chapter 1, such a proposal may sound outrageous. Surely, it might seem, we *need* inductive reasoning in science; we use such reasoning to confirm or support general propositions such as natural laws. That is how we learn from experience, fostering the growth of scientific knowledge.

Scientific knowledge does indeed grow, according to Popper. Understanding this growth is, for him, the central problem of the philosophy of science and of epistemology (the theory of knowledge) more generally. But he regarded the characterization of this growth in terms of induction as deeply mistaken. The belief that inductive arguments from the results of particular observations serve to confirm, support, or justify general empirical propositions is a dangerous illusion. That this belief – *inductivism* – is the wrong way to understand the growth of knowledge can be seen most clearly, according to Popper, if we consider an alternative view, which he calls *falsificationism*.

The key to falsification lies in a logical asymmetry that we noted in Chapter 1. Although the bulk of our story about Zig and the doughnuts was concerned with the relationship between Zig's many unpleasant observations of particular horseradish doughnuts and the proposition that all horseradish doughnuts are distasteful, the story began with an apparently quick and successful *refutation* of a proposition that Zig initially believed: All doughnuts are delicious (the Law of Doughnut Delectability – *LDD*). It took only a single foul-tasting horseradish doughnut to reveal the falsity of the *LDD*.

2.1.1 The Logical Structure of Generalizations

Such a falsification can even be represented in terms of a valid *deductive* argument. It will help us to see how such an argument works if we translate the *LDD* into a more precise logical form.

The logical form of *LDD* can be specified if we think of it as a generalization of a *conditional*. A conditional takes the form 'If p, then q,' where p and q should be understood as *variables* that stand in place of propositions that could be expressed by sentences all by themselves, like 'the snowman is melting.' The part (p) of the conditional that immediately follows 'if' is called the *antecedent*, and the part that follows 'then' (q) is called the *consequent*. *LDD* can be thought of as making the conditional 'if it is a doughnut, then it is delicious' completely general.

There is something a bit odd about statements like 'it is a doughnut,' 'it is delicious,' or 'if it is a doughnut, then it is delicious.' Absent any context to determine what it is that 'it' refers to, such statements lack any definite truth values. Whether the statement 'it is a doughnut' is true depends on what 'it' refers to when that statement is made.[1] To turn the conditional 'if it is a doughnut, then it is delicious' into a universal generalization, we preface it with a universal 'for any' quantifier:

Law 2 (LDD′) *For any thing whatsoever, if it is a doughnut, then it is delicious.*

LDD′ states exactly the same thing as *LDD*; the form of *LDD′* allows us to see more clearly that the law makes a statement not only about doughnuts but also about literally everything. What it states is this: Select, however you

[1] Logic therefore does not consider 'if it is a doughnut, then it is delicious' to be a sentence, and the same applies to its antecedent and consequent. In spoken language, this expression *could* be uttered so as to give it the meaning of its generalized form *LDD* (consider what is probably meant by someone who says something like, 'Hey, if it has cheese on it, then I like it'). Logicians attempt to tame these idiosyncrasies of natural languages by putting everything into symbolic forms like '$\forall(x)$ [Doughnut(x) $\supset$ Delicious(x)],' thus separating in symbols the form of the conditional – 'Doughnut(x) $\supset$ Delicious(x),' which is read as 'if x is a doughnut, then x is delicious' – from the operation of generalization – '$\forall$ (x),' which is read as 'for every thing x.'

Table 2.1 *Doughnuts and other things*

Name	Doughnut?	Delicious?
a	Yes (chocolate)	Yes
b	Yes (horseradish)	No
c	No (a piece of asparagus)	Yes
d	No (a submarine)	No

like, anything from any location in the universe. Is the thing selected a doughnut? If so (whether it is a chocolate, blueberry, or horseradish doughnut), then it is delicious. If not (whether it is an electron, a lemon, or Luxembourg), then it may be delicious, or it may be not.

To understand the behavior of logical conditionals like 'if it is a doughnut, then it is delicious,' consider how we might apply *LDD′* to individuals. Suppose we obtain four things, by whatever means. We can give these four things arbitrary names, so that we can talk about them without committing ourselves to any assumptions about what kinds of things they are. (Imagine, if you like, that you are presented with four boxes of arbitrary size and are simply told that each contains one thing to which *LDD′* will now be applied.) Let's call these things *a*, *b*, *c*, and *d*. Applying *LDD′* to them yields the following conditional sentences:

LDD'_a If *a* is a doughnut, then *a* is delicious.
LDD'_b If *b* is a doughnut, then *b* is delicious.
LDD'_c If *c* is a doughnut, then *c* is delicious.
LDD'_d If *d* is a doughnut, then *d* is delicious.

Conditionals like these are false when their antecedent is true and their consequent is false. Otherwise they are true. Suppose we open our four boxes and inspect their contents. We want to know of each thing whether it is a doughnut and whether it is delicious. Our data are given in Table 2.1.

Given these data, LDD'_b is false, and the other three conditionals are true.[2]

[2] It might seem odd to say that it is true that if *d* is a doughnut, then *d* is delicious, given that the name *d* refers to a submarine. Classical logic assumes, however, that all well-formed sentences, however peculiar their content, are either true or false, and treating LDD'_d, where *d* is a submarine, as *false* would lead to the unfortunate consequence that one could falsify a generalization about doughnuts by making observations of submarines.

2.1.2 Falsification Revisited: A *Modus Tollens* Argument

Now consider the import of our observation of *b* for the sentence *LDD*′. From the observation that *b* is a doughnut and *b* is not delicious, it follows that there is a thing such that it is a doughnut and it is not delicious. But it is a logical consequence of *LDD*′ that it is not the case that there is a thing that is a doughnut and is not delicious. We can therefore express the refutation of *LDD*′ in the following argument:

1. If *LDD*′ is true, then it is not the case that there is a thing that is a doughnut and is not delicious.
2. There is a thing that is a doughnut and is not delicious.
3. Therefore, *LDD*′ is not true.

In this argument, premise (1) must be true because it is a conditional whose consequent is a logical consequence of its antecedent. Premise (2) is taken to be true on the basis of observing a particular horseradish doughnut (more about this step below). These two premises suffice to show, using the deductively valid argument just given, that (3) *LDD*′ is false.

Karl Popper took this logical point to show that one may *falsify* a general empirical proposition without appeal to induction. For such falsification one can rely on a deductive form of argument known as *modus tollens*. Such arguments take the form: *If p then q. It is not the case that q. Therefore, it is not the case that p.*

2.1.3 Newton's Experiment as Falsification

It might seem that Popper's rejection of inductive arguments must lead to the rather odd conclusion that Newton's achievements in his optical experiments – an icon of the 'scientific revolution' – were an illusion. To draw this conclusion would, however, be to misunderstand the falsificationist view. Newton did indeed advance scientific knowledge with his prism experiments, but not by inductively confirming his claim about light. In Popper's view, Newton's contribution to the growth of knowledge has two facets: a *success* and a *failure*.

First, Newton *succeeded in falsifying* a claim about light. Recall how he noted that the 'Laws of Opticks vulgarly received' led to the expectation in his first experiment that a round image will be formed on the screen by the refracted light from the Sun. More precisely, from the hypothesis that all light is refracted by the same amount for a given refracting medium and a given angle of incidence, Newton derives the prediction that light refracted by the prism from the Sun will form a round image on the screen. The oblong image that Newton found when he performed his first experiment conflicts with this prediction, leading, via *modus tollens*, to the denial of the hypothesis that all light is refracted by the same amount for a given refracting medium and a given angle of incidence. So Newton succeeded in establishing the denial of a proposition that many had previously thought to be true.

Second, Newton *failed to falsify* his own hypothesis (*DR*, for Differential Refrangibility) about light: that light from the Sun consists of rays that differ in their refrangibility. According to the falsificationist, such a failure contributes to the growth of scientific knowledge not simply because a falsification did not happen (which would also be true if the experiment had never been performed, or if it were designed so poorly as to be irrelevant to the hypothesis in question), but because Newton's experiments were of a sort that might have falsified this hypothesis, yet they did not. In other words, Newton's experiments attempted, but failed, to falsify the hypothesis of *DR*.

Fair enough, you might be thinking, but Newton's experiment also failed to falsify other hypotheses about light's refrangibility, such as the hypothesis (DR_{NH}) that light from the Sun *that arrives in the Northern Hemisphere* consists of rays that are differentially refrangible. In the problem of induction, we saw that any finite set of individual observations was logically compatible with a multitude of different generalizations, and one might think that a similar problem is surfacing here.

2.1.4 Corroboration

Popper's response to this issue is crucial to his account of the growth of knowledge. To characterize such growth, we need to attend not only to those hypotheses that have survived potential falsification but also to the *corroboration* of hypotheses. Corroboration is a technical term, introduced by Popper, that is meant to be free of any of the inductive valences of terms such as confirmation, support, or verification. Popper stipulates that the degree of corroboration of a hypothesis is a function both of the extent to which the hypothesis has survived potentially falsifying tests and of the 'boldness,' or more precisely the *degree of falsifiability* of the hypothesis itself.

One theory is more falsifiable than another if, roughly speaking, it affords more opportunities for being falsified. Both *DR* and DR_{NH} could have been falsified by Newton's particular experiment, but *DR* – because it asserts something about light from the Sun everywhere – affords many *more* opportunities for falsification. Experiments similar to Newton's could be carried out in the Southern Hemisphere, on the Space Station, or on other planets. All these locations would provide opportunities to attempt to falsify *DR*, but such experiments would be irrelevant to DR_{NH}. By contrast, any experiment that tested DR_{NH} would necessarily also test *DR*. Thus, while it is true that both *DR* and DR_{NH} survived falsification in Newton's experiment, they are not equally corroborated by that experiment.

Theories that make broader, more general claims and theories that are simpler in their description of the world will tend to be more falsifiable because they have larger classes of potential falsifiers. A more falsifiable theory "*says more* about the world of experience" than one that is less falsifiable because it rules out more possible experimental outcomes (Popper, 1992, 113, emphasis in original). Popper's analysis of falsifiability centers not on what a hypothesis says *will* happen but on what it forbids, on the experimental results that, according to that theory, will never be produced. The more potential observations a theory rules out in this way, the more falsifiable it is.

Thus Popper endorses a methodology that advises us to prefer always those theories that are the most highly corroborated – that is, those that are the most falsifiable ones to have survived testing so far.

2.2 Falsificationism, Demarcation, and Conventionalism

2.2.1 Demarcation and the Aim of Science

By his own account, Popper began to formulate his philosophy of science in Vienna around 1919, when "the air was full of revolutionary slogans and ideas, and new and often wild theories" (Popper, 1965, 34). He noted a contrast among some of these "wild" theories. Einstein's General Relativity completely reconceptualized the phenomenon of gravity. The first experimental test of General Relativity was performed in 1919, a test that Popper regarded as precisely the kind of potential falsification that made for good science. By contrast, other widely discussed theories – for example, the Marxist theory of history and the psychoanalytic theory of Sigmund Freud – seemed to be so flexible as to be able to explain any potentially relevant observation whatsoever, even though the individual case might seem initially problematic.

In the case of Marxist history, Popper wrote that, although the theory in its early formulations made testable predictions, when these predictions were in fact falsified, the advocates did not respond by giving up the theory but instead "re-interpreted both the theory and the evidence in order to make them agree" (Popper, 1965, 37). He considered Freudian psychoanalysis not susceptible to falsification at all: "There was no conceivable human behaviour which could contradict [it]" (Popper, 1965, 37).[3]

Popper came to see the challenge posed by this contrast between Einstein's and Freud's theories in terms of what he called the 'demarcation problem': how to distinguish science from pseudoscience (or pre-science). He postulated falsifiability as a criterion for this demarcation: "[T]he criterion of the scientific status of a theory is its falsifiability, or refutability, or testability" (Popper, 1965, 37).

By stipulating this criterion, Popper's philosophy of science makes a claim not only about what makes a good theory but also about what science is *about* or, more precisely, what constitutes the *aim of science*. For Popper, to engage in science is to pursue knowledge by *considering only falsifiable theories* (i.e., theories that entail sentences denying the occurrence of certain possible experiences) and then *attempting to falsify those same theories* (by producing the circumstances under which those experiences might be produced). Popper sought in this way to articulate an ideal that would stand in opposition to the possibility that scientific theories might become matters of dogma, in which adherence to a theory is adopted simply as a matter of decision not subject to criticism. Popper referred to the latter danger – exemplified for him in the attitudes of some adherents to the Marxist theory of history and Freudian psychoanalytic theory – as a form of *conventionalism*.

[3] Popper did not conclude from this that Freud's ideas had no value. He allowed that, although Freud's theory is untestable, it "is of considerable importance, and may well play its part one day in a psychological science which is testable" (Popper, 1965, 37). He compared such 'pre-scientific' ideas to myths that could provide suggestions to later scientists.

As great as Einstein's General Theory of Relativity is, according to Popper, it would be a breach of good scientific conduct to accept that theory as a matter of convention, simply because of a prior decision to do so. Doing so would prevent the growth of knowledge.

It might therefore seem that Popper's philosophy of science rests on a certain normative claim (i.e., a claim about what one *ought to* do or believe): that it is *better* to engage in scientific inquiry by attempting to falsify theories than by attempting to confirm them through inductive arguments or by attempting to defend them against refutation. This yields a potential problem for the falsificationist, who cannot accept this general proposition on the basis of an inductive argument. Yet it also seems unfalsifiable, for what possible observation would falsify it?

2.2.2 Rules as Conventions

The first thing to be said in response to such an objection is that Popper denies that falsificationism is based on a proposition that one believes. Instead, falsificationism is constituted by a commitment to a certain set of rules. These are rules for doing science. The rules that govern any kind of activity are, according to Popper, conventions. A proposed set of rules for doing science can no more be established by logical arguments than can a proposed set of rules for playing chess. Popper chooses falsificationist conventions for their effectiveness in preventing scientific theories themselves from becoming conventions.

In other words, if science is to be a rule-governed activity, then one *must* be a conventionalist about methodology. Given that conventionalism about method is unavoidable, Popper advocates choosing conventions for the conduct of science that preclude conventionalism about the *content* of science.

A second important point is that, although Popper acknowledges that falsificationism is not falsifiable in the same sense that Einstein's General Theory of Relativity is falsifiable, it is susceptible to *criticism*. Popper offers susceptibility to criticism as a criterion more appropriate to the evaluation of conventional rules and philosophical positions more generally. In particular, one can ask of a proposed methodology for science whether the pursuit of science in keeping with its rules leads to desirable results. Popper writes:

> My only reason for proposing my criterion of demarcation is that it is fruitful: that a great many points can be clarified and explained with its help . . . It is

> only from the consequences of my definition of empirical science, and from the methodological decisions which depend upon this definition, that the scientist will be able to see how far it conforms to his intuitive idea of the goal of his endeavours. (Popper, 1992, 55)

We can thus see that Popper embraces (because he believes he must) a certain kind of conventionalism regarding science: namely, that scientific inquiry is directed at enabling the pursuit of *modus tollens* arguments, such as the one considered above, to be brought to bear on highly falsifiable hypotheses. Let us now consider the premises of such arguments.

2.3 The Empirical Basis

2.3.1 Avoidable Falsifications

You may recall that Zig was very quick to accept the falsification of *LDD* after tasting a single nondelicious horseradish doughnut, but very reluctant to accept the generalization that all horseradish doughnuts are distasteful. Suppose he had been inclined oppositely, however, and was loathe to give up his belief in the *LDD*. Does logic really force him to give it up? Of course, he has to admit that *LDD* really is a generalization and so it applies to the object he just tasted. But this falsification also depends on two further claims about that object: that it (1) *is a doughnut* and (2) *is not delicious*. Accepting these claims enables him to conclude that there is something that is both a doughnut and not delicious, in direct contradiction to the logical consequence of *LDD′* that no nondelicious doughnut exists. Although we might think it strange to dispute whether something really is a doughnut or not, Zig could dig in his heels on this point. Although it is a doughnut-shaped object sold in a doughnut shop, perhaps it is not really a doughnut. Perhaps it is a bagel disguised as a doughnut. If we show him the dough being made and then fried, rather than baked, he might insist that we did not use the kind of flour required for a doughnut. (Perhaps if it were made with rice flour, it would not be a proper doughnut.) When all else fails, he can resort to conspiracy theories or plead hallucination. Perhaps none of these responses would seem reasonable to us, but the point is that they could be made in a manner that is logically consistent with the truth of the *LDD*.

The consequence of this is that, although Popper jettisoned induction in favor of falsification as the engine of the growth of knowledge at least in part because he considered induction logically suspect, it remains the case that falsification is not achieved by logic alone. The decision to accept the premises of a falsifying *modus tollens* argument is not forced by logic.

2.3.2 Deciding to Accept a Falsifying Statement

Popper was forthright on this point. In order for scientific inquiry to proceed by enabling the pursuit of falsifications, scientists must *accept* certain statements as potentially falsifying premises, even though those premises are not forced upon them by logic. These accepted statements form what Popper calls *the empirical basis* (Popper, 1992, 93–111). Popper requires statements comprising the empirical basis to take the form of singular existential claims (such as 'there is a doughnut that is not delicious in Minot, North Dakota,' or 'there is a refracted image of the Sun that is not round in Cambridge, England').

Now one might suppose that, even though such a statement is not forced upon us by logic, it is nonetheless *justified by experience*. Newton did, after all, *observe* that the image of the Sun cast upon the screen was not round, and surely that experience of his constitutes support for the claim that there was, in his experimental setup, an image of the Sun that did not display the shape predicted by the "vulgar" theory of uniform refrangibility of light. Popper, however, considers this view a mistake. Statements, he insists, are not *supported* by anything except other statements that logically entail them. An experience is a psychological phenomenon, not a statement. We therefore cannot *apply* the rules of logic to experiences, but only to statements about them. Justification and support are the wrong categories for talking about the relationship between experiences and the statements we accept as describing those experiences.

Accepting a statement, according to Popper, should be regarded as a *decision* made in response to an experience. We might consider some decisions more appropriate than others in the light of a given experience, but these are practical – not logical – judgments. The decision to accept a statement as part of the empirical basis should be judged as either fruitful or not, and in this sense, the empirical basis has a conventional status, like rules of method. Decisions about the empirical basis must be judged with regard to how well they serve the aim of pursuing falsifying arguments to drive the growth of scientific knowledge. Popper's comments on this are worth quoting in full:

> The empirical basis of objective science has nothing 'absolute' about it. Science does not rest upon solid bedrock. The bold structure of its theories rises, as it were, above a swamp. It is like a building erected on piles. The piles are driven down from above into the swamp, but not down to any natural or 'given' base; and if we stop driving the piles deeper, it is not because we have reached firm ground. We simply stop when we are satisfied that the piles are firm enough to carry the structure, at least for the time being. (Popper, 1992, 111)

2.4 Conclusion

Popper's falsificationist ideas remain very influential. The criterion that a scientific theory must be falsifiable receives wide support among scientists, although they may not be aware of the full philosophical framework of falsificationism that Popper built around that criterion. Popper's attempt to characterize the growth of scientific knowledge in a manner that makes no appeal to induction has also confronted a number of challenges. For example, Wesley Salmon has argued that falsificationism cannot make sense of the apparently rational *use* of highly corroborated theories for the purpose of making practical predictions (Salmon, 1981).[4] In Chapter 3 we will consider a challenge that confronts both inductivists and falsificationists and that will enter into most of the philosophical debates that lie ahead: the problem of underdetermination.

[4] An example is the planning of trajectories of an interplanetary unmanned spacecraft, such as the Cassini–Huygens mission. These probes are able to travel to distant planets such as Saturn by using a 'gravitational slingshot' or 'gravity assist,' the execution of which demands the use of a theory of gravity, such as Newton's or Einstein's.

3 Underdetermination

3.1 Introduction

We might think of the problem of induction in terms of a logical *gap* between the premises of an inductive argument and its conclusion. So that scientific knowledge could be achieved without having to leap such gaps, Karl Popper proposed a theory of scientific method that focused on falsification rather than induction. In doing so, however, he had to confront the fact that, although falsifications can be *represented* with a gapless, deductively valid argument, logic does not require one to make the inference thus represented. Another gap, however, looms over both inductivism and falsificationism.

We saw in Chapter 2 how Zig could have avoided falsifying the Law of Doughnut Delectability by insisting that the nondelicious object he tasted was not, after all, really a doughnut. In such a case the logical gap exists because logic does not force on the observer a particular description of one's observation. Another logical gap arises because the logical connection between the hypothesis or theory being tested and the description of the data or observations is not direct. One needs additional premises (often called *auxiliary hypotheses* or *auxiliary assumptions*) to establish a logical connection between a theory and a description of a possible observation. Falsification starts with a determination of the falsehood of the observation that is expected assuming that *both* the hypothesis under test and the auxiliary assumptions are true. The resulting *modus tollens* argument, therefore, only shows that *it is not the case that both the hypothesis under test and the auxiliary assumptions are true*. But this is compatible with the hypothesis under test remaining true while one or more of the auxiliary assumptions is/are false.

The existence of such a logical gap is often expressed as the *underdetermination thesis*: Theories are underdetermined by data. The underdetermination thesis is often associated with the French physicist and historian/philosopher of science Pierre Duhem, who gave a prominent articulation of it in his book *The Aim and Structure of Physical Theory*, first published in 1906 (Duhem, 1991). In this chapter we will consider Duhem's underdetermination thesis, look at how his thesis was subsequently taken up and modified, and finally consider its bearing on our understanding of scientific knowledge. In later chapters we will discuss how certain approaches to scientific inference attempt to confront the problems posed by formulations of the underdetermination thesis.

3.2 Duhem's Thesis

Contemporary philosophers of science often associate the underdetermination thesis with Pierre Duhem, but distinctive features of Duhem's own thesis are often forgotten. Before we turn to the contemporary discussion of underdetermination, let us consider Duhem's original argument. First we will look at an example Duhem uses to illustrate underdetermination. Then we will note two significant features of Duhem's discussion: (1) he regards the thesis as having special importance for physics, as opposed to other sciences; (2) he takes the thesis to express a *problem* confronting the physical investigator and then discusses, in a general way, the resources needed for the investigator to *solve* that problem.

We might express Duhem's thesis as follows:

Thesis 1 (UD) Given that a set of premises $\{P_1, P_2, \ldots, P_n\}$ deductively entails a statement O describing a possible observation, and that from experimental observation one concludes that the statement O' is true, where O' entails that O is false, it follows as a deductive consequence that the conjunction 'P_1 and P_2 and … and P_n' is false. It does not follow from this that any particular member of $\{P_1, P_2, \ldots, P_n\}$, including the hypothesis one seeks to test, is false.

This thesis follows in a straightforward way from basic rules of deductive inference. Thus, the issue is not whether UD is true (which it certainly is) but what import it has for the conduct of scientific inquiry and our understanding of its results.

First, however, we should pause to note that, as articulated here, UD is really a thesis that applies in contexts in which a theory is being evaluated by *testing its consequences*. It will therefore be relevant to any methodology of science, inductivist or falsificationist, in which such testing plays a role.

3.2.1 A Test of Newton's Emission Theory of Light

To understand the argument Duhem makes, let us consider his own example involving another aspect of Newton's theorizing about light. You may recall from Chapter 1 that in defending his conclusions about the differential refrangibility of light against Hooke's objections, Newton argued that his conclusions did not depend on any substantive assumptions about the nature of light itself. Nonetheless, Newton did have a theory about the constitution of light. According to his 'emission theory,' light consists of very small projectiles emitted by bodies such as the Sun at very high velocities. These projectiles penetrate through transparent bodies, which in turn reflect and refract the light as a result of the action of the particles that make up the bodies on the light particles. In effect, what Newton proposed was that the matter of a transparent medium exerts an attractive force upon light particles, so that light passing into a more dense medium would speed up due to the greater attractive forces exerted. On this account, when light passes from one transparent medium into another (say, from air to water), the index of refraction (from which one can calculate the angle of refraction) is equal to the velocity of light in the second medium divided by the velocity of light in the first. That light should travel *faster* in a denser medium (like water) than in a less dense medium (like air) became an object of experimental interest when the French physicist Françoise Arago proposed a method to test this consequence of Newton's theory.[1]

The experiment itself was not performed by Arago, but separate measurements by Hippolyte Fizeau and Léon Foucault led to the conclusion that, contrary to the prediction from Newton's theory, light travels more slowly in water than in air.

Foucault's experiment is noteworthy for its technical precision and ingenuity of design (see Figure 3.1).[2] Sunlight entered the apparatus through

[1] Arago's proposal had its source in a similar idea due to the English physicist Charles Wheatstone, who had used a rotating mirror to measure the velocity of electric current.

[2] See Preston (1890) and Tobin (2003), on which the accompanying description is largely based, for a helpful discussion.

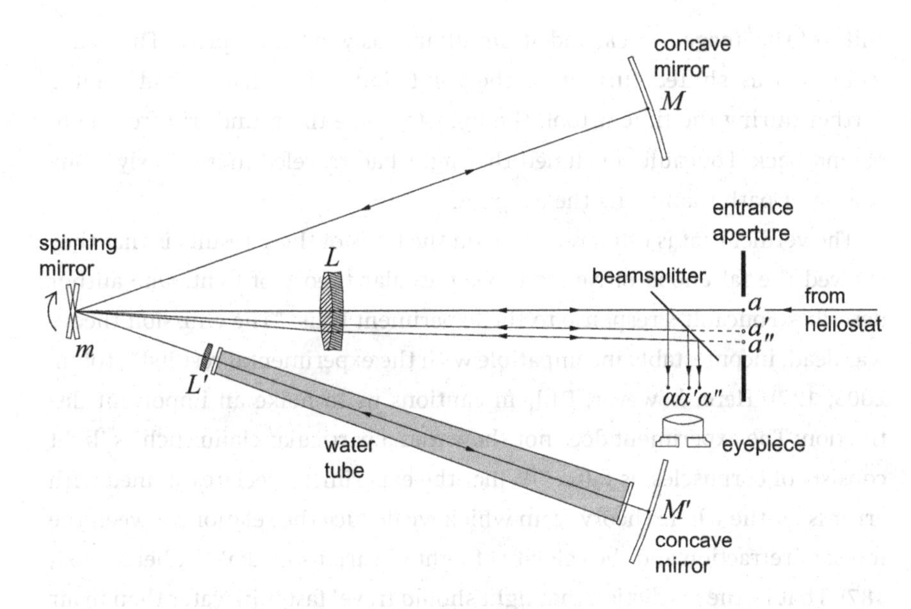

Figure 3.1 Foucault's experimental arrangement. From Tobin (2003, 125). Image courtesy of William Tobin. Reproduced with permission of Cambridge University Press through PLSclear.

a rectangular aperture containing a single vertical wire. Light then passed through an achromatic lens *L* on its way to a plane mirror *m*. The mirror *m* rotates (with the help of a small steam engine) around a vertical axis passing through its center. When *m* is appropriately positioned, light reflects from *m* to either of two concave spherical mirrors *M* and *M′*. An additional lens *L* and a 3 m tube of water are placed along the beam path between *m* and *M′* (the lens *L* corrects for refraction due to the water in the tube). The mirrors *M* and *M′* then reflect the image of the wire in the entrance aperture back to *m*, through *L*, and to a glass plate (a 'beamsplitter') that deflects the image to an eyepiece. With *m* stationary and oriented either towards *M* or *M′*, a rule was marked in the eyepiece to indicate the original position of the image of the wire.

When *m* is set spinning, however, given the finite velocity of light, it rotates through some small angle during the time it takes light to make the journey from *m* to *M* or *M′* and back. As a consequence, the image of the wire is shifted to the right in the eyepiece.[3] Foucault was able to compare the

[3] Quite apart from the comparison between the air-path and the water-path, this enables one to use the apparatus to measure the speed of light. But the short distance between *m* and *M* needed to maintain a bright and clear image in Foucault's setup limited the

shifts of the image from M and M' simultaneously in the eyepiece. The image from M' was shifted further to the right, indicating that m had rotated further during the time it took the light to make the round trip from m to M' and back. Foucault concluded that light had traveled more slowly along the water-path than along the air-path.

The verdict that is often rendered on the basis of these results is that they showed the falsehood of Newton's corpuscular theory of light. One author describes Foucault's response to the experiment thus: "The emission theory was dead, incontestably incompatible with the experimental results!" (Tobin, 2003, 127). Here, however, Duhem cautions us to make an important distinction: The experiment does not show that a particular claim such as 'light consists of corpuscles' is false. "[W]hat the experiment declares stained with error is . . . the whole theory from which we deduce the relation between the index of refraction and the velocity of light in various media" (Duhem, 1991, 187). That is, the prediction that light should travel faster in water than in air depends not only on the premise that light consists of corpuscles emitted by luminous bodies but also on additional assumptions, such as those about the forces exerted on such corpuscles by matter of a refracting medium and the reactions of the corpuscles to such forces. The conclusion that light travels more slowly in water than in air logically demands only that we acknowledge that not all of those premises are true. This, however, does not allow us to judge the truth or falsehood of the premises individually. "In sum," Duhem writes, "the physicist can never subject an isolated hypothesis to experimental test, but only a whole group of hypotheses," and if the group fails the test, "the experiment does not designate which one should be changed" (Duhem, 1991, 187).

3.2.2 Duhem on Physics

Duhem begins his discussion with a contrast between physiology and physics, noting with approval the advice of physiologist Claude Bernard that for the physiological investigator, "theory should remain waiting, under strict orders to stay outside the door of the laboratory" (Duhem, 1991, 181). In a

accuracy of this measurement. Michelson carried out a variation on Foucault's experiment with a different placement of the lens, a much greater separation of the mirrors, and hence greater accuracy (Michelson, 1879).

physics experiment, however, Duhem notes that "it is impossible to leave outside the laboratory door the theory that we wish to test, for without theory it is impossible to regulate a single instrument or to interpret a single reading" (Duhem, 1991, 182). Duhem concludes this from the fact that the theories of physics themselves describe the operation of instruments used by physicists to make measurements, produce phenomena, and manipulate entities and processes.

Because the use of instruments, such as thermometers, oscilloscopes, phototubes, and so on, and the interpretation of data resulting from these devices depend upon the theories of physics that explain how such devices work, performing an experiment in physics requires one to assume from the outset a substantial body of physical theory. He acknowledges that experimenters working in chemistry and physiology use some of these same instruments and thus rely on similar theoretical assumptions, but in these cases the theory assumed is concerned with a domain distinct from that of the theory being tested. Only the physicist is "obliged to trust his own theoretical ideas or those of his fellow-physicists" (Duhem, 1991, 183).

Duhem here makes an interesting concession: "From the standpoint of logic, the difference is of little importance; for the physiologist and chemist as well as for the physicist, the statement of the result of an experiment implies, in general, an act of faith in a whole group of theories" (Duhem, 1991, 183). As we continue our discussion, we will return to the question of whether 'an act of faith' is an apt description of what is needed from the experimenter. Setting that issue aside, it might seem that Duhem here admits that his distinction between the implications of underdetermination for physics as opposed to other fields is of little importance. As we will see, however, the qualifying expression 'from the standpoint of logic' is significant. Duhem does not think that the standpoint of logic alone matters for judging the import of an experiment.

3.2.3 Duhem's Problem and "Good Sense"

As noted previously, Duhem's UD thesis is unobjectionable. For the scientific investigator, however, UD cannot be the end of the discussion because the thesis burdens the investigator with a pressing *problem* that is both epistemological and practical. At some point one has to report what one has learned from the experiment performed. What should you say? If you report only that one of a whole set of premises you used to derive a predicted result has been shown to be false, the very next question you will be asked is, "Can you say anything about *which* premises we ought to suspect of being incorrect?"[4] You would like to answer positively, but logic does not give you any basis for doing so. Call this the 'underdetermination problem' or, if you like, 'Duhem's problem.'

Duhem notes that different responses to any given underdetermination problem may be not only possible but even defensible in some sense. One physicist, having encountered an experimental result at odds with the prediction from some theory, might choose to "safeguard certain fundamental hypotheses" in the theory by making adjustments to some of the less central assumptions (about causes of error, or about how the instruments work). Another, contemplating the very same result, may take it as counting against those same fundamental hypotheses protected by the first physicist (Duhem, 1991, 216–217). Duhem notes of these two physicists: "The methods they follow are justifiable only by experiment, and if they both succeed in satisfying the requirements of experiment each is logically permitted to declare himself content with the work that he has accomplished" (Duhem, 1991, 217).

Let us note two aspects of this statement. First, Duhem does require that each physicist must satisfy 'the requirements of experiment.' Although he does not elaborate on this point, this presumably means that one may not wantonly disregard the fact that certain kinds of error need to be controlled for, that measurements need to be made following a consistent procedure, that sources of bias in one's sampling procedure need to be taken account of, and so on. Second, both physicists are 'logically permitted' to declare themselves satisfied. But, as Duhem notes immediately following this passage, "Pure logic is not the only rule for our judgments; certain opinions, which do not fall under the hammer of the principle of contradiction are in any case perfectly unreasonable." Duhem attributes such judgments of unreasonableness to a faculty that he calls "good sense" (Duhem, 1991, 217).

By his own admission, Duhem's invocation of good sense as the means by which to solve underdetermination problems lacks precision and clarity. The

[4] Should the outcome of the experiment be consistent with the prediction from the hypothesis under test, one might be asked a somewhat different question: "Were that hypothesis to be false, is it likely that one would have gotten a result that did not agree so well with the predicted result?"

reasons for preferring one judgment to another that good sense provides are 'vague and uncertain;' they are apt to be a matter of (temporary) disagreement among different investigators, leading to controversies. But for precisely this latter reason, scientists have an obligation to *cultivate* their good sense, for by doing so they may shorten the time it takes for an inadequate hypothesis to be replaced by "a more fruitful" one (Duhem, 1991, 218). Here it is evident that, for Duhem, the underdetermination problem is no reason for despair about progress in science, for good sense (whatever that might really be) will eventually produce not simply the termination of disagreement but also agreement on what is in some sense the *right* decision. "[I]n order to estimate *correctly* the agreement of a physical theory with the facts, it is not enough to be a good mathematician and skillful experimenter; one must also be an impartial and faithful judge" (Duhem, 1991, 218, emphasis added).

3.3 Underdetermination Revisited: Quine

Duhem's discussion fails to provide any very concrete advice regarding 'good sense.' Aside from being impartial, can experimenters *do* anything that would help sharpen their experiments to point more definitively to a particular conclusion?

Although one might think that making progress on this kind of question would be the natural next step, the discussion of Duhem's thesis took a different direction in the mid twentieth century. In his influential 1951 essay "Two Dogmas of Empiricism," W. V. O. Quine took up the UD thesis and put it to a use that is quite different from Duhem's (Quine, 1980). But, because he cited Duhem in a footnote as having defended the doctrine that "our statements about the external world face the tribunal of sense experience not individually but only as a corporate body" (Quine, 1980, 41), readers came to associate both Duhem's and Quine's views. The thesis Quine defended came to be known as 'the Duhem-Quine thesis.'[5]

[5] This conflation is perhaps not so surprising. It seems likely that Quine's audience, like Quine himself, had not read Duhem (the reference is absent from the original version; Quine, 1951), being more acquainted with the Anglo-American variant of logical empiricism to which Quine was responding. See Quine (1991, 269) for his own account of how he came to insert the reference to Duhem, and Gillies (1993) for a useful comparison of Duhem and Quine.

Unlike Duhem, Quine was not primarily concerned with the role of logic and method in scientific experiment. Instead, Quine sought to address issues concerning language. In particular, Quine's argument attacked two widely held philosophical views about the meanings of words. The first of the two "dogmas" that he rejected holds that there is a fundamental and important distinction between *analytic* statements (such as 'every carnivore eats meat'), which are true or false in virtue only of the meanings of the words that were used to express them, and *synthetic* statements (such as 'an emu is wearing my favorite scarf'), the truth or falsehood of which depends on something more than the meanings of the words that were used to express them. The second dogma, which he calls *reductionism*, Quine describes as "the belief that each meaningful statement is equivalent to some logical construct upon terms which refer to immediate experience" (Quine, 1980, 20). (In our discussion of logical empiricism in Chapter 4 we will consider a view akin to what Quine calls reductionism.)

We will bypass Quine's arguments against these two dogmas and turn our attention to his conclusion. He claims that abandoning reductionism and the analytic/synthetic distinction leads one to the view that our knowledge, taken all together, "is like a field of force whose boundary conditions are experience" (Quine, 1980, 42). Experience, that is, constrains our beliefs about the world in some way, but only weakly. If the expectations we form about future experiences – based on the beliefs that make up the 'field' – conflict with our actual experience, we have to change our beliefs somehow. "But the total field is so underdetermined by its boundary conditions, experience, that there is much latitude of choice as to what statements to reevaluate in the light of any single contrary experience" (Quine, 1980, 42).

How much latitude, you ask? Quine's answer is striking: "Any statement can be held true come what may, if we make drastic enough adjustments elsewhere in the system." Moreover, we can make any adjustment we like because "no statement is immune to revision," not even the laws of logic (Quine, 1980, 43). As Quine well knows, we do not make or withhold revisions arbitrarily. We have *reasons* (in the appropriate circumstances) for taking our experience of strawberry flavor to result from having eaten an actual strawberry rather than from having a strawberry hallucination or being possessed by strawberry demons. But Quine insists that these reasons are ultimately 'pragmatic.' It would be difficult and costly to revise our beliefs to interpret strawberry experiences with a new set of beliefs about strawberry experience-inducing demons. Nonetheless, we *could* do so, and the fact that we do not reflects simply our reluctance to undertake such costly revisions to our beliefs when they are so easily avoided.

3.4 A Big Deal?

A comparison of Duhem's and Quine's discussions reveals that, although they share the UD thesis as their premise, they conclude rather different lessons for the scientific enterprise. According to Quine, were it not for 'pragmatic' considerations, our reasoning about the world would be unconstrained, since even 'statements of the kind called logical laws' may be revised. Duhem draws a somewhat more restrained conclusion, noting simply that quite different responses to an experimental outcome can be properly defended, provided that the investigators have met the 'requirements of experiment.'

Should we take the implications of UD to be a big deal? That human reasoning involves more than logic alone is hardly a surprise. As a purely formal science concerned with deductive relationships between one statement and another, logic cannot by itself solve the problem of differentiating between that which deserves belief and that which does not. But, as Duhem notes in a previously quoted passage, "certain opinions, which do not fall under the hammer of the principle of contradiction are in any case perfectly unreasonable" (Duhem, 1991, 217). The pressing question is not whether it is logically possible to hold any statement true 'come what may' but what standards beyond logical consistency we should apply when deciding what to believe and what status those standards have, given that they are not derivable from logic alone.

As Larry Laudan notes, Quine's claim that "any statement can be held true come what may" sounds a lot less methodologically significant if we recall that the word "can" expresses different kinds of possibility depending on how it is used (Laudan, 1996). In the present context we could understand Quine's claim to mean "given any particular statement that contradicts other beliefs a person holds there will be a strategy for maintaining belief in that statement without such contradiction by modifying other beliefs." This does not, however, mean that one may *justifiably* or *rationally* believe a statement regardless of the experience one has had.[6] Deciding what one *ought* to believe demands more than the answer to the question of what one *can* believe while remaining logically consistent. (Alternatively, we might say, the question for the philosophy of science might not really be about what one ought to *believe* at all but rather what theory is best corroborated, supported, or tested. Popper's falsificationism is one philosophy of science in which questions about belief are not central to scientific method. We will encounter another such approach in the error-statistical philosophy of Chapter 9.)

Consider Foucault's observation that the wire-image produced by light traveling the air-path shifted less than that produced by light traveling the water-path. Foucault confronted an underdetermination problem: whether to present his results as counting against the view of light as consisting of corpuscles, or as conflicting with some other assumption, since the result refutes not any isolated hypothesis about the nature of light but rather "the whole theory from which we deduce the relation between the index of refraction and the velocity of light in various media" (Duhem, 1991, 187). In truth, the situation is even worse, since we need not only theoretical assumptions to logically connect the corpuscular theory to the predicted shift; we also need statements describing the setup of the experiment. For example, were the distance from mirror m to mirror M sufficiently *greater* than the distance from m to M', then the result obtained by Foucault might come about even were the corpuscular theory to be correct.

Consequently, had Foucault been motivated to protect the emission theory, he could have done so in a logically consistent way, simply by denying that the two beams of light traveled the same distance in the manner required for the experiment to be probative. Even had Foucault personally measured the two paths to be of the same length, he could have insisted that the measuring device had changed length during the measuring process, or that he had been under the influence of powerful hallucinogenic drugs while recording the results of that measurement. These possibilities were also open to members of the scientific community who heard reports of Foucault's experiment, including those who, unlike

[6] As Laudan goes on to argue, however, one finds precisely these latter claims defended on the grounds of the UD thesis, including the claim that "[a]ny theory can be shown to be as well supported by any evidence as any of its known rivals" (Laudan, 1996, 39).

Foucault, actually *were* motivated to protect Newton's theory from refutation.

Foucault *could* have done any of these things, but it seems reasonable to expect that he would have encountered serious criticism for such a procedure. Failing to ensure that the light-path and air-path are of equal length would constitute a serious defect in experimental design and a failure to meet, in Duhem's words, "the requirements of experiment." Other changes in belief (pleading hallucination, assuming some change in the measuring device), apparently motivated *only* by a desire to protect the emission theory from falsification, seem "perfectly unreasonable," in Duhem's words. Moreover, describing Foucault's decision *not* to respond to his results in these ways as an 'act of faith' (as Duhem does) seems misleading if by that we mean he could not have had *reasons* (beyond convenience) for relying on the assumptions that he did. Acquiring such reasons would seem to result from, or even be the point of, careful safeguards aimed at securing the reliability of one's experimental procedure.

3.4.1 An Epistemology of Experiment

The strategies that experimentalists use to validate their conclusions are at the core of an *epistemology of experiment* advocated by Allan Franklin, based on his detailed studies of a wide variety of experiments in physics (Franklin, 1986, ch. 6, 2012).[7] Although a detailed discussion of Franklin's strategies would take us away from the focus of the present chapter, they include the use of calibration techniques (in which known phenomena are reproduced to test the functioning of an apparatus), reliance on independently well-tested theory (either to explain the phenomenon produced in the experiment or to warrant reliance on an apparatus whose proper operation that theory predicts), the use of statistical arguments, and – perhaps most importantly for our present concern – the elimination of plausible sources of error and alternative explanations.

[7] Although Franklin's strategies were drawn from studies of experiments in physics, David Rudge has applied them to an experiment in biology investigating industrial melanism (Rudge, 1998, 2001), and Franklin himself applied them to a discussion of Meselson and Stahl's experimental work on DNA replication (Franklin, 2005, ch. 3).

Although the logical core of the Duhem problem stands, these strategies provide important resources for scientists to resolve underdetermination problems in a piecemeal – yet reasoned – way. Even the conscientious application of them does not guarantee that one will not fall into error in one's conclusions, since they draw on resources beyond the truth-preserving strategy of deductive inference. In particular, they require investigators to rely on *judgments* of various kinds, such as the extent to which alternative explanations are plausible, or whether a theory has been sufficiently well tested to make its application in a particular instance trustworthy.

3.5 Conclusion

We have seen how both inductivist and falsificationist ways of thinking about scientific reasoning face the problem of a logical gap. What is less clear, however, is the significance of that gap for the results of scientific inquiry. If logic exhausted our resources for evaluating scientific theories, it would seem to follow that our decision to respond in one way – rather than another – to our observations of the world would ultimately be arbitrary, or at best driven by mere preference based on considerations of factors such as convenience. Something like this latter view does seem to emerge from Quine's discussion of underdetermination in 'Two Dogmas of Empiricism'.

But Duhem's discussion should give us pause before conceding the apparent assumption that we have only logic (and considerations of convenience) to go on when deciding what to make of our observations. Duhem points, if perhaps vaguely, to two additional resources: (1) methodology ('the requirements of experiment'), which provides standards for the competent execution and recording of experiments or observations, so as to both reduce our susceptibility to error and enhance our ability to learn about the hypotheses we subject to experimental tests, and (2) judgment ('good sense'), grounded in prior study of both phenomena and theories within a domain, which, we might suppose, allows the investigator to judge the extent to which a particular theory can be regarded as a plausible contender in light of all the relevant knowledge. To be sure, we would like to have a better, more clearly articulated explanation of these resources, and Franklin's epistemology of experiment represents one effort in that direction. We will consider further efforts involving ideas about *probability* in Chapters 8 and 9.

4 Logical Empiricism

4.1 Introduction

Chapter 3 explored the limitations on how far deductive logic alone can guide us in decisions about scientific theories based on experimental outcomes. In the early decades of the twentieth century, another group of philosophers and scientists in Europe pursued a philosophical program that attempted to exploit the resources of formal logic as far as possible in the articulation of a 'scientific philosophy.'

In this chapter we will take a brief look at the philosophical movement known as *logical empiricism* or *logical positivism*, paying particular attention to two related and central projects of the movement – an account of the structure of scientific theories, and a theory of empirical confirmation. The emphasis philosophers of science have placed on these projects reflects a commitment to the idea that scientific inquiry contributes to human knowledge through the production and confirmation of *theories*, while observation and experimentation serve a subsidiary role as sources of confirming or disconfirming evidence relating to theories. Later some philosophers challenged both this assumption – arguing that experimentation produces knowledge that is philosophically significant beyond simply its role in confirming theories (Ackermann, 1985; Franklin, 1986; Hacking, 1983) – and even the idea that *confirmation* of theories provides a helpful way of thinking about what we learn from scientific inquiry (Mayo, 1996). To appreciate the importance of such challenges, we will need first to consider the logical empiricist program itself, which exerted great influence on subsequent philosophical thinking.

4.2 The Vienna Circle and Logical Empiricism

4.2.1 The Vienna Circle and the Philosophy of Science

Although the questions pursued by contemporary philosophers of science have roots deep in the beginnings of Western philosophy, the emergence of the philosophy of science as a distinct discipline within philosophy, with its own professional organizations, its own scholarly journals, and its own programs of specialized training, was a twentieth-century phenomenon. Integral to the formation of the philosophy of science as a specialized field of study were the activities and eventual dispersion of a group of philosophers and scientists who met in Vienna, Austria, and have been loosely identified as 'the Vienna Circle.' In addition to fostering their own philosophical approach, the Vienna Circle influenced philosophers from other countries who joined them for study and discussion. When the Nazis seized power in Germany and Austria, many members of the group (which included quite a few Jews and leftist intellectuals) fled to other countries, many of them making their way, eventually, to the United States, where they became influential scholars and teachers.

Subsequent scholarship has shown that the philosophical views of those affiliated with the Vienna Circle were quite diverse (see, e.g., Friedman, 1999; Richardson & Uebel, 2007; Stadler, 2003), to the point of challenging the idea that one can articulate any unifying thesis or doctrine that can be reliably attributed to the movement. Nonetheless, a significant characteristic of the Vienna Circle was the pursuit of a philosophy that might articulate a scientific way of knowing that would be free of bias, free of dogma and superstition, free of nationalist ideology, and beneficial to humankind. Inspired by recent advances in physics (like their fellow Viennese Karl Popper, with whom they vehemently disagreed on many issues), they sought to achieve these lofty goals with the use of recent advances in formal logic and (for at least some members of the movement) with the guidance of an idea about what separated meaningful from meaningless discourse.

4.2.2 Meaning versus Metaphysics

The rough idea that meaningful discourse is distinguished from meaningless nonsense – by the fact that the former, but not the latter, consists of statements that can be *verified by experience* – is strongly associated with the

Vienna Circle. This idea came to be known as the verification principle, and the philosophical orientation in which it played a central role came to be called either *logical positivism* or *logical empiricism*. The verification principle is not merely stipulated but stems from a conviction that the clearest exemplars of meaningful discourse are to be found in such successful empirical sciences as physics. The membership of the Vienna Circle itself represented expertise in scientific disciplines, including physics (Philipp Frank), mathematics (Hans Hahn, Olga Hahn-Neurath), and social science (Otto Neurath), alongside philosophers (Moritz Schlick, Viktor Kraft, Rudolf Carnap). By contrast with successful empirical science, logical empiricists regard the history of metaphysical theorizing in philosophy as rife with examples of meaningless statements that could never be verified by any observation of any kind.

While physical science provides the Vienna Circle with their best example of a knowledge-generating enterprise, it also proves to be the source of their greatest challenge. It is all fine and well to *say* that the meaningfulness of physics is guaranteed by the fact that it consists of statements that can be verified by experience, but is that really true? For example, Einstein, writing in 1918, explained that he considered his General Theory of Relativity (regarded by logical empiricists as a signature achievement of modern science) to be based on principles entailing that the 'G-field' (the 'condition of space' described by a mathematical object called the 'fundamental tensor,' essential to the articulation and application of Einstein's theory) "is determined *without residue* by the masses of bodies" (Einstein, 1918, 241–242, quoted in Norton, 1993, 306, emphasis in original).

To those not familiar with General Relativity, this statement sounds opaque at best. Even for those well versed in the relevant physics, however, this statement does not obviously show itself to be one that can be verified by experience as required by the logical empiricists' principle, since the G-field itself is not directly observable. What set of observations could suffice to verify Einstein's statement about the G-field? To put the problem succinctly, logical empiricists have to articulate the verification principle so that the examples constituting – by their own reckoning – the clearest cases of meaningful discourse do actually *satisfy* the principle.

The most influential efforts by logical empiricists to tackle this problem invoke a linguistic distinction that divides the terminology employed by a scientific theory into three distinct categories: (1) the *logical* vocabulary V_L

(consisting of terms such as 'every,' 'some,' or 'not'), (2) the *observational* vocabulary V_O, and (3) the *theoretical* vocabulary V_T. Proposals for drawing a distinction between V_O and V_T varied somewhat, but the basic idea is that the truth or falsehood of a statement containing only terms in V_O (such as, perhaps, 'red' or 'to the left of' or 'coinciding with') can be decided directly by appeal to observations, while statements using theoretical terms (such as 'electrically charged,' 'acidic,' or 'G-field') cannot be. To vindicate the verification principle, the logical positivists seek to show that in successful empirical sciences, statements that include terms from V_T (like 'this sample is acidic') stand in some *appropriate relation* to statements using only terms from V_O (perhaps 'if this sample is applied to litmus paper, the litmus paper will turn red').[1] Statements expressing these relations are *correspondence rules* because they indicate what, in experience, corresponds to the theoretical terms invoked.

In seeking to make sense of the empirical character of scientific theories, logical empiricists take a linguistic approach focused on terms, the sentences in which they appear, and logical relations among those sentences. Theories, on this view, are collections of sentences having a certain structure. Applying a formal logic directly to theory and its relation to statements involving observational terms requires such a *syntactic* conception. More precisely, this conception holds that one can identify a theory by reference to a set of sentences that use a theoretical vocabulary to assert the basic principles of the theory. These are the *axioms* of the theory. A set of *correspondence rules* allows one to deduce the observable implications of the theory from the axioms. The theory thus consists of the axioms, the correspondence rules, and all the other sentences that can be deduced from the axioms, correspondence rules, and deductive consequences thereof.

If a statement using theoretical vocabulary bears cognitive meaning in virtue of the fact that it stands in an appropriate relation to statements using only observational vocabulary, then some account must be given of that relation. An appealingly simple proposal is that statements in the successful disciplines of empirical science that use theoretical vocabulary can simply be *defined explicitly and completely* with statements using only observational terms. This would require that for every such theoretical statement there

[1] For purposes of illustration we adopt the debatable proposition that 'litmus paper' is part of the observational vocabulary.

is some purely observational statement to which it is logically equivalent.[2] Thus, one might attempt to secure the meaningfulness of an assertion using the term 'acidic' by means of a correspondence rule that serves as a *definition* such as 'this sample is acidic if and only if it is the case that if this sample is applied to litmus paper, the litmus paper will turn red.' Such sentences have been called *reduction sentences* because they purport to show how statements involving theoretical terms can be reduced to statements involving only observational terms, which is to say that theoretical terms in the original statement could be replaced by constructions involving only observational terms without losing or changing what the original statement says.

This simple approach will not work, however. First, even a seemingly simple statement such as 'this sample is acidic' says much *more* than any one observation-language statement such as 'if this sample is applied to litmus paper, the litmus paper will turn red.' Being acidic has implications for many different kinds of chemical reactions into which the sample might enter, and not simply its interaction with litmus paper. One might counter this objection by insisting that 'this sample is acidic' needs to be analyzed using a more complicated observation-language statement describing other observational tests of acidity as well as the litmus-paper test, but the defender of reduction sentences now clearly bears the burden of showing that such a reduction can be completed, which there is reason to doubt. Scientists are always discovering new ways to reveal what is ordinarily unobservable.

That brings us to a second and more fundamental problem. As Rudolf Carnap, arguably the most prominent and influential of the Vienna Circle logical empiricists, acknowledged (Carnap, 1936, 1937), a statement like 'this sample is acidic' *cannot* be equivalent to another of the form 'if this sample is subjected to conditions *x*, then observable condition *y* will be realized' simply because the latter will inevitably be true in some circumstances in which the former is false. As we noted in Chapter 1, a conditional statement of the form 'if *p* then *q*' is considered by logicians to be true whenever *p* is false, regardless of whether or not *q* is true (in the case where *p* is true, 'if *p* then *q*' is true in case *q* is also true, false otherwise). For the present example, this has the consequence that 'if this sample is applied to litmus paper, the

[2] Two sentences S_1 and S_2 are *logically equivalent* just in case S_1 logically entails S_2 and S_2 logically entails S_1.

litmus paper will turn red' is true for any sample that is not applied to litmus paper, regardless of its chemical properties or composition. Thus, it will turn out that many things that are certainly not acidic (like the milk in my refrigerator) will make the conditional 'if the sample is applied to litmus paper, the litmus paper will turn red' come out true, simply in virtue of never being applied to litmus paper.

Logical positivists sought to get around this problem by weakening the relationship between statements using V_T and those using V_O but struggled to avoid weakening the requirement so far as to allow 'metaphysical' statements (which they intend to eliminate as meaningless) to satisfy the criteria of cognitive meaningfulness thus developed.

Here is an example of such an attempt: Given that asserting the logical equivalence of a statement including vocabulary from V_T with a statement using terms only from V_O and V_L proved problematic, one might use a definition sentence with a form different from one asserting logical equivalence. Compare the following two sentences:

1 'This sample is acidic if and only if, if it is applied to litmus paper, then the litmus paper turns red.'
2 'If this sample is applied to litmus paper, then it is acidic if and only if the litmus paper turns red.'

Sentence (1) is a reduction sentence asserting an explicit definition of 'acidic' that falls prey to the objection under consideration. Sentence (2), however, makes no positive or negative commitment regarding the acidity of samples that are never applied to litmus paper, and thus gives no explicit definition of acidity in general; yet, it does establish a logical connection between the theoretical term 'acidic' and the terms 'litmus paper,' 'applied to,' and 'red' (which we are here supposing to belong to the observational vocabulary). Such more 'liberal' developments in the articulation of the logical empiricist approach to scientific theorizing were prominent through the middle decades of the twentieth century (Carnap, 1936, 1937, 1955; Hempel & Oppenheim, 1948; see the introduction to Suppe, 1977 for an overview).

One more important modification is worth noting. Although statements about particular samples and slips of litmus paper might be verifiable, scientific discourse also relies on statements of generalizations, and these clearly cannot be verified, if by that we mean definitively proven to be true. (Carnap, in fact, went further and affirmed that no synthetic statement at all could be "verified completely and definitively" (Carnap, 1936, 427).) The verification principle requires modification in terms of some weaker notion, leading the logical empiricists to focus on *confirmation*, the subject of a considerable body of technical work. We will not pursue the details of these efforts, but some general features of the project deserve attention insofar as confirmation theory constitutes a significant undertaking of some philosophers of science, within which logical empiricist ideas remain influential even for those who do not subscribe to many of the characteristic doctrines of the movement.

4.3 A Theory of the Confirmation of Theories

One of the ways in which logical empiricism appealed to scientifically minded philosophers was in its power to open research areas that retained philosophical significance while providing opportunities to apply formal tools of logic and mathematics to the problems that arose in those areas. The development of confirmation theory provides an illustration of the power of this aspect of the logical empiricist project as well as its limitations.

Confirmation theory as a branch of the logical empiricist project is associated most strongly with the efforts of Rudolf Carnap, whose *Logical Foundations of Probability* (Carnap, 1950, 1962) opened the way for further work. Carnap presents his project in that book as an *explication* of concepts relating to confirmation. He did not start with an assumption that there is some truth of the matter about how theories are confirmed or are supposed to be confirmed by evidence, which he as the philosopher must figure out. Rather, he considered confirmation to be a concept already in use but lacking an exact formulation. To explicate is to transform "an inexact, prescientific concept, the *explicandum*, into a new exact concept, the *explicatum*" and provide for the latter "explicit rules for its use" (Carnap, 1962, 3). The adequacy of the *explicatum* is judged by its similarity to the explicandum, exactness, fruitfulness, and simplicity (Carnap, 1962, 5–8).

Carnap distinguishes three concepts of confirmation: (1) the *classificatory* concept, expressed in a sentence like 'That the mean Northern Hemisphere temperature of the past 30 years exceeds that of any thirty-year period of the past 800 years is confirmed by the evidence from proxy data such as tree ring patterns and chemical concentrations in ice core samples'; (2) the *comparative* concept expressed in a statement like 'The effectiveness of vaccines against

COVID-19 infection is better confirmed by randomized controlled treatment studies than by anecdotal evidence'; (3) the *quantitative* concept, which Carnap also calls the 'degree of confirmation.' Carnap considers it an open question whether this third, quantitative, concept "occurs in the customary talk of scientists" (Carnap, 1962, 23). Specifying an explicatum for this quantitative concept becomes central to the project of developing a theory of confirmation.

Carnap identifies the quantitative concept degree of confirmation with the first of two concepts of probability that he distinguishes, using the notational device of a subscript. *Probability*$_1$ corresponds to degree of confirmation, while *probability*$_2$ corresponds to relative frequency (to be discussed in detail in Chapter 9). A degree of confirmation then consists of a relation between sentences that constitutes the subject matter of *inductive logic* in the same manner that logical consequence is a relation between sentences that is the concern of deductive logic. Both relations are, in Carnap's view, "objective and logical" in the sense that "if a certain probability$_1$ value holds for a certain hypothesis with respect to a certain evidence, then this value is entirely independent of what any person may happen to think about these sentences, just as the relation of logical consequence is independent in this respect" (Carnap, 1962, 43). The project of developing a theory of confirmation thus belongs to the inductive branch of logic.

We will forego here any attempt to summarize the theoretical apparatus that Carnap proposes, which requires a rather elaborate formal scheme. Instead, let us consider some prominent, philosophically significant features of Carnap's approach.

Confirmation theory is to be applied to an 'object language' that is formal, rather than a natural language like Italian or Yoruba. Among other things, this means that the language used to express sentences to which we can apply inductive logic lacks features such as connotation, ambiguity, metaphor, or vagueness. Each name singles out a unique individual, and each predicate (which will specify either a property applicable to individuals or a relation among two or more individuals) can be strictly affirmed or denied of any individual (or group of individuals in the case of a relation). Predicates in the object language either apply or they do not; they do not apply 'to some degree.' A *state description* is a sentence in the object language that affirms or denies – of each individual or group of individuals that can be named in the language – every predicate in the language. A state description says everything there is to say, in that language, about everything in the world! Even in fairly sparse languages with only a few named individuals and a few predicates, there will be many possible state descriptions, although perhaps only one is true.

Most sentences, of course, are not as exhaustive as a state description. A sentence about individuals will typically mention at most a few, and only a sparse number of facts about them. A generalization will typically be about no more than a small number of properties or relations. But for any given sentence, one may in principle specify the class of possible state descriptions that are compatible with that sentence. Consider a sentence *s* in some language *L* that simply asserts, 'Individual *a* is round.' Every state description in *L* will specify whether or not *a* is round. Any state description that denies *a* is round is incompatible with *s*, while any state description that affirms *a* is round is logically compatible with *s*. The class of all state descriptions in the latter category constitutes the *range* of *s*.

Carnap then relates the inductive logical relation of degree of confirmation to the deductive logical relation of entailment as follows: Suppose we have two sentences *e* and *h*. To say that *e* deductively entails *h* (in Carnap's terms, '*e* L-implies *h*') is to assert that the range of *e* is entirely contained within the range of *h*. In other words, every possible state description in *L* that affirms *e* also affirms *h*. An analogous statement of inductive logic would assert the degree to which *e* confirms *h*. For example, the statement that *e* confirms *h* to the degree of three-fourths (in Carnap's notation, $c(h, e) = \frac{3}{4}$) means that three-fourths of the range of *e* is contained in the range of *h*, a situation represented in Figure 4.1.

To assign a number to this ratio of ranges, however, requires assigning numbers (more precisely a *measure function*) to the ranges themselves. This introduces the problem of deciding what that measure function should be. Thus we face a problem for inductive logic that does not arise for deductive logic. The range of one sentence either is or is not contained within the range of another, and that is all that is needed to determine whether the relation of deductive entailment holds between two sentences. No measure is required. Choosing a measure function as the basis of a confirmation measure is among the problems of confirmation theory that are simultaneously 'technical' (producing a usable mathematical device) and philosophical, insofar as the kinds of reasoning involved in addressing the problem go beyond strictly

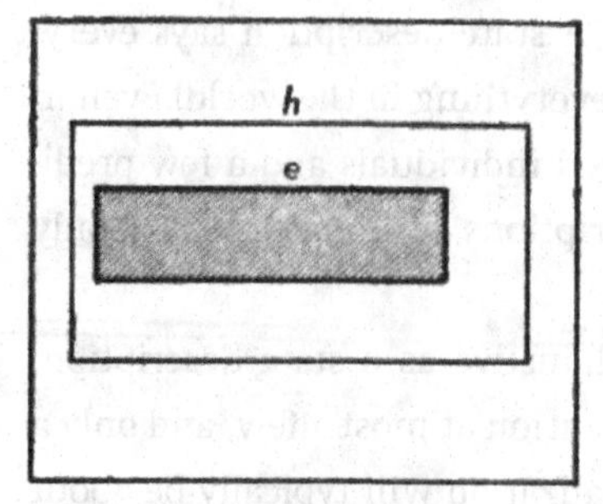
h
e

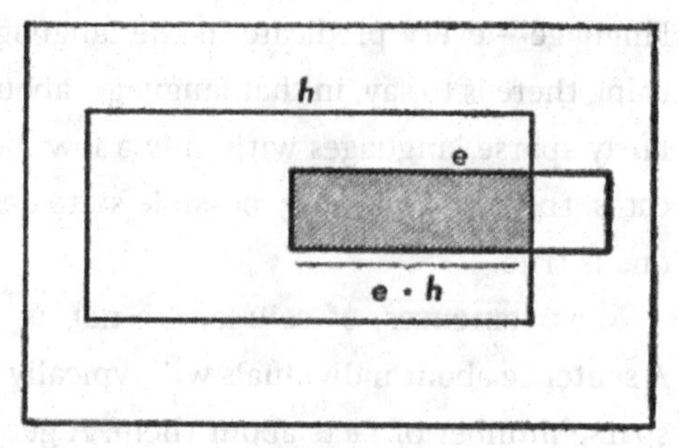
h
e
e · h

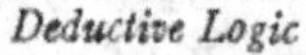
Deductive Logic
'e L-implies h' means that the range of e is entirely contained in that of h.
Inductive Logic
'c(h,e) = 3/4' means that three-fourths of the range of e is contained in that of h.

Figure 4.1 A visual representation of the relationship between '*e* entails *h*' and '*e* confirms *h* to the degree ¾.' Used with permission of University of Chicago Press - Books, from Carnap (1962, 297); permission conveyed through Copyright Clearance Center, Inc.

mathematical questions. Carnap advocates, somewhat tentatively, a specific choice of measure, leading to a specific confirmation function (Carnap, 1962, 562ff). The details of his chosen function need not detain us, but Carnap's approach to the problem illuminates the limitations of purely formal considerations in solving philosophical problems of scientific inquiry.

Carnap refers to the choice of a confirmation function as a choice of inductive method and notes that this "is fundamentally not a theoretical question" but rather a "practical decision" not to be "judged as true or false but only as more or less adequate, that is, suitable for given purposes" (Carnap, 1952, 53). Theoretical results regarding the various confirmation functions among which one must choose are relevant to this choice, but neither deductive logic nor inductive logic contains sufficient resources to determine the choice, which might draw upon considerations of "performance, economy, aesthetic satisfaction, and others ... Here, as anywhere else, life is a process of never ending adjustment; there are no absolutes, neither absolutely certain knowledge about the world nor absolutely perfect methods of working in the world" (Carnap, 1952, 55).

In an earlier paper dealing with the question of whether abstract entities exist (Carnap, 1950), Carnap had distinguished between *internal* and *external* questions, where the former are addressed within a framework of language and rules governing its usage and the latter concern the framework itself.

Arthur Burks (Burks, 1963) noted that this distinction could be applied in a parallel way to questions within and about inductive logic. Accepting Burks's suggestion as "illuminating," Carnap notes that "important questions of the philosophy of induction belong to the external questions of induction" and that he regards external questions, such as "whether or not to accept a certain *c*-function" as the basis for probability assignments, as "practical questions." Carnap, nonetheless, emphasizes that philosophical work such as explication importantly helps to "specify the factors which are relevant for a rational decision of a practical external question." Such practical decisions, when made rationally, must in turn involve deliberations over theoretical questions like those Carnap pursues in developing his confirmation theory (Carnap, 1963, 982).

This framing of the choice as a matter of deciding on a method rather than a matter of accepting a proposition as something to hold or believe reflects an element of pragmatism in Carnap's thought that became more prominent among some logical empiricists as their views developed. Pragmatism, as a way of approaching philosophical issues, began in the US with the contributions of Charles Sanders Peirce, William James, and John Dewey, among others. Although pragmatism resists easy definition, it constitutes an approach to philosophical issues that emphasizes problems that arise for people in their conduct of life and not only in the seminar room, and that invites clarification by focusing on the practical consequences of adopting one point of view rather than another, where "practical" is to be construed quite broadly. (See Richardson, 2002 for an illuminating comparison of logical empiricism and pragmatism that treats them both as (in part) attempts at a 'scientific philosophy' but differing on some important questions regarding the scope of their revolutionary undertakings.)

4.4 Conclusion

The logical empiricists, while far from being the first philosophers to engage problems related to scientific knowledge and theorizing, were historically significant in their efforts to articulate a far-reaching approach to science. They drew upon the formal apparatus of logic to serve a broader philosophical agenda. Much of the discourse as well as professional infrastructure of current philosophy of science owes some kind of debt to the Vienna Circle and those it engaged as logical empiricism came to dominate the philosophy of science in the English-speaking world in the middle of the twentieth century.

Thus far I have said rather little about the significant problems that faced efforts of logical empiricists in their pursuit of formal frameworks for confirmation theory and other elements of their 'scientific philosophy' approach. In particular, here I will not undertake to detail the various 'paradoxes of confirmation' that have been widely construed as revealing limitations on how far one can rely on formal considerations in characterizing the extent to which given evidence lends support to a particular hypothesis. (See, e.g., Goodman, 1955; Hempel, 1945; Rinard, 2014; Stalker, 1994). As we have seen, even Carnap himself acknowledged such limitations to a degree. The project also faced W. V. O. Quine's challenge, discussed in Chapter 3: because of underdetermination, there can, in principle, be no unique way of connecting statements about our sensory experience of the world with statements that involve 'theoretical' language, however we might construe that distinction.

As we will see in Chapter 5, further challenges to logical empiricism took aim at the possibility of achieving its broader aim: to articulate a 'scientific philosophy' in which inquirers could resolve disagreements over matters of substance, at least to a significant extent, by appeal to the neutral ground of empirical observation.

5 Postpositivist Views on Scientific Progress and Rationality

5.1 Introduction

The logical empiricists' pursuit of 'scientific philosophy' aimed at an approach to scientific inquiry that would produce solutions to various philosophical problems relating to scientific inquiry: What is a scientific theory? What makes a statement cognitively meaningful? What language is appropriate for the articulation of the empirical content of scientific knowledge? How can the confirmation of theories by evidence be evaluated and quantified? Solutions to these problems couched in terms of formal logic promised to provide a systematic approach to understanding how scientific inquiry could yield results that one could *rationally* embrace (i.e., results supported by good reasons) and that over time would constitute *progress* (i.e., would lead to continual improvements in our knowledge about the world). In this chapter we examine responses to logical empiricism that cast doubt upon the prospects for achieving these aims. Some of these responses pose problems for the approach taken by logical empiricists to achieve these aims. Some suggest that the aims are not achievable by any means. The term 'postpositivist' has been used as a loose designation referring to these responses found in work from the late 1950s into the 1980s, especially work that relies on historical analyses of developments in science. Particular logical empiricist claims (or claims attributed to logical empiricists by their critics) targeted by the postpositivists include the following: that the acceptability of competing scientific theories can be judged within a single logical framework; that one can divide the vocabularies used by scientists into distinct observational and theoretical components; that one can judge how well the evidence supports a hypothesis by applying a mathematical measure to a logical relationship between the evidence and the hypothesis; that the pursuit and achievement of scientific progress can be depicted or captured in terms of rules; that scientific inquiry can result in a unified body of knowledge; and that a scientific theory can be understood as a set of sentences with a certain logical structure.

5.2 Scientific Revolutions as Changes in Paradigm

Between 1938 and 1970, the University of Chicago Press published the first two volumes of a project initiated by the logical empiricists: the *International Encyclopedia of Unified Science.* Vienna Circle member Otto Neurath – a social scientist, philosopher, and socialist reformer – articulated the way in which this project connected philosophical and humanitarian aims in the first volume of the *Encyclopedia*: "To further all kinds of scientific synthesis is one of the most important purposes of the unity of science movement, which is bringing together scientists in different fields and in different countries, as well as persons who have some interest in science or hope that science will help to ameliorate personal and social life" (Neurath, [1938] 1955, 1). *The Structure of Scientific Revolutions* by Thomas Kuhn appeared as number two in the second volume of the *Encyclopedia* (Kuhn, 1962, hereafter *SSR*).[1] This book came to be seen as contributing to the demise of the logical empiricist movement from which the *Encyclopedia of Unified Science* was born (but see Reisch, 1991 for a corrective to this received view).

Kuhn sought to understand scientific revolutions from the perspective of a *historical* study rather than through logical analysis. In this way, Kuhn's project fundamentally differs from that of the logical empiricists. Some of Kuhn's apparent conflicts with the logical empiricists thus arise from the different questions they pursue. Nonetheless, we will note some genuine points of disagreement. Kuhn's historical studies convinced him that scientific knowledge develops through two quite distinct patterns or modes. In the *normal science* mode, change accumulates slowly and research is "firmly based upon one or more past scientific achievements, achievements that some particular scientific community acknowledges for a time as supplying the foundation for its further practice" (Kuhn, 1996, 10). Normal science, as Kuhn describes it, proceeds amidst a general agreement within

[1] *SSR* went through three editions. The second edition (1970) added a lengthy postscript, but left the text otherwise nearly unaltered. The third edition (1996) added an index. All page citations here are to the third edition.

a scientific community about what problems most need to be solved and what standards are to be used in the evaluation of efforts to solve those problems. Kuhn emphasized that such consensus is best understood as a byproduct of the acceptance by the community of certain 'past scientific achievements' as *exemplars* of good scientific work.

Kuhn called such exemplars *paradigms*, borrowing a term that describes examples of grammatical forms in the learning of languages. But he used the word in other ways as well. The term *paradigm* took on a special importance in Kuhn's work, and he used it throughout *SSR*, denoting sometimes a kind of theoretical framework, sometimes something more pervasive like a "worldview," sometimes a scientific community united by its commitment to such a worldview, sometimes a line of research aimed at a particular set of problems, and so on. One critic, citing Kuhn's "quasipoetic style," counted "not less than twenty-one different senses" in which Kuhn used the term in *SSR* (Masterman, 1970, 61–65). In a postscript to the second edition of *SSR*, Kuhn sought to bring order to this thicket of paradigm locutions by distinguishing two primary senses of the term. First, there is the paradigm as the "entire constellation of beliefs, values, techniques," as well as generalizations and models that are shared by a scientific community, for which he now introduced the term "disciplinary matrix" (Kuhn, 1996, 175). The second paradigm notion is the exemplar: "the concrete problem-solutions that students encounter from the start of their scientific education" and that "show them by example how their job is to be done" (Kuhn, 1996, 175).

The stability of a paradigm in both senses is needed in order to sustain normal science. Kuhn emphasizes that we can expect to find paradigms wherever a group of researchers regularly produces solutions to scientific *puzzles* and reaches broad consensus about whether particular efforts succeed in solving those puzzles. Solving such puzzles constitutes progress within the paradigm. Paradigms vary greatly with regard to their size, longevity, and degree of specialization. Neo-Darwinian evolutionary biology is a sprawling discipline with many subfields, while molecular phylogenetics constitutes a more narrowly focused (though still quite ambitious) endeavor.

No paradigm could be expected to be sufficiently rich and stable to serve as the basis of solutions to all of its problems. Postpositivist authors emphasized that any scientific theory, in the words of Imre Lakatos, "at any stage of its development, has unsolved problems and undigested anomalies. All theories, in this sense are born refuted and die refuted" (Lakatos, 1978, 5). As we will explore later in this chapter, even an enterprise as successful and long-lasting as Newtonian celestial mechanics encounters problems that it cannot solve. As a problem resists solution and vexes researchers, it becomes an *anomaly*. A persistent anomaly stands as an obstacle to paradigmatic scientific progress and, when sufficiently significant and resistant to solution, gives rise to a second mode of scientific change: *revolutionary science.*

One of Kuhn's own examples of revolutionary science comes from the history of quantum physics. *The Old Quantum Theory* was inaugurated through Niels Bohr's 1913 'planetary' model of the hydrogen atom. Bohr's model featured a nucleus around which the electron travels in well-defined orbits. The model allowed only certain orbits, corresponding to specific discrete energy levels for the atom as a whole. If the electron underwent a transition from a higher-energy state to a lower-energy state, it would emit electromagnetic radiation carrying energy equal to the difference between the two, but it did so *without* passing through the continuum of energies between its initial and final states (a 'quantum leap,' you might say). In this way, Bohr sought to explain the fact that hydrogen gas emits light in a discrete spectrum.

Bohr and others applied the Old Quantum Theory successfully to numerous problems, particularly those concerning spectra emitted by certain elements. The Old Quantum Theory served well for calculating spectra for hydrogen atoms with one proton and one electron (and the same treatment was successfully extended to 'hydrogenic' atoms, such as sodium, with an inner core of electrons closely bound to the nucleus, which together with the nucleus could be treated as one body, with a single additional orbiting electron). But the theory did not work for three-body systems. It also made incorrect predictions regarding the splitting of spectral lines in a magnetic field, known as the *Zeeman effect.* Such problems led to a situation that Kuhn characterizes as a *crisis* for the Old Quantum Theory.

A Kuhnian *crisis* occurs when members of a research community find that some anomaly (or set of anomalies) has proven so resistant to solution within the current paradigm that they begin to doubt that resolving the anomaly is possible. Members of the community increasingly feel called upon to do the impossible. Emphasizing the psychological dimension of such crises, Kuhn quotes Wolfgang Pauli, writing to the young physicist Ralph Kronig during the crisis of the Old Quantum Theory: "At the moment physics is again terribly confused. In any case, it's too difficult for me, and I wish I had been a movie comedian[2] or something of the sort and had never heard of physics" (Kuhn, 1996, 84, quoting from Kronig, 1960).

Revolutions, according to Kuhn, begin when doubts about the ability of the paradigm to resolve its own anomalies drive members of the community to consider routes to a solution outside of the paradigm's constraints. In the case of Quantum Theory, Werner Heisenberg made the first breakthrough in a 1925 paper that laid the groundwork for a radically new way of doing quantum mechanics, making use of a new mathematical apparatus and eliminating Bohr's planetary orbits. Heisenberg's theory predicted that quantum particles like electrons would have well-defined positions or momenta when those properties were measured but did not describe determinate trajectories for quantum particles at other times. Heisenberg's paper (Heisenberg, 1925), together with subsequent joint efforts involving Max Born and Pascual Jordan (Born, Heisenberg, & Jordan, 1926; Born & Jordan, 1925), laid the foundations for the New Quantum Theory.

Motivated by a roughly positivist idea (but see Bokulich, 2008), Heisenberg saw Bohr's 'unobservable' atomic orbits as a theoretical defect. Albert Einstein objected that the theory by which Heisenberg sought to remove this defect gave an incomplete account of mechanics at the atomic scale. On Kuhn's account, such controversy is typical of revolutions. The community does not convert all at once. Some will resist, others will grasp immediately for the relief from anomaly offered by a new proposal. Kuhn quotes Pauli, writing again to Kronig, shortly after the appearance of Heisenberg's 1925 paper: "Heisenberg's type of mechanics has again given me hope and joy in life. To be sure it does not supply the solution to the riddle, but I believe it is again possible to march forward" (Kuhn, 1996, 84, quoting from Kronig, 1960).

5.3 Observation as "Theory-Laden" and Incommensurability

Kuhn's historical analysis leads him to a picture of scientific change in which the evidential support for holding a theory is tied to the paradigm in which that theory developed. After a revolution, evidential significance alters. Such paradigm dependence marks one of the clearest points of disagreement

[2] Remarks that he resembled the film actor Edward G. Robinson delighted Pauli (Enz, 2002).

between Kuhn and his logical empiricist predecessors. The logical empiricist approach requires drawing a distinction within the language of science between its observational vocabulary and its theoretical vocabulary, so that one can assess the cognitive meaningfulness of scientific language. Kuhn's paradigm dependence of evidence calls into question the very possibility of drawing this distinction. Moreover, Kuhn argues that the very language used to describe evidence from observations draws its meaning from the theoretical context in which that evidence was assessed. Observations are "theory-laden."

Norwood Russell Hanson introduced the term 'theory-laden' in 1958 while defending a similar view. Hanson asks whether Tycho Brahe, whose theory of the cosmos had the Sun orbiting the Earth,[3] and Johannes Kepler, whose theory placed the Earth in orbit around the Sun, "see the same thing in the east at dawn" (Hanson, 1958, 5). The question serves as "the beginning of an examination of the concepts of seeing and observation" (Hanson, 1958, 5).

Both affirmative and negative answers can be defended depending on how one resolves an ambiguity in the question. Is there a single physical object that both Kepler and Tycho see? Hanson affirms that there is: the Sun. It is precisely because this is true that further pursuit of the question is philosophically interesting, for "seeing is an experience," and one can then ask whether, or in what sense, the experiences of Kepler and Tycho are the same. Hanson defends the claim that Kepler and Tycho, in a sense, do not see the same thing: Kepler's and Tycho's seeing of the Sun are different. Seeing is not only a matter of the pattern of activation of the retina – "there is more to seeing than meets the eyeball" (Hanson, 1958, 7). Neither can the experience of seeing be reduced to the physical object that is seen. "Seeing is not only the having of a visual experience, it is also the way in which the visual experience is had" (Hanson, 1958, 15). Hanson also rejects the idea that Kepler and Tycho observe the same thing but simply interpret or use their observations differently. Kepler sees the Sun coming into view as the Earth rotates on its axis. Tycho sees the Sun orbiting into view from a fixed Earth. Tycho and Kepler do not first do the same thing (see the Sun) and then two different things (interpret their seeing of the Sun). Each act of seeing is a unified whole

[3] Tycho Brahe's model places the Sun in orbit around the Earth, but all of the other planets orbit around the Sun. In this way, his theory combined elements of the geocentric (earth-centered) model of Ptolemy and the heliocentric (sun-centered) model of Copernicus.

and each is different. To say that different scientists "all make the same observations but use them differently ... does not explain controversy in research science. Were there no sense in which they were different observations they could not be used differently" (Hanson, 1958, 19). Hanson concludes, "There is a sense, then, in which seeing is a 'theory-laden' undertaking. Observation of *x* is shaped by prior knowledge of *x*" (Hanson, 1958, 19).

For Kuhn, Hanson's theory-ladenness of observation becomes one aspect of a central commitment of his philosophy of science: the *incommensurability* of rival paradigms. Literally, incommensurability means 'lacking a common measure,' a phrase indicative of the problem faced by scientists in a period of Kuhnian revolution: One needs to choose a paradigm but lacks a common measure for determining which of the competing candidates is *better*. Incommensurability may be understood, however, in both a *methodological* and a *semantic* sense (Bird, 2000).

Suppose that in a period of crisis for paradigm P_1, a rival paradigm P_2 takes shape, such that members of the community must make a choice to continue working within P_1 or switch their allegiance to P_2. One might suppose that this would be a matter of evaluating the extent to which the theoretical content of each of the paradigms is supported by the available data. According to the thesis of *methodological incommensurability*, however, the methods and standards for deciding the level of support are not neutral regarding the paradigms in question: "[T]he choice is not and cannot be determined merely by the evaluative procedures characteristic of normal science, for these depend in part upon a particular paradigm, and that paradigm is at issue" (Kuhn, 1996, 94).

Semantic incommensurability adds to the difficulty. According to Kuhn, terms may be used in competing paradigms so differently that they have different meanings. Kuhn's example in *SSR* involves the term 'mass' as used in Newton's dynamics and Einstein's dynamics. For Newton, mass is an invariant quantity, that is, the mass of a given body has the same value regardless of the state of motion of that body relative to a given observer. In Einstein's theory, however, two observers who are in motion relative to one another will disagree in their measurements of the mass of a body. (The difference will be negligible if the relative velocities are small compared to the speed of light; unless you are an experimental physicist, it is practically guaranteed that you will never have to worry about this effect!) Kuhn argued that such a scientifically important difference entails that 'mass' does not mean the same thing in Newton's physics as it does in Einstein's.[4]

Semantic incommensurability points to perhaps the clearest disagreement between Kuhn's account and logical empiricism. A logical empiricist's interpretation of this state of affairs would insist upon the distinction between the theoretical and observational vocabulary of a theory. Since Newton's dynamical theory and Einstein's dynamical theory say such different things about mass, this must be regarded as a theoretical term, not an observational one. Each theory would then use distinct bridge principles to connect statements involving mass to statements in the observational vocabulary, and the choice between theories would be based on seeing which theory entailed observation-vocabulary statements in better agreement with the results of actual observations.

This reply assumes that we can draw a principled and well-defined distinction between observational and theoretical vocabularies – in short, that observation vocabulary is *theory-neutral*. Kuhn, however, rejects any such distinction. He not only rejects the idea of a theory-free observational vocabulary; he suggests that in some sense *there is not even a paradigm-independent world that we observe*: "[T]hough the world does not change with a change of paradigm, the scientist afterward works in a different world" (Kuhn, 1996, 121).

Kuhn, like Hanson, draws this conclusion from the ways in which scientists are trained to make observations of a certain kind and incorporate apparently theoretical concepts and languages into their observations:

> Looking at a contour map, the student sees lines on paper, the cartographer a picture of a terrain. Looking at a bubble-chamber photograph the student sees confused and broken lines, the physicist a record of familiar subnuclear events. Only after a number of such transformations of vision does the student become an inhabitant of the scientist's world, seeing what the scientist sees and responding as the scientist does. (Kuhn, 1996, 111)

[4] Kuhn's point about mass in relativistic dynamics has been the subject of a controversy among physicists involving a mixture of theoretical, experimental, historical, philosophical, and pedagogical concerns. The controversy broke into the open in 1989 when Lev Okun argued that the concept of relativistic mass was fundamentally mistaken (Okun, 1989). Jammer (2000, 51–61) provides a useful survey of the debate that also addresses the way that Kuhn's thesis of incommensurability itself has been entangled in this dispute.

But this training takes place *within* a paradigm (a disciplinary matrix), so that when a revolution occurs "the scientist's perception of his environment must be reeducated – in some familiar situations he must learn to see a new gestalt" (Kuhn, 1996, 112). How precisely to understand such passages is not perfectly clear, but they convey at least this much: One cannot hope to resolve disputes over rival paradigms by adopting a description of the relevant data that is free of assumptions belonging to the very paradigms at issue.

5.4 Incommensurability: A Problem for Progress and Rationality?

As a consequence of methodological incommensurability, a scientific revolution changes the exemplars that serve as the basis for the articulation of standards. Correspondingly, a revolutionary change involves the rejection of the old exemplars *and the accompanying standards of adequacy* by which problem solutions are to be evaluated. A new paradigm comes with its own exemplary solutions. Inevitably these will fall short of the old standards, but will succeed by the new standards (for which they are the basis).

The result is one of Kuhn's most controversial theses: During periods of revolution, choosing between paradigms cannot be guided by *accepted* standards at all. The choice between the old paradigm and a potential new one "is not and cannot be determined merely by the evaluative procedures characteristic of normal science, for these depend in part upon a particular paradigm, and that paradigm is at issue. When paradigms enter, as they must, into a debate about paradigm choice, their role is necessarily circular. Each group uses its own paradigm to argue in that paradigm's defense" (Kuhn, 1996, 94).

Critics allege that Kuhn's depiction of scientific revolutions in SSR amounts to the abandonment of experimental reasoning and logical inference. If "[e]ach group uses its own paradigm" to defend its commitment to that paradigm, then those arguments can only appeal to those who already accept the paradigm in question. Acceptance comes first, arguments come later. Arguments over paradigm choice then seem more like armed conflict, or perhaps political campaigns, than rational deliberation on the basis of empirical evidence, since the arguments cannot do any work until one has somehow gotten one's opponent to adopt one's own point of view. And then,

Welche Thiere gleichen einander am meisten?

Kaninchen und Ente.

Figure 5.1 A duck (*Ente*) that is also a rabbit (*Kaninchen*). The German caption “Welche Thiere gleichen einander am meisten?” means “Which animals resemble one another the most?”

what work is left for the argument to do? As depicted in an influential critical review of *SSR*, Kuhn’s account entails that “the decision of a scientific group to adopt a new paradigm is not based on good reasons; on the contrary, what counts as a good reason is determined by the decision” (Shapere, 1964, 392). Kuhn’s own language and choice of analogies for explaining his view encouraged the perception that his account of scientific revolutions is radically subjective. He compares a scientific revolution to a political revolution: The latter occurs when crisis becomes so acute that “political recourse fails”; likewise the former occurs in a crisis by methods going beyond the “evaluative procedure characteristic of normal science” (Kuhn, 1996, 93–94). He compares changes in paradigm to visual “gestalt shifts”: “What were ducks in the scientist’s world before the revolution are rabbits afterwards” (see Figure 5.1).[5] He uses language to describe changes in paradigm that highlight similarities with religious conversion: “The transfer of allegiance from paradigm to paradigm is a conversion experience that cannot be

[5] As Kuhn himself noted, N. R. Hanson had gone much further in the discussion of gestalt shift images as elucidating some similar views about how theoretical commitments affect one’s empirical evidence (Hanson, 1958).

forced" (Kuhn, 1996, 151), and "[a] decision of that kind can only be made on faith" (Kuhn, 1996, 158).

This suggestion of subjectivity creates a challenge for Kuhn's account of scientific progress. Suppose we start with the rough intuition that scientific progress is constituted by some process of *improvement*. Improvement can happen in various ways: Less accurate data is replaced by more accurate data; unreliable predictions are replaced by more reliable predictions; confusion is replaced by understanding; worse theories are replaced by better theories. Such evaluative claims presume standards by which we draw the distinction between worse and better. If all standards are specific to paradigms, then we can invoke no independent standards to judge successive changes in paradigm as progressing in the manner suggested by our rough intuition.

Kuhn's later admission of some paradigm-transcending scientific values, such as simplicity, accuracy, fruitfulness, consistency, and scope (Kuhn, 1977), does not significantly alter these problems of progress and rationality. Although he acknowledges that these values are embraced by scientists regardless of the specific paradigms to which they are committed, he at the same time insists that applying these values requires invoking paradigm-specific commitments to guide their interpretation and prioritization. (In this way, a scientist's choice of theory or paradigm is underdetermined by these values insofar as they are shared and paradigm-independent.)

Kuhn's treatment of scientific progress, nonetheless, deserves attention for its suggested reconceptualization of scientific inquiry itself. In the final chapter of *SSR*, Kuhn invokes an analogy with another revolutionary idea in science: Darwinian evolution. What most shocked Darwin's contemporaries, according to Kuhn, was "neither the notion of species change nor the possible descent of man from apes," since evolutionary ideas had been in circulation for some time. What really distinguished Darwin's theory of evolution from the ideas developed by others was that "[t]he *Origin of Species* recognized no goal set either by God or nature." Nonetheless, it described a process of "gradual but steady emergence of more elaborate, further articulated, and vastly more specialized organisms" – a process that Darwin himself understood as one of advancement or progress (Kuhn, 1996, 171–172).

Likewise, Kuhn suggests, we should regard the development of science as progressive: "The net result of a sequence of such revolutionary selections, separated by periods of normal research, is the wonderfully adapted set of instruments we call modern scientific knowledge," and the process is "marked by an increase in articulation and specialization." Yet the process need not be understood as directed toward "a set goal, a permanent fixed scientific truth" (Kuhn, 1996, 172–173). Science does progress, according to Kuhn, but the incommensurability of earlier and later paradigms poses an obstacle to the idea that there is a "coherent direction of ontological development" in a revolutionary transition. On the contrary, "the notion of a match between the ontology of a theory and its 'real' counterpart in nature now seems to me illusive in principle." Kuhn here articulates a version of *antirealism* about science (Kuhn, 1996, 206; McMullin, 1993).

We will take up debates between realists and antirealists in Chapter 10. For now, note the tension between Kuhn's rejection of science as aimed at truth and his embrace of the idea that science progresses. If science does not progress toward truth and is marked by transitions between incommensurable paradigms, how can we meaningfully consider it progressive? One possibility is that Kuhn's notion of progress is precisely that of increasing "articulation and specialization." K. Brad Wray argues that the increasing specialization of science should be understood not merely as a sociological phenomenon but as serving important epistemic functions, enabling scientists to enhance the accuracy of their work by focusing on ever-narrower domains (Wray, 2011).

5.5 Imre Lakatos: The Methodology of Scientific Research Programs

Postpositivist treatments of progress and rationality give prominence to an historical analysis of scientific change but vary significantly in how they analyze history and the aims they pursue. In the work of Imre Lakatos, we find an integration of historiographical method and philosophical aims that significantly differs from the approach of Thomas Kuhn. Lakatos sought to demonstrate the rationality of scientific change and the close connection between that rationality and a certain conception of progress.

Lakatos's life story deserves a longer telling than I can give here.[6] Born into a Jewish family in Hungary in 1922 (his family name was Lipsitz), he

[6] See Larvor (1998) and Kadvany (2001) for more on Lakatos's life and the development of his philosophy.

adopted a false identity when Hungary entered the Second World War on the Axis side, first to avoid forced labor by the Hungarian government, then to escape deportation to an extermination camp when the Nazis occupied Hungary in 1944. (His mother and grandmother were killed in Auschwitz.) He adopted the name Lakatos after the Russians invaded Hungary. After studying mathematics, physics, and philosophy, he served in the postwar Hungarian Ministry of Education but was then imprisoned for nearly four years on charges of 'revisionism.' When anti-Soviet resistance swelled up in 1956, Lakatos helped write a statement proclaiming the freedom of scientific inquiry on behalf of the National Committee of the Hungarian Academy of Sciences. Shortly after Soviet tanks rolled into Budapest to put down the uprising, Lakatos emigrated, first to Vienna, then to Cambridge, where he wrote a PhD thesis in the philosophy of mathematics. He joined Karl Popper in the Department of Philosophy, Logic, and Scientific Method that Popper had founded at the London School of Economics. He was appointed Professor of Logic in 1969, but died in 1974 at the age of fifty-one.

Lakatos codified his view of scientific change in a long essay titled "Falsification and the Methodology of Scientific Research Programmes," first published in a collection of essays arising from a symposium held in 1965 (Lakatos & Musgrave, 1970) that features critical responses to Kuhn's work, including that of Karl Popper, as well as an essay by Kuhn that illuminatingly contrasts his philosophy of science with Popper's falsificationism. Accordingly, Lakatos's essay features contrasts with both Kuhn and Popper. Importantly, Lakatos seems to have conceived his approach from the outset as an improvement on Popper's falsificationism.

Lakatos thought falsificationism had failed to deal adequately with the problem of the underdetermination that we encountered in Chapter 3. He sought to remedy this by specifying criteria by which to judge whether a decision to retain an apparently falsified theory is rational.

Lakatos begins his critique of Popper's falsificationism by distinguishing it from what he calls 'dogmatic falsificationism,' in contrast to the 'methodological falsificationism' he attributes to Popper. For the dogmatic falsificationist, a statement describing empirical results simply reports the results of *observation*, which is to be regarded as a distinct 'psychological' process from *theorizing*. An observational statement, according to this view, is *proved* (or at least justified) by the observation of facts.

The methodological falsificationist (such as Popper himself) rejects this untenable claim. As discussed in Chapter 2, Popper denied that experiences such as the observation of a fact could justify statements because the only thing that could do that would be another *statement* that logically entailed the first. Neither psychological states ('in the mind') nor states of affairs ('out in the world') can play this role. Contrary to the views of the logical empiricists (and consistently with the thesis of theory-laden observation endorsed by Kuhn and Hanson), statements used to describe experimental or observational results – and potentially falsify theories – are articulated in language already bearing theoretical commitments. Their acceptance as part of the empirical basis results from the *decision* of the investigator.

But Lakatos emphasizes an important difference between *naïve* and *sophisticated* methodological falsificationism. For naïve methodological falsificationists (NMFs), a theory is "'acceptable' or 'scientific'" if it is falsifiable by experiment. The sophisticated methodological falsificationist (SMF) considers a theory scientific (acceptable) only if it "leads to the discovery of novel facts" (Lakatos, 1970, 116). The NMF considers a theory falsified when they decide that a statement accepted as part of the empirical basis conflicts with that theory. For SMFs, a theory T is falsified only if some *other* unrefuted theory T' has been proposed that predicts *novel* facts (facts that T predicts will not hold, or at least that it presents as very improbable), at least some of which have been corroborated, and that explains facts previously explained or predicted by T.

Lakatos's methodology seeks to improve upon NMF and SMF by doing two things: First, it replaces individual theories with sequences of historically related theories (*scientific research programs*) as the unit of methodological appraisal. Second, it articulates criteria by which to *evaluate* the responses of scientific research programs to potential falsifications. Lakatos intends the resulting 'methodology of scientific research programs' to answer questions about what 'intellectual honesty' requires in response to a potential falsification of a given theory.

A scientific research program is, for Lakatos, a historical entity with four main components. These include (1) a *hard core* of "assumptions which, by methodological decision, as it were, are kept unfalsified" (Zahar, 1973, 100). To obtain a theory, one must conjoin the assumptions from the hard core with (2) auxiliary assumptions constituting the *protective belt*; these assumptions are expendable, and when a potential falsification or anomaly arises for a theory thus constructed within the research program, investigators direct falsification at the protective belt rather than the hard core. Methodologically, the construction and subsequent modification of a scientific research program is guided by (3) a *negative heuristic*, which directs the investigator to deflect potential falsifications away from the hard core and toward statements in the protective belt, and (4) a *positive heuristic*, a kind of guideline or policy for the initial construction and subsequent modification of the protective belt. Thus, a research program comprises a series of theories constructed by investigators who – guided by the negative and positive heuristics – combine the unchanging hard core with the changing contents of the protective belt.

An example from the history of astrophysics illustrates how we might apply these ideas.

5.5.1 A Tale of Two Planets

The planet Uranus was discovered by William Herschel in 1781. Over the following years, observations of Uranus proved difficult to reconcile with predictions drawn from Newtonian celestial mechanics (NCM); essentially, this is Newton's laws of motion and law of gravity, mathematically elaborated through the subsequent century and applied to the known bodies of the solar system). In 1845, Françoise Arago, director of the Paris Observatory, encouraged the young astronomer-mathematician Urbain Le Verrier to work on the anomalies that observations of Uranus presented for NCM.

Other astronomers had speculated that the problem might arise from the gravitational field of "a new planet whose elements would be known according to its action on Uranus" (Wilhelm Bessel, writing in 1840, quoted in Hanson, 1962, 361). Using NCM, Le Verrier set to work to calculate: (1) the precise effect of Jupiter and Saturn on Uranus's orbit; (2) the exact disagreement between the predictions and the observations of Uranus's orbit, accounting for the perturbations determined in his first step; and (3) what hypothetical properties of a possible new planet would introduce a further perturbation in Uranus's orbit sufficient to reconcile prediction and observation.

Le Verrier could not persuade the Paris astronomers to search for the planet that his analysis predicted, so he wrote to Johann Gottfried Galle at the Berlin Observatory. Galle's student Heinrich Louis d'Arrest suggested making a comparison with a recently constructed chart of the stars, and just after midnight on September 24, 1846, the Berlin astronomers found the planet. It lay within 1° of Le Verrier's predicted location and had an apparent diameter in very close agreement with the predicted value.

Independently of Le Verrier, John Couch Adams had taken up the problem tackled in part (3) of Le Verrier's calculations. He was even less successful than Le Verrier in persuading others to take his calculations seriously. He communicated his results to James Challis, director of the Cambridge Observatory, and to George Airy, the Astronomer Royal.[7] Only when Airy got wind of Le Verrier's Paris presentations did he set Challis's observatory to work on a search for the planet, which began on July 29, 1846. Challis did not have as good a star map as Galle, and only after the announcement of the Berlin observation did the English astronomers realize that they had observed the planet in August without realizing it.[8]

Note that by the criteria of NMF, these astronomers made the wrong move. When they 'observed' the conflict between the data on Uranus's orbit and the prediction from NCM, they should have *abandoned* Newton's theory as falsified. Instead, they proposed to save the theory by adopting a new auxiliary assumption. SMF does not condemn this response, but does not tell us why it was reasonable to introduce a new auxiliary hypothesis *rather than* trying to come up with a replacement for the core commitments of NCM. The methodology of scientific research programs means to provide us with the rationale that SMF fails to deliver.

Applying Lakatos's methodology to our example, Le Verrier and Adams regarded NCM as the hard core. The anomalous features of Uranus's orbit presented a problem to be solved, but – in keeping with the negative heuristic – not a reason by themselves to reject NCM. Furthermore, they regarded statements about the bodies to be found in the solar system as lying within the protective belt. That belt had been modified before (when Uranus

[7] One aspect of the multifaceted controversy surrounding this episode concerns why neither Airy nor Challis saw fit to follow up on Adams's prediction, (See Sheehan, Kollerstrom, & Waff, 2004; Standage, 2000 for two engaging accounts).

[8] Accounts of the episode sometimes assert that Le Verrier and Adams independently "made the same calculations." In fact, Adams only tackled one part of Le Verrier's three-part project. Their resulting calculations generally disagreed with each other and with the actual orbit of Neptune, but all three happened to coincide for a brief period, during which the observations in Berlin and Cambridge were made (Hanson, 1962).

was added to the solar system) and could be modified again. The positive heuristic they followed allows one to hypothesize new bodies with orbital parameters that explain otherwise anomalous orbits of celestial bodies.

A central motivation for the methodology of scientific research programs (MSRP) is that, in a certain sense, every theory is 'born falsified' (in the sense that every theory has unresolved anomalies throughout its historical development), and yet theoretical principles can be rendered unfalsifiable by simply redirecting the inference that an error has occurred onto some statement in the protective belt. Sometimes it is reasonable to do this (as in the case of the NCM anomalies posed by Uranus's orbit). However, it is not *always* reasonable to protect a theory from falsification in this way. Sometimes the lesson to draw from an apparent failure of a theory is that the theory is wrong! We would like a principle that allows us to differentiate these two possibilities. Another episode from the history of astronomy illustrates why.

5.5.2 A Tale of Another Planet?

Urbain Le Verrier, the hero of the discovery of Neptune, also worked on the orbit of Mercury. In the 1840s, he set out to determine the theoretical orbit of Mercury, for which very precise observational data were available. But he could not make the data fit the theory. Norwood Russell Hanson comments that for Le Verrier "the theory of Mercury was like a frayed garden hose conveying high-pressure steam. When he adjusted it here, it sprang a leak there; he hadn't enough hands, or ideas, to plug every hole at once" (Hanson, 1962, 366). In an 1849 report to the Paris Academy, he traced the difficulties to a mismatch regarding the precession of Mercury's perihelion (the point in the orbit at which the planet comes closest to the Sun). According to NCM, the perihelion of Mercury's orbit should advance 527″ of arc over the course of a century, but the observational data indicated an additional 38″ of advance. (In 1882 Simon Newcomb corrected the magnitude of the anomaly to approximately 43″ per century.)

Le Verrier attempted in 1859 to reconcile theory and observation by adjusting the planet's orbital parameters (such as its mass), but these were already very narrowly specified by existing data and theory. If he changed them by very much, it ruined the fit between theory and data elsewhere. So he turned to the trick that had worked so well in the case of Uranus:

He considered the postulate of a new planet (or "group of planetoids") inside Mercury's orbit, previously unobserved because of its (or their) proximity to the Sun. A new telescopic search was needed.

Many observational reports of an intra-Mercurial planet followed. (These described the observation of *transits*, in which a dark spot appeared in front of the Sun; such spots had to exhibit planetary motion, lest they turn out to be sunspots.) Le Verrier received one report in December 1859 from a provincial doctor and amateur astronomer named Edmond Modeste Lescarbault. Lescarbault's report was sufficiently credible that Le Verrier, by now the director of the Paris Observatory and a scientific dignitary, traveled to the town of Orgères-en-beauce to query Lescarbault directly. The doctor was convincing. Le Verrier announced the finding to the Acadèmie des sciences in January 1860 and arranged for Lescarbault to receive the Legion of Honor. Thus was the planet that came to be known as *Vulcan* discovered (Baum & Sheehan, 2003; Hanson, 1962). Or would have been, if it existed.

In the ensuing decades, in Hanson's words, "Vulcan's forge cooled" as various astronomers failed to observe Vulcan. Even had Lescarbault's planet been real, Le Verrier's determinations of its features based on Lescarbault's data indicated that it could not reconcile Mercury's anomalous orbit with NCM. Ultimately, NCM proved unable to explain the anomalous precession of Mercury's perihelion. Instead, a radically new theory of gravity – Einstein's General Theory of Relativity – would explain this phenomenon. But that is another story ...

5.5.3 Progress

Lakatos's MSRP is primarily a theory of scientific *progress:* It draws its principal distinction between *progressive* and *degenerating* research programs. Whether a research program is progressive depends on its ability to generate predictions of *novel* facts, that is, "facts which had been either undreamt of, or have indeed been contradicted by previous or rival programmes" (Lakatos, 1978, 5). A research program is *theoretically progressive* when it generates predictions of novel facts. It becomes *empirically progressive* when such predictions are corroborated by subsequent investigation. A research program in which responses to anomalies are not motivated by the positive heuristic of the research program or do not result in corroborated novel predictions is *degenerating*.

NCM displayed progress in its encounter with the anomalous orbit of Uranus. By adding a heuristically motivated hypothesis about a new planet, Le Verrier constructed a new theory with novel content, which was then corroborated. The attempt to supplement NCM with a hypothesis about an intra-Mercurial planet was theoretically progressive, but NCM failed to convert this into empirical progress. As Lescarbault's Vulcan proved illusory, astronomers explored other means to solve the problem of Mercury's orbit.

One such effort reveals the challenge of applying Lakatos's MSRP. In 1894 Asaph Hall suggested a revision of Newton's law of gravity itself. He proposed to change the parameter that dictates how the strength of the gravitational force depends on the distance between the two gravitating masses. The constraints on viable revisions were significant, because any such change would have to preserve the very good *agreement* between other astronomical data and the standard formulation of Newton's law (in which gravitational force F_G is proportional to the inverse of the distance r squared, i.e., $F_G \propto r^{-2}$). Hall determined that revising Newton's law such that $F_G \propto r^{-2.00000016}$ would yield the right value for the precession of Mercury's perihelion. As William Harper relates, in 1895, Simon Newcomb, "by then the major doyen of predictive astronomy, cited Hall's proposal as 'provisionally not inadmissible' after rejecting all other accounts he had considered" (Harper, 2007, 937).

Was this a theoretically progressive change within the research program, or was NCM at this stage degenerating? That depends on whether modifications to the distance dependence in Newton's law of gravity adhere to NCM's positive heuristic. Research programs do not, however, wear their heuristics on their sleeves. Newton's law of gravitation asserts that $F_G \propto r^{-2}$. That same mathematical relationship can be expressed as $F_G \propto r^{\chi}$, where $\chi = -2$. The latter expression separates the *form* of the distance dependence from its *parameter values*. Do both statements belong to the hard core of NCM, or only the form $F_G \propto r^{\chi}$? This question lacks a clear answer; worse, it is not clear what kind of evidence we would need to answer it. But whether Hall's proposal violates the negative heuristic of NCM depends entirely on the answer to this question.

In any case, Hall's proposal did not survive. In 1903 Ernest William Brown used data on the lunar orbit to show that Hall's proposed change to the distance parameter in Newton's law exceeded by two orders of magnitude the maximum allowable difference from –2. And a departure small enough to be allowed could not account for the anomalous orbit of Mercury (Harper, 2007). Something else had to give, and that something was the very conception of gravitational force that Newton had introduced, and that had survived for so long.

5.6 Novelty and Rationality

We have seen that Lakatos's distinction between progressive and degenerating research programs turns on their ability to predict *novel facts*. The novel prediction requirement calls to mind a widely held intuition that it is more impressive when a theory can predict something previously unknown than when it can merely explain what is already known.

This idea has a long history in scientific thought. It is clearly expressed, for example, in Christian Huygens's 1678 *Treatise on Light*. He describes how scientific principles may be established with "a probability which is little short of certainty" when the "assumed principles" yield predictions that are "in perfect accord" with observations, the observations are numerous, and "above all when one employs the hypothesis to predict new phenomena and finds his expectations realized" (Matthews, 1989, 126–127). Nineteenth-century scientist and philosopher William Whewell called the ability of a theory to predict new kinds of phenomena the "consilience of inductions" and held that a consilient theory such as NCM enjoys a higher grade of support than a theory that merely predicts new facts of a kind that theory is already known to explain (Whewell, 1847).

Getting clear about what makes a fact *new* or *novel*, and *why* such predictions should render theories preferable, proves difficult. For example, Le Verrier's prediction of the existence and location of Neptune strikes us as an impressive success not only for Le Verrier himself but also for the theory of NCM on which he relied. Once the existence of Neptune was established, one could then use the knowledge of the parameters of Neptune's orbit to predict where it would be in the night sky on some date in the future. This does not impress us as much. Yet in both cases one derives from the theory a statement about some state of affairs that is not known already to obtain.

Both predictions would be novel in the *temporal* sense, according to which a fact is novel if, at the time that it is predicted, it is not among the facts known to science. But the difference in the import of the two predictions calls for a new distinction. What impresses us in a theory is not just the ability to generate a statement about some state of affairs that is in the future but rather the prediction of some new *kind* of fact, or phenomenon, previously unknown to science (as required for Whewell's "consilience"). Le Verrier's prediction of Neptune seems to meet this stronger requirement, whereas the prediction of the location in the night sky of an already known planet on some future date does not.

To implement this idea systematically, we need some way of determining which predictions involve new kinds of facts rather than new facts of the same kind. The status of a research program also becomes dependent upon biographical matters regarding scientists working on that program. Would Le Verrier's prediction not count as novel if another astronomer had already observed Neptune? What if numerous astronomers had seen the planet? How many would it take to disqualify the prediction from being novel? If only Le Verrier's state of knowledge mattered, would we have to revise our evaluation of NCM upon the discovery of a diary entry revealing that Le Verrier had secretly already observed Neptune before his calculations? It is not clear *why* such biographical details should matter to the status of a theory.

Alan Musgrave (1974) notes that Lakatos articulated a nontemporal conception of novelty. According to Lakatos's account, as Musgrave interprets and defends it, scientists do not use novel predictions for the confirmation of a theory T in isolation, but evaluate T in comparison to some prior or 'background' theory T_0. For Le Verrier this background theory T_0 would perhaps be constituted by the 'hard core' principles of NCM in conjunction with, among other things, existing knowledge of the first six planets in the solar system. Le Verrier derived his prediction from a new theory (call it 'NCM+Neptune') combining all those elements with a 'theory of Neptune' specifying its hypothetical orbital parameters. Regardless of whether anyone in particular, or even the entire astronomical community, already knew of the existence of such a planet, the theory NCM+Neptune has novel predictive content in comparison to T_0, which makes the episode progressive.

In Elie Zahar's account, a fact is novel in the heuristic sense relative to a given hypothesis "if it did not belong to the problem-situation which governed the construction of the hypothesis" (Zahar, 1973, 103). John Worrall has articulated this idea in terms of what he calls "use-novelty": Some fact E that is entailed by a theory T cannot serve as a test or "potential falsifier" (in Popper's terms) of *T* if *E* was used in the construction of *T*, that is, if *T* was "engineered to entail" *E* (Worrall, 1989a, 148; see also Worrall, 1985; for a subsequent exchange on the issue, see Mayo & Spanos, 2009, 125–169).

One can see in this discussion about novelty how philosophers of science have grappled with challenges posed by the postpositivist elevation of the history of science as a crucial touchstone for the philosophy of science. Historical descriptions of how certain scientific claims have come to be accepted do not immediately yield conclusions about whether there were good reasons to accept those claims, or what those reasons might be. History may tell us when and by whom a theory was formulated and a fact predicted by that theory became known, but philosophers are often interested in the rationality of accepting a theory on the basis of a predicted fact.

Lakatos's approach explicitly subordinates historical fact to philosophical theories about rationality. Scientific methodology is important, according to Lakatos, because it "concerns our central intellectual values." Lakatos would have us value *progress* and regard scientific decisions as rational when they promote progress.

Let us, for the sake of argument, accept this point. What follows from it for the decisions that scientists must make, such as whether a theoretical claim can safely be assumed, or whether a theory confronting anomalies is worth trying to save? Can Lakatos's methodology of scientific research programs help the scientist make rational decisions on such matters?

By Lakatos's own insistence, the determination of a research program's status as progressing or not *cannot* by itself answer any such questions. It can only show in retrospect the rationality (or failure of rationality) of such decisions, through what Lakatos calls the *rational reconstruction* of the history of science. A rational reconstruction starts with a certain view of what is rational (for Lakatos, this would be the methodology of scientific research programs, but for a Popperian falsificationist or an inductivist, it would be something different) and then uses that theory of rationality to show how scientists were rational in a particular scientific episode, under the tentative presumption that those scientists were rational. Lakatos acknowledges that one must then compare the rational reconstruction of the episode with "actual history" in order to "criticize one's rational reconstruction for lack of historicity, and the actual history for lack of rationality" (Lakatos, 1978, 53).

Such a method will be of no direct help for decisions regarding current, ongoing science. Lakatos states explicitly that he does not intend anyone to use his criterion to execute a "quick kill" of a scientific theory, as one might use a Popperian falsification: "One must treat budding programmes leniently: programmes may take decades before they get off the ground and become empirically progressive" (Lakatos, 1978, 6). Neither is it a rule, according to Lakatos, that one should stay with a research program as long as it *is* progressive: "It would be wrong to assume that one must stay with a research programme until it has exhausted all its heuristic power" (Lakatos, 1978, 68). It is starting to seem as though there are no clear, objective rules for scientists to follow.

On the contrary, Lakatos insists, his methodology provides an objective basis for the decision to reject some programs outright: "[S]uch an objective reason is provided by a rival research programme which explains the previous success of its rival and supersedes it by a further display of heuristic power," that is, the ability to generate empirical progress (Lakatos, 1978, 69). *Finally,* something clear-cut and definitive . . .

. . . Except! Lakatos immediately recomplicates things by noting that empirical progress is often not susceptible to quick and unanimous appraisal: "*[T]he novelty of a factual proposition can frequently be seen only after a long period has elapsed*" (Lakatos, 1978, 69, emphasis in original). What appears to be an established fact predicted by existing theory can be regarded as novel if a new theory provides us with a new *interpretation* of that fact. Hence we should be tolerant of "budding" research programs, but even an "old, established and 'tired' programme, near its 'natural saturation point' . . . may continue to resist for a long time and hold out with ingenious content-increasing innovations even if these are unrewarded with empirical success" (Lakatos, 1978, 72).

Lakatos summarized these nuances thus: "[T]he methodology of scientific research programmes does not offer instant rationality" (Lakatos, 1978, 6). His friend in life and foe in philosophical debate, Paul Feyerabend, took Lakatos's denial of instant rationality to be a way of evading a deeper truth to which Lakatos's insights pointed, but which Lakatos would not accept: There is no universal, exceptionless methodology for distinguishing the rational from the irrational, the progressive from the degenerating, good science from bad science. Feyerabend espoused an "epistemological anarchism" according to which the only universally valid and exceptionless rule of method is "anything goes," and he considered Lakatos to be a "closet anarchist." Not surprisingly, Lakatos rejected this characterization of his own views.

5.7 Feyerabend: Epistemological Anarchism

In 1975, Paul Feyerabend published *Against Method* and became notorious among philosophers of science. Like Imre Lakatos, Feyerabend came of age in the midst of the turmoil enveloping Central Europe in the 1930s and 1940s.[9] Born in Vienna in 1924, he was drafted into the German army in 1942. After serving several tours of duty on the Russian front, including commanding a company of 'seasoned soldiers,'[10] he was shot in the spine, leaving him paralyzed from the waist down. He used a wheelchair for a while until he learned to walk with crutches and, throughout his life, suffered bouts of intense pain requiring medication. After the war, Feyerabend returned to Vienna, where his studies included physics and mathematics. After completing a PhD in philosophy in 1951 he traveled to London, where he studied with Karl Popper. His teaching career took him from Bristol to Berkeley to wandering scholar, lecturing widely and sometimes holding three positions simultaneously on two continents (Berkeley, London, and Berlin).

Against Method advocates a view that Feyerabend calls *epistemological* (or sometimes *theoretical*) *anarchism.* He proclaims at the outset, "Science is an essentially anarchic enterprise: theoretical anarchism is more humanitarian and more likely to encourage progress than its law-and-order alternatives" (Feyerabend, 1988, 9).

The term 'essentially' deserves emphasis. The anarchic character of science is not, Feyerabend emphasizes, a mere historical contingency. Although

[9] Feyerabend's autobiography is very readable and conveys his humor and breadth of interests (Feyerabend, 1995). Paul Hoyningen-Huene's obituary for Feyerabend includes a biographical sketch and appreciation of Feyerabend's contributions to the philosophy of science (Hoyningen-Huene, 2000).

[10] "There I was, a dedicated bookworm, with no experience, the symbols of authority on my shoulders, confronted with a bunch of skeptical experts. It happened to me again, twenty years later, when I was supposed to teach the Indians, blacks, and Hispanics who had entered the university as part of Lyndon Johnson's educational programs. Who was I to tell these people what to think?" (Feyerabend, 1995, 46–47).

Feyerabend will invoke the history of science in his arguments, he is not simply *observing* something about how science has, as a matter of fact, been pursued; he is *advocating* anarchism as an integral and value-enhancing part of the scientific undertaking. What this means for Feyerabend is that science should not be pursued by always following a fixed set of rules. The idea of a theory of rationality like Lakatos's MSRP is, on this view, not only doomed but also dangerous.

In his opening salvo, Feyerabend claims *two* distinct advantages of an anarchistic approach to science. First, it is "more humanitarian . . . than its law-and-order alternatives." Feyerabend's epistemological anarchism does not rest only on epistemological considerations, but is part of a more general view of the conditions for human flourishing, which Feyerabend associates with nineteenth-century British philosopher John Stuart Mill. Second, epistemological anarchism is "more likely to encourage progress than its law-and-order alternatives." Feyerabend's advocacy of epistemological anarchism thus relies on two main lines of argument: *the argument from progress* and *the argument from human flourishing.*

5.7.1 The Argument from Progress

This argument seeks to establish the following: Whatever rules you might think of as constitutive of the scientific method, there will be occasions when the best way to advance scientific knowledge will be to break those rules. "To those who look at the rich material provided by history," Feyerabend writes, "and who are not intent on impoverishing it in order to please their lower instincts, their craving for intellectual security in the form of clarity, precision, 'objectivity,' 'truth,' it will become clear that there is only *one* principle that can be defended under *all* circumstances and in *all* stages of human development. It is the principle: *anything goes*" (Feyerabend, 1988, 19, emphases in original).

This expression, "anything goes," has become a kind of icon of Feyerabend's philosophy of science. Plucked out of context, however, this expression can create misunderstanding. In particular, one might think that with this expression Feyerabend tells us that one may advance scientific knowledge by whatever means one likes. This would be a very odd position to take; it is not hard to see that most things that one *could* do in science (setting your apparatus on fire, pouring maple syrup into the spectrometer) will not work. Indeed, finding even *one* thing to do that will significantly advance the state of knowledge in a given field requires great ingenuity and effort.

Feyerabend's position is more nuanced than just "anything goes." Instead, he argues that any rule we might regard as a 'universal' guide to scientific inquiry will, under some circumstances, prevent scientists from contributing to the progress of science (whatever we think of as progress). The only rule guaranteed to never prescribe an action that prevents progress is "anything goes," but this is because it is a rule that neither prescribes nor prohibits any action at all.

To support this claim, Feyerabend employs two argumentative strategies. The first presents general arguments regarding the relationship between rules and their context of application and regarding the conditions under which arguments succeed in changing opinions. The second strategy employs episodes drawn from the history of science. Feyerabend's writings weave these two strategies together, using history to lend plausibility to his philosophical arguments, and refracting his interpretation of historical events through his own philosophical lenses. For the sake of understanding, I will discuss the strategies separately, beginning with the philosophical arguments.

If we knew exactly what the world is like, then perhaps we could come up with a system of rules that would constitute an effective procedure for determining its structure, its constitution, the general principles that best describe how the world works. Of course, in that case we would not really need these rules, because we would already know everything that the rules were supposed to help us discover. Feyerabend argues that the effectiveness of a rule for pursuing science will always, in one way or another, depend on what the world is like, which is just what we do not know. We must "keep our options open." "Epistemological prescriptions may look splendid when compared with other epistemological prescriptions, or with general principles – but who can guarantee that they are the best way to discover, not just a few isolated 'facts,' but also some deep-lying secrets of nature?" (Feyerabend, 1988, 12).

We encountered this idea in Chapter 1: Whether it is reasonable to generalize from the unpleasant taste of a sample of horseradish doughnut to horseradish doughnuts in general depends, in part, on the extent to which the taste attributes of doughnuts are uniform, a fact about which we are imperfectly informed. Similarly, the reasonability of Newton's generalization about the variable refrangibility of the rays in white light depends on the extent to which light behaves the same, regardless of its location and source.

Replacing induction with falsification will not solve the problem. For their success, falsificationist rules also depend on contingent and uncertain facts. A falsificationist response to the anomalies in Uranus's orbit might call for the rejection of NCM on the basis of that theory's failed predictions. That is the right response only if there are no other celestial bodies exerting gravitational forces that render Uranus's orbit consistent with Newton's theory, within the limits of accuracy available at the time. This is so even if one appeals to Lakatos's criteria of progressive research programs to guide one's response.

Feyerabend concludes that whether an application of any rule of inference leads to success will be a contingent matter, depending on additional facts not known in advance.

Alright, so science depends on *fallible* methods. Even when we apply the methods correctly (whatever we think correctness might require), the conclusions we draw might be mistaken. That a rule is fallible, however, does not entail that we should not follow it. In particular, if following a methodological rule yields *reliable* results – or results more reliable than those obtained by alternative methods – then it is reasonable for scientists to follow that rule.

Were Feyerabend asserting merely the *fallibility* of widely endorsed methodological rules in this way, we would have little reason to disagree, but also little reason to consider him an original and provocative philosopher. But Feyerabend goes further. He claims that given any rule that we might consider to be (partly) constitutive of the scientific method, there will be occasions when we can *better* advance scientific knowledge by following a rule that points us in the *opposite* direction.

For example, Feyerabend recommends the use of *counterinduction*, which "advises us to introduce and elaborate hypotheses which are inconsistent with well-established theories and/or well-established facts" (Feyerabend, 1988, 20). This bit of advice will strike many as counterintuitive, to say the least. Isaac Newton certainly seems to have disagreed. The last of his four "Rules for the Study of Natural Philosophy" states: "In experimental philosophy, propositions gathered from phenomena by induction should be considered either exactly or very nearly true notwithstanding any contrary hypotheses, until yet other phenomena make such propositions either more exact or liable to exceptions." Newton goes on to explain that the point of this rule is "so that arguments based on induction may not be nullified by hypotheses" (i.e., alternative theories introduced for no other reason than their possible ability to explain the observed phenomena) (Newton, [1687] 1999, 796). Newton's rule, by making the conclusions of inductive arguments provisional, implicitly acknowledges the *fallibility* of inductive reasoning that we just highlighted, but clearly proscribes counterinduction when a claim has strong inductive support.

Newton notwithstanding, Feyerabend believes it is reasonable to introduce alternatives that are incompatible with both well-established theories and well-established facts.

Even a well-established theory may have weaknesses that will only be discovered if one has some other incompatible theory with which to compare it, since "some of the most important formal properties of a theory are found by contrast, and not by analysis" (Feyerabend, 1988, 21).

In a certain sense, such a statement seems eminently reasonable, since distinctive features of a given theory can be highlighted by comparison with another theory with which it competes for our acceptance. In the *Origin of Species*, Charles Darwin devotes a significant amount of attention to comparisons between the kinds of explanations of traits of species offered by his theory of evolution by means of natural selection and those offered by the theory of 'special creation.' He thus helps his readers appreciate the differences between the two accounts – a comparison that remains illuminating long after the scientific evidence for evolution by natural selection has been well established. That arguments offered in support of one theory often involve a comparison with another (either an already existing competitor or a foil contrived just for the occasion) is not really a controversial claim.

As Feyerabend elaborates his point, however, he evidently intends to be making a point that is controversial. Consider the following:

> A scientist who wishes to maximize the empirical content of the views he holds and who wants to understand them as clearly as he possibly can must therefore introduce other views; that is, he must adopt a *pluralistic methodology* . . . Proceeding in this way he will retain the theories of man and

> cosmos that are found in Genesis, or in the Pimander,[11] he will elaborate them and use them to measure the success of evolution and other 'modern' views. He may then discover that the theory of evolution is not as good as is generally assumed and that it must be supplemented, or entirely replaced, by an improved version of Genesis. Knowledge so conceived is not a series of self-consistent theories that converges toward an ideal view; it is not a gradual approach to the truth. It is rather an ever increasing *ocean of mutually incompatible (and perhaps even incommensurable)* alternatives. (Feyerabend, 1988, 21, emphasis in original)

Feyerabend's choice of examples seems outrageous. Is he seriously claiming that we can improve our understanding of biological species by replacing Darwinian ideas with those found in the biblical text of Genesis? Feyerabend often plays devil's advocate, defending views to which he has no real commitment. This makes it hard to gauge the seriousness of the specific example he gives here. We should not, however, let that distract us from the seriousness of the main point, which concerns the very aim of science itself. Science aims, he tells us, not at a single comprehensive account of the phenomena in some domain but at an 'ocean' of accounts that are not even compatible with one another.

He concludes – and it is a bit of a leap – that we should not discard any idea: "There is no idea, however ancient and absurd, that is not capable of improving our knowledge" (Feyerabend, 1988, 33).

The second argument for counterinduction, which defends the value of introducing theories inconsistent with "experiments, facts, observations," begins with the point – familiar to us by now – that "no single theory agrees with all the known facts in its domain" (Feyerabend, 1988, 39). (Recall Lakatos: "[E]very theory is born refuted.") This makes the "demand to admit only those theories which are consistent with the available and accepted facts" fatal to scientific advance. "Hence, a science as we know it can exist only if we drop this demand" (Feyerabend, 1988, 51).

But Feyerabend's argument does not stop there: We should not merely tolerate such contradiction of the "facts": we should *seek it out.* "Counterinduction is thus both a *fact* – science could not exist without

[11] The Pimander, which describes a mysterious cosmological revelation, constitutes one chapter of a collection of second- and third-century CE Greek texts presenting eclectic and opaque teachings of a spiritual nature called the *Corpus Hermeticum.*

it – and a legitimate and much needed *move* in the game of science" (Feyerabend, 1988, 54, emphasis in original). Such a counterinductive move benefits scientists by allowing them to criticize the facts themselves (i.e., to show that they are not facts after all), in ways that might not otherwise be possible. Feyerabend's most developed historical exemplar of this proposal can be found in his treatment of Galileo's arguments in defense of Copernicus's heliocentric theory of the cosmos.

5.7.2 Galileo on the Motion of the Earth

In 1632, after a great deal of maneuvering to obtain the needed license from the Roman Catholic Church (the *imprimatur*), Galileo Galilei published his *Dialogue Concerning the Two Chief World Systems*. The title referred to two competing accounts of the nature and constitution of the cosmos. The *geocentric* (earth-centered) system was rooted in Aristotle's natural philosophy. Claudius Ptolemy had developed a mathematical model of it in the second century CE, which subsequent astronomers had elaborated into a rather complex system. Nicolaus Copernicus had devised a mathematical model (also rather complicated) of a *heliocentric* (sun-centered) system, which he described in a book titled *De Revolutionibus*, published just as Copernicus died in 1543.

We will not concern ourselves here with the details of the two systems.[12] For our purposes, the significant difference between them is just this: According to the first system, the Earth remains stationary and the motions that we observe of the planets, the Moon, the Sun, and the stars across the sky result from those bodies following orbits centered on the Earth. According to the second system, the Earth has a twofold motion consisting of its *diurnal* rotation about its own axis and its *orbital* motion around the Sun.

Galileo's defense of the Copernican system embroiled him in a theological dispute that led ultimately to his trial and to the Church pronouncing him "vehemently suspected of heresy." The Church had put *De Revolutionibus* on the index of prohibited books in 1616, and around the same time, Galileo had been personally warned by the Vatican's chief theologian, Cardinal Robert

[12] Thomas Kuhn's *The Copernican Revolution* remains a worthwhile resource for understanding these developments (Kuhn, 1957).

Bellarmine, against teaching or defending Copernicus's system.[13] Galileo's *Dialogue* uses a transparent artifice to evade Bellarmine's proscription: a conversation between the figures Salviati, Sagredo, and Simplicio. Officially Salviati does not advocate Copernicanism; he intends "only to adduce those arguments and replies, as much on one side as on the other . . . and then to leave the decision to the judgment of others" (Galilei, 1967, 107). In practice, Salviati presents a relentless stream of arguments directed against the Aristotelian, geocentric viewpoint and in defense of the Copernican system. Galileo uses Sagredo as a neutral arbiter between Salviati and Simplicio, the staunch and sometimes obtuse defender of Aristotelian geocentrism.

Feyerabend discusses many aspects of the *Dialogue*'s extended argumentation. Here I will confine this discussion to one line of argument that illustrates especially vividly Feyerabend's strategy of arguing from historical cases for the value of counterinduction.

Imagine, *if you can*, that you live in the pre-Copernican era. Perhaps you have never even heard of the idea that the Earth might be in motion. You might well find the suggestion of a moving Earth positively ludicrous. Indeed, the Copernican system requires the Earth to have two quite rapid motions. We now know that the Earth travels about the Sun at approximately 30 km/s, while the Earth's daily rotation about its own axis entails that locations on the equator travel at about 0.5 km/s. The magnitudes of these speeds alone suggest that we are in a constant state of considerable motion, and surely, it might seem, such motion would manifest itself to us in *some* noticeable ways.[14]

Opponents of the Copernican system accordingly pointed to a very simple, readily observed phenomenon as conclusive evidence that the Earth *cannot* be

[13] A large literature has grown up around the Galileo affair. Finocchiaro (1989) provides a useful resource, while Blackwell (2006) presents an intriguing development in scholarship.

[14] The first figure would have been estimated much lower in Galileo's day, when estimates of the distance from Earth to Sun were systematically low. In the second century BCE, Aristarchus of Samos had used an ingenious method to arrive at the result that the distance from the center of the Earth to the Sun is about 1528 earth radii (roughly 11,000,000 km). The ninth-century CE Arab astronomer Al Fargani, using a different method, placed the Sun at 1220 Earth radii's distance (about 9,000,000 km) (Kuhn, 1957, 80–81, 274–278). Modern measurements put the figure at about 150,000,000 km. Even with the smaller values, pre-Copernican intuitions would lead one to conclude that the speed at which the Earth would have to orbit the Sun must surely be *noticeable*.

moving as required in a heliocentric model. They pointed out that objects that are dropped *fall in a straight line* toward the surface of the Earth. The Aristotelian natural philosophy explains this phenomenon naturally: Objects such as rocks, bricks, books, and Fabergé eggs fall because they primarily consist of the *earth* element, the *natural motion* of which is toward the center of the cosmos (which is also why we find the planet Earth situated just there). But, the argument goes, a simple experiment would show such motion to be incompatible with the motion of the Earth about its own axis.

Consider a stone dropped from the top of a tall tower. According to the Copernican theory, the tower is moving very rapidly, along with the rest of the Earth, from west to east. Suppose the stone is dropped from the western side of the tower. We should then observe the stone to fall away *at an angle* from the tower, for once it has been released the tower will continue its motion from west to east, but there is nothing to make the stone keep up with the tower. It will fall behind while it also falls to the ground. But, of course, this is not what we see at all. The dropped stone falls in a straight line, perpendicular to the surface of the Earth.

According to Feyerabend, Galileo's defense of Copernicanism requires him to both *replace* the fact 'the dropped stone falls in a straight line, perpendicular to the surface of the Earth' with another fact – 'the *apparent* motion of the stone is described by a straight line, perpendicular to the surface of the Earth' – and then show how this new fact is powerless to refute the Copernican theory, since one can interpret it in terms of either a stationary or a moving Earth. To do this, Galileo must get his reader to take a single fact and, in light of a theory with which that fact is incompatible, introduce a new distinction between the sensations experienced and the interpretation given to those sensations. Feyerabend uses the term *natural interpretations* to refer to mental operations that operate so immediately upon our sensations that we do not easily notice them as interpretations of those sensations.

In the present case, the natural interpretation is the move from our visual impressions of the stone falling parallel to the walls of the tower to the claim that 'the stone's *actual* motion is described by a straight line, perpendicular to the surface of the Earth.' Galileo's problem is that to his Aristotelian opponents this is not a move at all. They identify the actual motion of any object with the motion they observe it to exhibit with the unaided eye. To get his opponents to admit the distinction that renders that observation powerless against the Copernican system requires that Galileo *first* get them to adopt a new perspective on the observation different from and incompatible with Aristotelian epistemology. (Compare this move with Hanson's discussion of Tycho's and Kepler's observations of the setting Sun.)

Galileo's new perspective introduces two new principles. First is the principle of inertia (more accurately, a precursor to the more modern principle of inertia articulated as the first of Newton's Laws of Motion – a body with a given state of motion will remain in that state of motion if no forces act upon it). Galileo does not state this principle explicitly, but rather has Salviati lead Simplicio through a thought experiment involving a ball rolling on an unbounded frictionless plane, securing Simplicio's admission that such a ball, once set in motion, would have no reason to stop moving. Later, Simplicio is required to apply the same reasoning to the diurnal motion imparted to the stone by the moving Earth.[15] The second principle is the Principle of the Relativity of Motion. Galileo states this principle as follows: "Motion, in so far as it is and acts as motion, to that extent exists relatively to things that lack it; and among things which all share equally in any motion, it does not act, and is as if it did not exist" (Galileo, 1967, 116).

Neither of these principles has a place in the dominant Aristotelian natural philosophy of his day; so how can Galileo get his audience to accept the distinction he requires? Rather than pursue a lengthy excursus, suffice it to say that he relies on thought experiments like the one previously mentioned and his own expert manipulation of the dialogue format to have Salviati lead Simplicio (and the rest of his audience) to the following realization: If the Earth does rotate on its axis as the Copernican system requires, then the stone dropped from the tower will not be left behind, but will continue to move with the tower subsequent to being dropped because of inertia. This results in the stone having an actual motion (i.e., the motion that would be observed from a viewpoint external to the Earth) that is curved. But, because the diurnal motion is shared by the stone, the tower, and the observer, the Principle of Relativity entails that the motion will not

[15] The fact that we are now dealing with a circular motion means that the principle of inertia does not strictly apply. Galileo either regarded at least some circular motions as satisfying the principle of inertia, or else declined to state explicitly that the principle applies to the stone's 'horizontal' motion as an *approximation* of straight-line motion (Hooper, 1998).

appear to be curved; only the straight downward motion of the stone toward the Earth will be observed by an earth-bound observer.

Seemingly, Galileo's argument shows that what Galileo's opponents thought was a fact was no fact at all and reminds us to always confine ourselves to arguing on the basis of the sense impressions themselves. But, Feyerabend argues, this interpretation fails to adequately consider both Galileo's historical and argumentative situation and the role of theories in observation. Natural interpretations are unavoidable, according to Feyerabend. Galileo's opponents were not being unreasonable in taking themselves to have observed the straight-line motion of falling stones; they could hardly have understood things differently, given their cognitive situation. Telling them to attend only to sense impressions would have been useless because they took themselves already to be doing just that. Thus, the logical empiricist's prescription to separate the theoretical and observational vocabularies and purify our descriptions of any theoretical language will not work. Like Hanson and Kuhn, Feyerabend rejects the ideal of a purified, theory-free description of observations as unattainable. More than any of the other post-positivists, he emphasized that pursuing this unattainable ideal not only would be futile but also would have prevented many of the scientific discoveries we celebrate. Feyerabend concluded that a leap like Galileo's requires the scientist to *multiply interpretations* by introducing new theories that challenge the established facts of any given time. Because any candidate for becoming an established fact can only be relevant through such a theory-based interpretation, Feyerabend accepted a version of Kuhn's incommensurability thesis: There is no neutral ground available from which to adjudicate disputes between rival theories. Commitment to a theory comes before evaluation of the evidence.

5.7.3 The Argument from Human Flourishing

The other main line of argument pursued by Feyerabend is that a rule-bound approach to science is not conducive to human flourishing on either the individual or societal level. Educating people about science by advocating strict rules for the pursuit of knowledge "cannot be reconciled with a humanitarian attitude" (Feyerabend, 1988, 12).

Feyerabend regards humanitarian considerations as the primary *motivation* for his epistemological anarchism. He states in the third edition of

Against Method, "Anger at the wanton destruction of cultural achievements from which we all could have learned, at the conceited assurance with which some intellectuals interfere with the lives of people, and contempt for the treacly phrases they use to embellish their misdeeds was and still is the motive force behind my work" (Feyerabend, 1988, 272).

Feyerabend often cited John Stuart Mill's 1859 essay "On Liberty" as a precedent for his own views. Mill cowrote this essay with his wife Harriet Taylor Mill to discuss "the nature and limits of the power which can be legitimately exercised by society over the individual" (Mill, [1859] 1963, 213).[16] The central thesis of "On Liberty" is that "the sole end for which mankind are warranted, individually or collectively, in interfering with the liberty of action of any of their number, is self-protection. That the only purpose for which power can be rightfully exercised over any member of a civilized community, against his will, is to prevent harm to others" (Mill, [1859] 1963, 223). Mill aims to defend this thesis only on considerations of *utility*, that is, what conduces to the happiness of humanity as a whole.

An especially important consequence of Mill's thesis, for Feyerabend, concerns the freedom of thought and expression. Because one's thoughts never – and one's speech only in rare circumstances – threaten to harm another person, Mill espouses near-complete liberty of thought and expression. (He emphasizes that to offend a person is *not* to harm them.) Moreover, Mill aims to show (in keeping with his emphasis on utilitarian considerations) the great *benefit* to society of the expression of opinions that run counter to popular belief or what is encouraged by those wielding social or cultural power and influence.

As Mill summarizes:

> [T]he peculiar evil of silencing the expression of an opinion is, that it is robbing the human race; posterity as well as the existing generation; those who dissent from the opinion, still more than those who hold it. If the opinion is right they are deprived of the opportunity of exchanging error for truth; if wrong, they lose, what is almost as great a benefit, the clearer perception and livelier impression of truth, produced by its collision with error. (Mill, [1859] 1963, 229)

[16] The evidence regarding Harriet Taylor Mill's role in the authorship of this and other works of John Stuart Mill is scant, equivocal, and contested. My attributions of claims and arguments to "Mill" here should be read as similarly equivocal. See Miller (2022) for discussion.

Mill considers separately the cases in which an opinion that is subject to suppression is true and those in which it is false. That true opinions ought not to be suppressed seems plausible enough, but one might think that we are well justified in using the power of 'moral coercion' against views that are clearly incorrect. But, Mill notes, given human fallibility, we cannot be certain that what we regard as the truth really *is* the truth. Suppressing what we *believe* to be false might inadvertently amount to suppressing the truth.

But Mill goes further: Even when an opinion *really is* false, we benefit from its expression, for two reasons. First, we gain a better appreciation for the *reasons* that support a true opinion when we have to defend it against arguments in favor of a contrary view. Second, we gain a better understanding of the *meaning* of an assertion when we encounter opposing arguments that we then must defeat.

Feyerabend writes that Mill advocates the proliferation of contrary points of view "as the solution to a problem of *life*: how can we achieve full consciousness; how can we learn what we are capable of doing; how can we increase our freedom so that we are able to decide, rather than adopt by habit, the manner in which we want to use our talents?" (Feyerabend, 1981, 67, original emphasis).

Insofar as Mill advocates the expression of views contrary to those held by many or by the powerful, on the grounds that active engagement with such views benefits all, Feyerabend's citation of Mill seems apt. But has Feyerabend overreached in assimilating Mill's views to his own? Laura Snyder has pointed out that Mill, unlike Feyerabend, does not embrace the proliferation of views as an end in itself but rather as a *means* to the discovery of truths that might otherwise not be appreciated (Snyder, 2006, 214). He also regards it as a means to the appreciation of the reasons supporting those truths and even the very meaning of those truths.[17]

[17] Snyder – disputing Feyerabend's, as well as my own (Staley, 1999b) and Elisabeth Lloyd's (1997) views – rightly points out how Feyerabend and Mill differ on the proliferation of opinion as an *end*, citing passages in which Mill states that "cessation ... of serious controversy" is a necessary concomitant of human progress. However, she does not mention the immediately following sentences in which Mill states that, although such consolidation of opinion is "at once inevitable and indispensable, we are not therefore obliged to conclude that all its consequences must be beneficial." Mill's argument for the value of the devil's advocate role *follows* this comment (Mill, [1859] 1963, 250–251).

This last point, however, is crucial for understanding an important aspect of Feyerabend's appreciation of Mill. Mill goes so far as to claim that when an opinion has fallen so far out of favor that it no longer has supporters, a 'devil's advocate' ought to be assigned the task of defending it nonetheless, so that the rest of us may benefit from having it brought to our attention. Feyerabend himself often assumes this very role, by his own admission. As Elisabeth Lloyd has pointed out, reading Feyerabend in the role of Mill's devil's advocate can help us understand some of his more outrageous claims (Lloyd, 1997).

Recall how Feyerabend suggests that "the theory of evolution is not as good as is generally assumed and that it must be supplemented, or entirely replaced, by an improved version of Genesis." On the face of it, this appears to be scientifically irresponsible. Feyerabend offers no argument to indicate a defect in the 'modern synthesis' (of the Darwinian idea of evolution by means of natural selection and genetic theory) that could be remedied by this 'improved Genesis.' The bare possibility that there *might* be some such flaw is insufficient to make his point. Elsewhere, Feyerabend defended astrology, when the *Humanist* published a statement denouncing it, signed by 186 'leading scientists,' including eighteen Nobel laureates. Feyerabend gives a few reasons to think that it is plausible that "celestial events such as the positions of the planets, of the moon, of the sun influence human affairs" and accuses the denouncers of being ignorant of the history of astrology and even of the relevant science (Feyerabend, 1978, 91–96).

Can Feyerabend really mean to advocate for astrology? Well, yes and no. In a later work he acknowledges, "Astrology bores me to tears," but goes on to note that "it was attacked by scientists, Nobel Prize winners among them, without arguments, simply by a show of authority and in this respect deserved a defense" (Feyerabend, 1991, 165). He considers the attack on astrology to be ill informed, but, on the other hand, he does not really think much of astrology. To advocate on behalf of views for which one has no deep conviction is the mark of a 'devil's advocate.'

If the main point of Feyerabend's philosophy of science were simply that for any proposed rule of method, it is always *possible* that one might better advance science by breaking it than by adhering to it; then it is not clear why anyone should get very excited about it. In particular, this point is not strong enough to support any stronger claim about the benefits of proliferating theories than the claim that by doing so it is *possible* that one might come up with a better theory than one already has. The fact that a certain strategy *might* work is not a good-enough reason to pursue that strategy. If my car won't start, I *might* fix it by randomly loosening or tightening bolts and clamps, but that does not make this an advisable strategy for fixing cars. Peter Achinstein argues that if we try to interpret Feyerabend's proliferation thesis in a stronger form, it becomes 'absurd.' Specifically, Achinstein proposes the following way of giving Feyerabend's position "some teeth": "Believe or accept an 'invented and elaborated' theory as true, or probable, or as a good predictor, or as good in some way, if it is inconsistent with the accepted point of view, even if the latter should happen to be highly confirmed" (Achinstein, 2000, 38).

As Achinstein notes, we cannot take this proposal seriously. It asks us to believe or accept *too many* theories (there is no limit to the number of theories one can cook up if one does not care whether they have any evidence or plausibility). Moreover, the mere fact of being inconsistent with an accepted theory or point of view does not by itself constitute any *reason* to believe or accept a theory.

We might better appreciate Feyerabend's point – and its limitations – by revisiting his frequent citations of Mill. Mill did not advocate believing or accepting theories just because they disagree with accepted points of view. He advocated *attending* to arguments for claims that conflict with what is accepted, even if that requires having those arguments given by people who do not either believe or accept them. Similarly, I read Feyerabend as recommending simply that scientists (and others) be willing to try out *some* ideas that conflict with accepted theories (even well-supported ones), so that one can find out how good those ideas can be made. However, this bit of methodological advice is at best suggestive, leaving the question of *which* ideas to pursue completely open. This leaves Feyerabend's philosophy of science with a large gap in its *positive* recommendations.

This gap brings us back to Mill. An important aspect to Mill's philosophy that Feyerabend neglected was his positive contributions to scientific methodology. Although "On Liberty" figures among J. S. Mill's most influential writings, he devoted much more of his life to a weighty tome titled *A System of Logic*, which went through eight editions between 1843 and 1872. Prominent in this work is J. S. Mill's articulation of a set of methods for discovering causal relationships. ("Mill's methods" are widely taught to this day.) Although Mill's advocacy of methods of discovery might seem incompatible with the 'against method' position Feyerabend finds in "On Liberty," I have argued elsewhere that a careful reading of *A System of Logic* reveals Mill's nuanced understanding of the limitations on the rules of method that he advocates (Staley, 1999b). In short, Mill saw no conflict between recognizing that *no rule of method is universal and exceptionless* (just as Feyerabend insists) and advocating nonetheless that *some rules of method can be helpful and usually trustworthy*.

Ultimately the value of Feyerabend's contributions to the philosophy of science may lie as much in the way he made them as in their content. His provocateur's stance reflected his personality and his approach to philosophy. He wrote less to express his convictions than provoke others to question theirs. Consider the following passage from a dialog he wrote:

A: Are you an anarchist?
B: I don't know – I haven't considered the matter.
A: But you have written a book on anarchism!
B: And?
A: Don't you want to be taken seriously?
B: What has that got to do with it?
A: I do not understand you.
B: When a good play is performed the audience takes the action and the speeches of the actors very seriously; they identify now with the one, now with the other character and they do so even though they know that the actor playing the puritan is a rake in his private life and the bomb-throwing anarchist a frightened mouse.
A: But they take the writer seriously!
B: No, they don't! When the play gets hold of them they feel constrained to consider problems they never thought about no matter what additional information they may obtain when the play is over. And this additional information is not really relevant . . .
A: But assume the writer produced a clever hoax . . .
B: What do you mean – hoax? He wrote a play didn't he? The play had some effect, didn't it? It made people think, didn't it? (Feyerabend, 1991, 50–51)

Institutionalized science carries a great deal of cultural authority, and Feyerabend, while valuing scientific inquiry a great deal, sought to make people think critically about the basis for that authority, and about how beliefs that lack the backing of scientific authority get devalued. A danger of such criticism, if taken too far, is that it may lead us to fail to realize the potential benefits of scientific knowledge. But Feyerabend insisted that beliefs could be good in ways not recognized by the standards adopted by scientific institutions and that people might legitimately value such beliefs highly. He warned against failing to understand such beliefs simply because one has not taken the time to give them a fair and knowledgeable hearing.

5.8 Conclusion

What makes it rational to accept a scientific claim? What constitutes progress in science? How are these two concepts – rationality and progress – related? We can now see how challenging it is to answer the first two questions, on which the answer to the third question crucially depends.

Although the authors considered in this chapter all approach these questions in different ways, they share a conviction that the kind of formal approach advocated by logical empiricists – and the accompanying attempts to legislate what kind of content is compatible with an acceptably empirical science – is irreconcilable with the historical record of scientific change.

In attempting to articulate an idea of scientific change that allows for a greater absorption of theoretical elements into the empirical results constituting scientific evidence, they risk running afoul of other aspects of scientific practice. This holds particularly for advocates of incommensurability, when we consider scientific practices that seem designed to accomplish what these advocates consider impossible: the securing of a stable basis for the assessment of evidence that can survive changes – even revolutionary changes by Kuhn's standards – in a theoretical framework.

Consider an example from the beginning of evolutionary phylogenetics. In *Origin of Species*, Charles Darwin noted the structural commonalities between a bat's wings and (for example) the hands of humans as evidence that these distinct species share a common ancestry (Darwin, 1859, 434). Darwin was able to make this inference in spite of the fact that he had no good theory of the physical basis of inheritance. A revolution in biology brought together Darwin's theory of evolution by natural selection and the genetic theory of inheritance that emerged from the work done by Gregor Mendel in the nineteenth century and developed by an extended experimental and theoretical undertaking during the first two decades of the twentieth century. After this revolution in our understanding of genetics and evolution, that same shared morphology between bats and primates continued to serve as evidence of common ancestry, although just what that involves had been given a much deeper level of explanation.

Moreover, scientists, at least sometimes, *anticipate* significant changes in theorizing (without knowing just what those changes will be) and take measures in their experimental work to secure their evidence against such changes. For example, when he was performing experiments that provided crucial insights into the nature of electromagnetism in the middle of the nineteenth century, Michael Faraday took careful measures to describe his experimental results in a manner that would make them independent of any of the theories of electricity or magnetism of his day (Cobb, 2009). Isaac Newton, in the course of arguing for his Universal Law of Gravitation, ensured that his inference, though extremely bold in its scope, was free of any commitment to particular views about the cause of gravity, an issue that he regarded as unsettled (Harper, 2002, 2011).

Making sense of these historical episodes need not appeal to a theory-neutral observation language of the sort espoused by logical empiricists. Mary Hesse, for example, accepts Kuhn's thesis of theory-ladenness of observation, while defending the possibility of paradigm-neutral theory testing in her *network model* of scientific theories (Hesse, 1980). Hesse's argument relies on the observation that even where a term used in reports of observations does have different meanings in different theoretical networks, there will often be an area of "*intersection* of predicates and laws between the theories," the exploitation of which can yield paradigm-independent judgments about competing theories (Hesse, 1980, 97, original emphasis).

The thesis of methodological incommensurability faces another challenge: Scientific inquiry relies, to a large extent, on appraisal methods that survive changes in particular theories in particular disciplines. Moreover, when such methods do change, it is often unrelated to any particular theory to which those methods might be applied.

Consider the developments in theoretical statistics that have provided scientists in essentially every discipline with new tools both for evaluating the reliability and probative power of their testing procedures and for making the most efficient use of the data they produce. These developments have their own fascinating history (Hacking, 1990; Porter, 1988; Stigler, 1990), even as debates over the philosophy and practice of statistics continue (Mayo, 2018). Even if we regard these developments as Kuhnian revolutions, methods arising from advances in statistical theory contribute to our means for building up and evaluating knowledge from experimental data independently of the reigning paradigms in particular disciplines, whether epidemiology or cognitive neuroscience or astrophysics.

If some of the more sweeping claims of the postpositivists are hard to square with the facts about scientific practice, their skepticism about the possibility of rendering scientific progress and rationality as the outcome of any very strictly rule-governed process is not easily dismissed. To say that some methodological practices and standards survive changes in general theories – and that strategies are available to increase the chances that evidence does the same – does not suffice to restore the viability of a framework for evaluating the degree of confirmation for any theory, given any observation – without regard to any substantive claims – guided only by universal rules of method.

Interlude: Robert Boyle's Experiments with the Air Pump

In 1644, Evangelista Torriccelli (1608–1647) proposed an experiment that Vincenzo Viviani (1622–1703) then performed. Viviani filled a glass tube – sealed at one end – with mercury, then inverted that tube into a dish that also contained mercury, so that the open end of the tube was submerged in the mercury in the dish. A space appeared at the closed top of the tube, but a column of mercury remained in the tube, elevated at a height of 29 inches above the level of mercury in the dish. What, if anything, was in the space above the column of mercury? What kept mercury in the tube from falling down all the way into the dish? Natural philosophers across Europe were divided about whether a truly empty space, a void, was even possible. The dominant view, drawn from Aristotle, was that it was not, but interest in alternatives to Aristotle was growing, and the experiment of Torricelli and Viviani encouraged the idea that one might investigate this 'Torricellian space' and the elevation of the column of mercury beneath it experimentally.

Indeed, Robert Boyle (1627–1691) developed experimental techniques that allowed him to do just that. In particular, he developed an experimental technique to test the idea that the height of the column of mercury in Torricelli's tube was to be explained as the effect of an equilibrium between the cylinder of mercury in the tube and "the cylinder of air supposed to reach from the adjacent mercury to the top of the atmosphere" (Boyle, [1660] 1965, 33; see Figure I.1). The weight of the atmospheric air presses down on the mercury in the dish, exerting pressure on it that sustains the column at a height of about 29 inches. Although the mercury also pushes back against the atmospheric air, the air resists being compressed (Boyle calls this property the 'spring' of the air). Thus, at a height of 29 inches the weight of the column of mercury is in a state of equilibrium with the weight of the atmospheric air (the height of the mercury column serves to measure the weight of atmospheric air). Boyle sought to use his newly invented air pump to test this explanation. This apparatus had a pumping mechanism and a large glass receiver with an opening at the top that could then be sealed shut. (The glass-blowing techniques of the time could not produce a usable receiver larger than about 30 quarts – 28 liters – in volume.) Once sealed, the pump could be activated to suck air from the receiver.

To test the equilibrium explanation for the height of the mercury column in Torricelli's tube, Boyle placed into the receiver of his air pump a 3-foot-long tube filled with mercury and inverted into a dish containing mercury, similar to Torricelli's setup. The tube was too tall to fit inside the receiver. Where the tube extended out of the receiver, Boyle filled and sealed the space around it with a plaster from a substance called diachylon. When the pump was activated, the column of mercury in the tube fell. Continued operation of the pump lowered the mercury column further for about 'a quarter of an hour,' after which time the mercury in the tube remained slightly above the

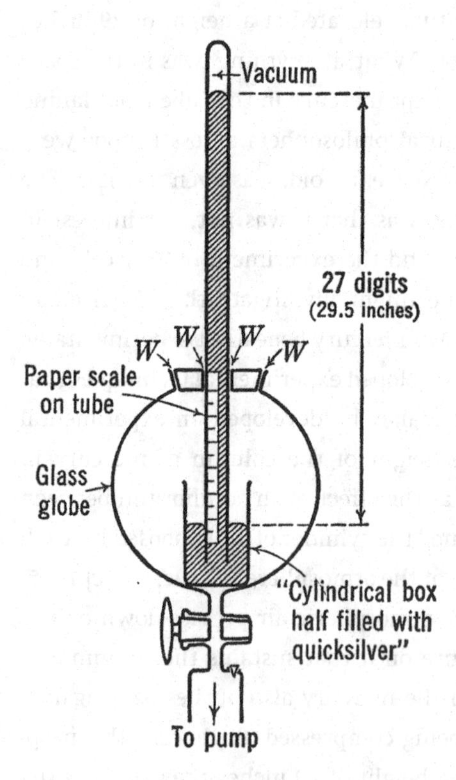

Figure I.1 Boyle's air pump setup to test his explanation of the 'Torricellian column.' The *W*s indicate spaces where diachylon is used to seal up spaces near where the tube extends above the receiver. From Conant (1957). Copyright © 1948, 1950, 1952, 1953, 1954, 1957 by the President and Fellows of Harvard College. Used by permission. All rights reserved.

mercury in the dish. Boyle explained this remaining very short column of mercury in the tube by supposing that, although "the receiver was considerably emptied" of air, some would "press in at some little avenue or other," in an amount that, although small, "was sufficient to counterbalance the pressure of so small a cylinder of quicksilver, as then remained in the tube" (Boyle, [1660] 1965, 33–34).

Boyle followed up this test by confirming that the column returned to nearly its original height when air was allowed to return to the receiver, and by repeating his experiment in the presence of "those excellent and deservedly famous Mathematic Professors, Dr. Wallis, Dr. Ward, and Mr. Wren"[1] (Boyle, [1660] 1965, 34). In addition, having pumped air out of the receiver and observing the descent of the column of mercury, he noted that the column would not only rise again to its original height when 'external air' was allowed to return to the receiver but also rise 'much above' its original height when additional air was pumped into the receiver. When that additional air was released, it "would fall again to the height it rested at before" (Boyle, [1660] 1965, 36).

Significantly, Boyle treats this experiment as a test of the equilibrium explanation of the Torricellian phenomenon, but *not* as a test regarding claims about the existence of a vacuum as debated by natural philosophers since at least Aristotle. As a test of the equilibrium explanation, the experiment works by creating conditions that approximate the *absence* of the column of atmospheric air that stands in an equilibrium relationship to the column of mercury in the Torricellian tube (the Torricellian tube being contained within a vessel that is nearly empty of atmospheric air): "[I]f we could perfectly draw the air out of the receiver, it would conduce as well to our purpose, as if we were allowed to try the experiment beyond the atmosphere" (Boyle, [1660] 1965, 33). But this does not suffice to draw conclusions regarding the existence of a vacuum, and Boyle declines to proclaim regarding 'that famous question.' He observes that "notwithstanding the exsuction of the air," light passes through the space of the receiver, enabling things inside of it (like the mercury column) to be seen. He also notes the

[1] John Wallis (1616–1703) was a clergyman and mathematician. Seth Ward (1617–1689) was an astronomer, mathematician, and bishop. Christopher Wren (1632–1723) was an architect, physicist, and mathematician. All were active in the early years of the Royal Society. Boyle and Wren were both founding members.

results of a previous experiment that showed that magnetic forces could act on things inside of the receiver (Boyle, [1660] 1965, 36–37). In the seventeenth century, both light and magnetism were generally assumed to involve some corporeal processes, though their exact natures were subject to unresolved disputes.

Boyle's first publication of results from experiments using the air pump drew criticism. In response, he performed further experiments, some leading him to assert a quantitative relationship between pressure and volume of air, close to what is today known as Boyle's law. Boyle's law states that at constant temperature, the pressure and volume are reciprocally related. This law is only approximately true (it strictly applies only to 'ideal gases'), but remains a standard element in the introductory courses in chemistry.

The criticisms that prompted these additional experiments came from Thomas Hobbes (1588–1679) and Franciscus Linus (1595–1675). (We will discuss Hobbes's criticisms in Chapter 6.) In a 1661 publication, Linus, 'a committed Aristotelian,' regards Boyle as defending the existence of the vacuum (Shapin & Schaffer, 1985, 156). Linus had noticed that if one performed the Torricellian experiment with a tube open at both ends but closed at the upper end using one's finger, they would experience the sensation of the finger being pulled down into the tube. Linus thinks this contradicts Boyle's explanation that the column of mercury is sustained by the pressure of atmospheric air pushing down on the mercury in the dish. He postulates an alternative explanation that does not involve a vacuum: There is, in the space above the column of mercury, a thread (or *funiculus*, perhaps consisting of rarified mercury) with its one end attached to the surface of the mercury and extending upward from there. If the top end of the tube is closed with one's finger, then the top end of the thread attaches to the finger and the weight of the mercury pulls the finger down while the tension in the thread pulls the column of mercury up, sustaining it in the tube (Shapin & Schaffer, 1985, 157–159).

Boyle describes his "adversary's hypothesis" of the funiculus as "needless" and proposes to "manifest by experiments purposely made" that the properties of spring (resistance to compression) and weight (or pressure) of air are fully capable of producing the phenomena in Torricelli's experiment (Boyle, [1662] 1965, 156). The experiment he then describes uses a J-shaped tube, closed at the end of its shorter leg, to which is attached a paper scale. Boyle filled the tube with mercury until the height of the mercury column in both

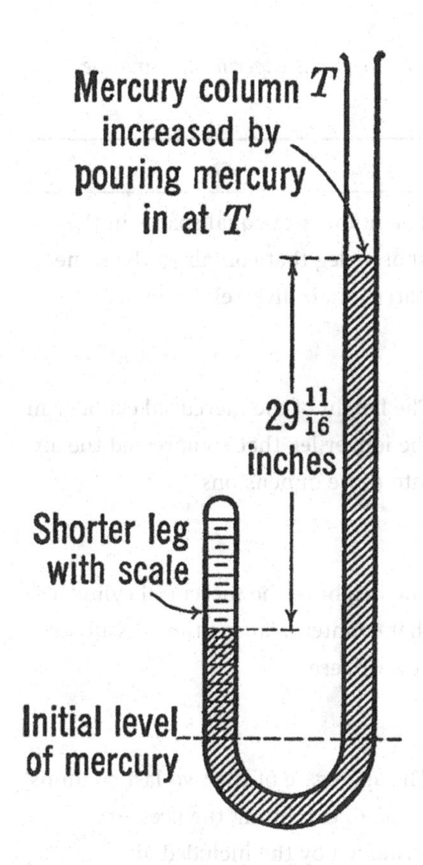

Figure I.2 Boyle's J-shaped tube. From Conant (1957). Copyright © 1948, 1950, 1952, 1953, 1954, 1957 by the President and Fellows of Harvard College. Used by permission.

legs was equal (see Figure I.2). Using the paper scale attached to the short leg, he divided the space above the mercury in that leg into forty-eight equal spaces. He then proceeded to add more mercury to the long open leg of the tube in increments. As he added more mercury, the column in the short leg rose, decreasing the space above the mercury there. Each time that space shrank by two of the original forty-eight intervals, he recorded the height of the mercury in the long leg of the tube, above the level of the mercury in the short leg. Boyle takes this to be the measure of the mercury that compressed the enclosed air (see Table I.1). By adding to that height the 29⅛ inches of mercury that would be equivalent to the weight of the atmospheric air, he calculates – for each volume of enclosed air as it grows smaller and smaller – the "pressure sustained" by that air (Boyle, [1662] 1965, 158). He recorded these numbers in a table and in a final column calculated the pressure that would be

Table I.1 *Boyle's data on the compression of air from his experiment using the J-shaped tube*

A	*B*	*C*	*D*	*E*
48	00	Added to 29 1/8 makes	29 2/16	29 2/16
46	01 7/16		30 9/16	30 6/16
44	02 13/16		31 15/16	31 12/16
42	04 6/16		33 8/16	33 1/7
40	06 3/16		35 5/16	35
38	07 14/16		37	36 15/19
36	10 2/16		39 5/16	38 7/8
34	12 8/16		41 10/16	41 2/17
32	15 1/16		44 3/16	43 11/16
30	17 15/16		47 1/16	46 3/5
28	21 3/16		50 5/16	50
26	25 3/16		54 5/16	53 10/13
24	29 11/16		58 13/16	58 2/8
23	32 3/16		61 5/16	60 18/23
22	34 15/16		64 1/16	63 6/11
21	37 15/16		67 1/16	66 4/7
20	41 9/16		70 11/16	70
19	45		74 2/16	73 11/19
18	48 12/16		77 14/16	77 2/3
17	53 11/16		82 12/16	82 4/17
16	58 2/16		87 14/16	87 3/8
15	63 15/16		93 1/16	93 1/5
14	71 5/16		100 7/16	99 6/7
13	78 11/16		107 13/16	107 7/13
12	88 7/16		117 9/16	116 4/8

A. The number of equal spaces in the shorter leg that contained the same parcel of air diversely extended.

B. The height of the mercurial cylinder in the longer leg that compressed the air into those dimensions.

C. The height of the mercurial cylinder that counterbalanced the pressure of atmosphere.

D. The aggregate of the two last columns *B* and *C*, exhibiting the pressure sustained by the included air.

E. What that pressure should be according to the hypothesis that supposes the pressures and expansions to be in reciprocal proportion.

Source: Conant (1957, 53, table 1).

sustained by the enclosed air "according to the hypothesis that supposes the pressures and expansions to be in reciprocal relation" (Boyle, [1662] 1965). Of course, this last number represents the value of the pressure that would be predicted by applying what we know today as Boyle's law. Although Boyle performed no statistical analysis on his data, the agreement between this predicted value and the number obtained by adding

29⅛ inches to the measured height of the mercury column in the long leg of the tube is reasonably good, given the limited accuracy of the available means of measurement (and given that Boyle's law itself holds only approximately).[2]

Boyle took this experiment not only as supporting his hypothesis regarding the reciprocal relationship between volume and pressure but also as serving to demonstrate, *contra* Linus, the adequacy of air's properties of spring and weight to explain the Torricellian phenomenon. The crucial point that he drew from his experiment was the potency of the spring of air. When the column of mercury in the long leg of the tube stood 29 inches above that in the short leg, the pressure sustained by the enclosed air was double that when the mercury was at the same level in both legs. At its greatest height of 88 inches above the mercury in the short leg, the pressure sustained was four times its original. Boyle comments, "And there is no cause to doubt, that if we had been furnished with a greater quantity of quicksilver and a very strong tube, we might, by a further compression of the included air, have made it counterbalance the pressure of a far taller and heavier cylinder of mercury" (Boyle, [1662] 1965, 159). The spring of the air is thus able to resist at least a weight (including that of the atmospheric air) of well over 100 inches – nearly 10 feet – of mercury, "and that without the assistance of [Linus's postulated] Funiculus, which in our present case has nothing to do" (Boyle, [1662] 1965, 159).

Linus hypothesized the funiculus to explain why the Torricellian column stays up in the tube. His fiber does its work by means of tension. But in the J-tube experiment, the phenomenon involves compression; appeal to the funiculus is not needed.

This establishes that the funiculus hypothesis is not *necessary* to explain the results of the J-tube experiment. But Boyle goes further and adds a test that counts against Linus's alternative hypothesis by producing a phenomenon that the funiculus is not *sufficient* to account for: "[W]e took care, when the mercurial cylinder in the longer leg of the pipe was about an hundred inches high," to suck on the open end of the tube. This caused the mercury in

[2] Boyle comments that although "in our table some particulars do not so exactly" agree with the predicted vales, "yet the variations are not so considerable, but that they may probably enough be ascribed to some such want of exactness as in such nice experiments is scarce avoidable" (Boyle, [1662] 1965, 159).

the long leg "to notably ascend" (Boyle, [1662] 1965). To explain the Torricellian column in the original experiment, the funiculus must not be strong enough to lift up the mercury column to a height of over 29 inches, but here, by partially evacuating air from above the column using suction produced by his mouth, Boyle is able to lift up a column of nearly 100 inches of mercury, which Linus's funiculus could manifestly not do. An explanation in terms of pressure and spring is readily available:

> [T]he pressure of the incumbent air being in part taken off by its expanding itself into the sucker's dilated chest; the imprisoned air was thereby enabled to dilate itself manifestly, and repel the mercury, that compress it, till there was an equality of force betwixt the strong spring of that compress air on the one part, and the tall mercurial cylinder, together with the contiguous dilated air, on the other part. (Boyle, [1662] 1965)

In this way, experimental support both for what we now know as Boyle's law and for the potential for air's properties of spring and weight to explain experimental phenomena went hand in hand, forming an important precedent for future experimental investigations.

6 Relativism and Social Constructionism

6.1 Introduction

A Kuhnian account of the discovery of Boyle's law would begin with the paradigms in place or in dispute at the time, perhaps dividing them into the plenist opponents of Boyle and those mechanical philosophers who accepted the possibility of a void. Exactly what Boyle had discovered with his air pump and with his experiment using the J-tube would, on such an account, differ depending on the paradigm from which the experiments were assessed. The apparent result is that we must evaluate whether Boyle's explanation of the changes observed in these experiments counts as scientific knowledge *relative to* some relevant paradigm.

Philosophers use the term *relativism* to refer to a broad range of views that have in common their susceptibility to being characterized in part by a claim of the form '*x* is relative to *y*' (Kusch, 2020). The variable *x* in this formula might refer to properties of things, moral values, or knowledge. The variable *y* might refer to individuals, cultures, social identities, or (as in the case of Kuhn) scientific paradigms.

Philosophers of science debate relativism as it applies to the status of knowledge (or sometimes truth) and whether it is relative to social conditions, interests, culture, or individual beliefs or perspectives. But debates over relativism about knowledge inevitably draw in other modes of relativism. Larry Laudan identifies three relativistic theses shared among the group that he identifies as postpositivists,[1] in a critique directed at both them and

[1] Laudan focuses on Kuhn and Feyerabend and also mentions "the later Wittgenstein, the later Quine, the later Goodman, Rorty, and dozens of lesser lights" (Laudan, 1996, 4). Laudan's main thesis is that the presuppositions that led the postpositivists to their relativist positions are *shared* with the logical empiricists; the key to overcoming the mistakes of both camps is to reject these presuppositions. In short, Laudan argues that

their logical empiricist antagonists: "(1) that evidence radically underdetermines theory choice – to the extent that virtually any theory can be rationally retained in the face of any conceivable evidence (*epistemic relativism*); (2) that the standards for theory evaluation are mere conventions, reflecting no facts of the matter (*metamethodological relativism*); and (3) that one conceptual framework or worldview cannot be made intelligible in the language of a rival (*linguistic relativism*)" (Laudan, 1996, 5, emphasis in original).

We have previously encountered some aspects of this kind of relativism. What Laudan refers to as epistemic relativism is clearly rooted in the problem of underdetermination discussed in Chapter 3, as that problem is formulated by W. V. O. Quine, who concluded that "[a]ny statement can be held true come what may, if we make drastic enough adjustments elsewhere in the system" (Quine, 1980, 43). You may also recall Laudan's criticism of this claim: Deciding what one can conclude from given evidence is not simply a function of what is logically permissible. Scientists rely on methodological norms that go beyond the principles of deductive logic.

The argument over relativism must then turn to the second relativistic thesis – metamethodological relativism. The relativist who holds this position will not be persuaded by Laudan's appeal to methodological norms that transcend deductive logic, for any such norms are, according to metamethodological relativism, merely conventions and thus contingent and arbitrary. To really confront relativism, then, requires consideration of the source and status of the methodological norms on which scientists rely.

Constructionism (also called constructivism) constitutes a closely related philosophical category to relativism that provides one kind of answer to our question about methodological norms. To describe a philosophical account as constructionist is to invoke a metaphor at odds with the metaphor of discovery. Describing Boyle's discovery of the reciprocal relation between pressure and volume as a *discovery* connotes that this relationship existed prior to and independently of Boyle's experiments. 'I discovered a huge spider living in my dresser drawer' is an inapt thing to say if you

the postpositivists accept the same unrealistic and algorithmic conception of the rationality of science that the logical empiricists embrace. Upon finding that the historical record of scientific inquiry could not be reconciled with that view, they conclude that scientific inquiry is not rational when they should reject the conception of rationality that logical empiricists rely upon.

assembled a spider habitat in your dresser drawer, purchased a tarantula from a pet store, and placed it in your dresser drawer. ('I made a home for my pet spider in my dresser drawer' would be more accurate.[2]) Constructionist views about knowledge reject the idea that when we gain some new knowledge of the world, we come into contact with a truth that precedes us and is independent of us, and which simply awaits our forming the right belief in the right way to make the leap from being true to being known. We make knowledge by a certain kind of process, and different views about the process yield different sorts of constructionist positions. The term *social constructionist* applies to those who regard the construction of knowledge as a social process.

6.2 Explaining Knowledge, Explaining Consensus

The social constructionist approach to scientific knowledge emerged from attempts by sociologists to apply their theories and research methods not only to the social structures of the scientific community but also to the results of scientific inquiry. Such efforts were pioneered by a few researchers in the 1970s and expanded into a significant approach in the sociology of science in the 1980s and 1990s. When Ian Hacking published a critical assessment of the idea of social construction in 1999, he opened with a nonexhaustive list of twenty-four things, the construction of which was mentioned in twenty-nine book titles in the library catalog – from authorship to Zulu nationalism. Notable entries from the middle of the alphabet included facts, illness, knowledge (four titles), nature, quarks, and reality (Hacking, 1999, 1).

Among the most influential early works in this tradition, David Bloor's *Knowledge and Social Imagery* is noteworthy for its expression of the social constructionist viewpoint on scientific knowledge as a challenge to the approaches taken by philosophers of science:

> The sociologist is concerned with knowledge, including scientific knowledge, purely as a natural phenomenon. The appropriate definition of knowledge will therefore be rather different from that of either the layman or the

[2] I'm no expert, but I suspect that a dresser drawer is a poor choice of habitat for a pet tarantula.

> philosopher. Instead of defining it as true belief – or perhaps, justified true belief – knowledge for the sociologist is whatever people take to be knowledge. (Bloor, 1976, 5)

Bloor goes on to lay out a program for the sociological study of knowledge, how and why it changes, how it is distributed, maintained, disseminated, organized, and so on. In pursuing this study, the sociologist will operate "in the same causal idiom as any other scientist" (Bloor, 1976, 5). Such a scientific approach to the study of science commits the sociologist to exemplifying four values that "are taken for granted in other scientific disciplines." These four values constitute the heart of the "strong program in sociology of knowledge." This approach, Bloor writes, will be (1) concerned with the causes (social but also other kinds) of beliefs; (2) impartial in regard to what requires explanation between beliefs that are true or false, or rational or irrational; (3) "symmetrical in its style of explanation," giving the same kind of explanation, for example, for true and false beliefs; and (4) reflexive in applying the same approach to the sociology of knowledge itself that it applies to other sciences (Bloor considers such reflexivity necessary to prevent sociology from becoming a "refutation of its own theories") (Bloor, 1976, 7).

The strong program in sociology of knowledge rests on two steps. First, the investigator defines knowledge in terms of consensus or shared belief. Second, they explain knowledge thus defined by examining the causes of consensus without regard to any putative status of that consensus as true, justified, rational, or not. Bloor emphasizes the contrast between such an approach and the 'teleological' approach of philosophers of science who treat 'successful' or 'rational' inquiry separately from that which leads to error or failure. Teleological approaches (such as Lakatos's rational reconstructions) explain research outcomes by appeal to social or psychological conditions in cases judged to result in drawing the wrong conclusion, or failing to accept the right conclusion. When a scientist succeeds, it is because their beliefs have been formed on the basis of good reasons and no social or psychological conditions need to be invoked. The teleological view regards Boyle's opponents, such as Linus and Hobbes, as having failed to appreciate his discoveries regarding the spring of air because their prior commitments to plenism and the idea that 'nature abhors a vacuum' prevent them psychologically from being able to see clearly the plain evidence offered by the air pump and J-tube experiments. Boyle himself, on the other hand, accepted these conclusions simply because the evidence supported them. (Exactly what philosophies of science are truly committed to the approach that Bloor characterizes here is not clear, and the possibility of a 'straw man' account of one's opponents should be kept in mind.)

The strong program promises a more 'naturalistic' alternative to teleological explanations of scientific knowledge. By describing the approach as naturalistic, advocates of the strong program refer to its reliance on the empirical methods of sociology to answer questions about how scientific knowledge comes into being. Such an approach, however, requires both of the steps just mentioned. In order to explain knowledge by means of the methods of sociology, knowledge must already be defined as the sort of thing that the social processes studied by sociologists are capable of producing. Treating knowledge (and consequently truth) as a matter of consensus – or, perhaps more to the point, convention – allows such a treatment.

Bloor's relativism is explicit: The "methodological relativism" of the strong program is "summarised in [its] symmetry and reflexivity requirements" (Bloor, 1976, 142). The fact that "the foul Pit of Relativism" is conventionally regarded as something mightily to be resisted in favor of the "high Peaks of Truth" should not, he argues, deflect the sociology of knowledge from its naturalistic path: "Relativism is simply the opposite of absolutism, and is surely preferable. In some forms it can at least be held authentically in the light our social experience" (Bloor, 1976, 142). Indeed, Bloor argues that both Popper's falsificationism and Kuhn's evolutionary view of scientific 'progress' require a relativistic understanding of scientific knowledge: Popper's insistence on the conjectural status of knowledge claims and acknowledgment that "[n]othing is absolute and final" entail relativism, just as Kuhn's denial that science progresses toward a fixed goal that exists independently of its own products does.

6.3 Underdetermination Arguments for Social Constructionism

6.3.1 Contending Views of Social Order

To get a more concrete view of the disputes between social constructionists and their critics, let's return to the example of the air pump.

An influential but highly contested social constructionist account of Boyle's discoveries is given by Steven Shapin and Simon Schaffer (Shapin & Schaffer, 1985). In their account, Boyle seeks not only to develop and apply experimental techniques to answer questions about the spring of air and establish laws regarding the relationship between air pressure and volume. Boyle is advocating, they argue, a "form of life" involving "technical, literary, and social practices" that enable the production of "experimental matters of fact" that could become the subject of consensus, shielded from "items of knowledge that were thought to generate discord and conflict" (Shapin & Schaffer, 1985, 18). One such source of conflict was an alternative conception of natural philosophy advocated by Thomas Hobbes (1588–1679), best known for his views regarding political authority as laid out in his 1651 publication *Leviathan*. Hobbes disputed the sufficiency of Boyle's air pump for establishing the experimental facts that Boyle claimed to have found through its use in experiments. Hobbes's objection to Boyle's claim rested in part on specific weaknesses regarding the adequacy of the apparatus (in short, the air pump leaked) and in part – Shapin and Schaffer claim – on the basis of a conception of philosophy itself, which Hobbes understood to be concerned with the causes of phenomena and governed by an ideal of demonstrative proof, such as one finds in geometry. This contrasts with Boyle's acceptance of experimental results on the basis of imperfect but sufficient evidence.

Shapin and Schaffer connect the dispute between Boyle and Hobbes over natural philosophy to a dispute over the political organization of the community of natural philosophers and the political organization of the broader society. They shared a concern about the problem of *dissension* (with the British Civil Wars of the mid seventeenth century still fresh in memory). The solutions they favored were quite different, however.

Boyle sought a "calm space" in which natural philosophers could form consensus, lay the foundations for philosophical knowledge, and "establish their credit in Restoration culture"[3] (Shapin & Schaffer, 1985, 76). To create such a space, he considered it crucial to distinguish between matters of fact (experimental facts, and the kinds of laws that could be established by them) and metaphysical claims. Boyle did not take a position regarding the debate

[3] The 'Restoration' period of British history refers to the restoration of the Stuart monarchy in England, Scotland, and Ireland in 1660 and some number of years following that event.

over the conceptual possibility of a void in nature in the sense of a space containing no material thing whatsoever. He did not, therefore, claim to have produced a void or "true vacuum" in the receiver of the air pump, but only "a space . . . such as is either altogether, or almost totally devoid of air" (quoted in Sargent, 1995, 39).

In Shapin and Schaffer's portrayal, this strategy of eschewing "first causes," or metaphysical principles, in experimental inquiry is part of a "language-game that Boyle was teaching the experimental philosopher to play" (Shapin & Schaffer, 1985, 51). The community of experimental philosophers could achieve consensus around matters of fact in virtue of their ability to witness the experiments performed (either actually or "virtually," by reading descriptions of the experiments). Performing experiments at the Royal Society, for example, provides both multiple witnesses, and witnesses with the social and moral standing to serve as trustworthy sources. The promotion of the experimental philosophy coincides with the establishment of a social category of competent witnesses. Shapin and Schaffer conclude that "the matter of fact was a social as well as an intellectual category" (Shapin & Schaffer, 1985, 69).

Hobbes's ideas about knowledge and about society are similarly linked in the Shapin and Schaffer account. His model of knowledge is based on the example of geometry, "the only science that it hath pleased God hitherto to bestow on mankind" (quoted in Shapin & Schaffer, 1985, 100). In short, philosophical knowledge requires beginning with the assigning of definitions and proceeding to demonstrations that are capable of producing universal assent. Hobbes builds a natural philosophy on such a model, beginning with definitions and proceeding with demonstrations. The main points of Shapin and Schaffer's account of Hobbes's natural philosophy are: (1) Hobbes does not admit the existence of any noncorporeal things, including spaces empty of bodies; his ontology is monist (there is just one kind of thing) and materialist (that one kind of thing is matter). (2) Hobbes uses his monist materialism to provide an alternative account of the phenomena that Boyle understands in terms of spaces being entirely or approximately empty of air; for example, the 'Torricellian space' above the column of mercury in the inverted tube inserted in the dish of mercury is filled with air that is propelled up through the column of mercury in the tube, having been displaced from its previous location by the rise of the mercury in the dish (Shapin & Schaffer, 1985, 363–364). (3) Hobbes considers alternatives to his natural philosophy that allow the existence of nonmaterial things to be *politically dangerous*. Whereas Boyle was concerned to create a space in which one could safely dispute physical causes of motion, Hobbes sought to bring an end to public expressions of dissent and held that philosophy could only begin from *agreed-upon* definitions. He associated belief in nonmaterial entities (whether immaterial souls or spaces empty of bodies) with political forces opposed to the absolute political authority of the Sovereign. These forces included priests, religious dissenters, and the "Greshamites."[4]

In Shapin and Schaffer's telling, this broader context is vital for understanding how Boyle's experiments with the air pump served to establish facts about air pressure such as Boyle's law. Emphasizing the underdetermination problem, they argue that alternative conclusions (whether of Linus or Hobbes) could be drawn from Boyle's observations, by relying upon different assumptions than Boyle's. To understand why Boyle's conclusions were accepted rather than those of his opponents, we need to understand why Boyle's assumptions were accepted. Those assumptions concern not only how the air pump works but also how knowledge of the natural world is constituted. The triumph of Boyle over Hobbes regarding air pressure, then, must be understood as the triumph of Boyle's vision of the social order (in particular the establishment of an intellectual community with the authority to decide matters of experimental fact) over Hobbes's vision of a unified earthly authority (the Sovereign) without contending and autonomous authorities, whether they be priests or experimentalists ('priests of nature'). In Shapin and Schaffer's words, "In the course of offering solutions to the question of what proper philosophical knowledge was and how it was to be achieved, Hobbes and Boyle specified the rules and conventions of differing philosophical forms of life" (Shapin & Schaffer, 1985, 332). The knowledge produced by Boyle's experiments was also a product of the contest between these competing forms of life.

6.3.2 The Experimenters' Regress

An example from more recent science will illustrate another social constructionist argument. The sociologist Harry Collins began studying the community

[4] 'Greshamites' refers to the group of natural philosophers, including Boyle, whose meetings began on the occasion of a lecture by Christopher Wren at Gresham College in London in 1660, and which would become the Royal Society of London for Improving Natural Knowledge in 1663.

of physicists working on the detection of gravitational waves in the early 1970s. In 2016, that community reached a consensus that such waves (a phenomenon predicted by Einstein's General Theory of Relativity) had been observed by LIGO (the Laser Interferometer Gravitational Wave Observatory). In the early 1970s, though, physicists working on gravitational wave detection were sharply divided. Joseph Weber of the University of Maryland had built a gravitational wave detector and found an excess of vibrations beyond what could be attributed to background sources ('noise'). But the magnitude of his findings also greatly exceeded expectations of what his detector should have been able to register, given its sensitivity and predictions from relativity theory. So Weber enhanced his experiment by making improvements to the apparatus itself and also by looking for coincidences between signals from two detectors separated by 1,000 miles (such coincidences would be expected from the massive wavefronts of detectable gravitational waves). Other physicists who built their own devices based on the same principles as Weber's did not find the same excess, although Weber continued to find an excess even when using multiple detectors separated by great distances. When Collins interviewed physicists working on this problem, beginning in 1972, the question was unresolved: Had Weber's experimental apparatus detected gravitational waves?

Collins saw this case as exemplifying a general problem that he called *the experimenters' regress*. Whether we should accept Weber's claims about gravitational wave depends on trusting the apparatus he used. But our judgment about that apparatus depends on our judgment about whether it produces the correct outcomes. In this case, "[w]hat the correct outcome is depends upon whether there are gravity waves hitting the Earth in detectable fluxes. To find this out we must build a good gravity wave detector and have a look. But we won't know if we have built a good detector until we have tried it and obtained the correct outcome! But we don't know what the correct outcome is until … and so on *ad infinitum*" (Collins, 1992, 84).

Through interviews with physicists involved in debates over Weber's claim, Collins documented extensive disagreements not only over the credibility of that claim but also over the various arguments and evidence offered against it. Weber's critics cited different reasons why his claim should be rejected. Collins categorized some of the reasons for disbelief as "nonscientific," including: "Faith in experimental capabilities and honesty, based on a previous working partnership," "Personality and intelligence of experimenters," "Reputation of running a huge lab," "Style and presentation of results," "Size and prestige of university of origin," and "Nationality" (Collins, 1992, 87). Collins concludes that "scientific criteria" are insufficient to decide whether Weber's results establish his claim about gravitational waves and thus "the experimenters' regress leads scientists to reach for other criteria of quality" (Collins, 1992, 88). Settling the question of what counts as a good gravity wave detector, and settling the question of whether gravity waves exist (at the level claimed by Weber), are "congruent social processes" (Collins, 1992, 89).

The experimenters' regress constitutes a particular kind of underdetermination problem. Experimenters use an experimental setup to arrive at a result, but that result cannot on its own provide an answer to the experimenters' question. Determining an answer to the question requires in addition a judgment that the experimental setup is adequately trustworthy for its intended purpose. In instances of the experimenters' regress, Collins argues, the scientist cannot make that judgment 'scientifically' unless they already know the answer to the question. Of course, if they did know the answer, they would not need to perform the experiment.

In Collins's study of Weber's claims about high fluxes of gravitational waves, the resolution of this underdetermination problem is presented in a manner that exemplifies a social constructionist treatment. Collins argues that the controversy over Weber's results was resolved not by the presentation of reasons and evidence but also by the exertion of social forces of influence and persuasion. He emphasizes in particular the efforts of the leader of one research group (identified only as 'Q' in keeping with the sociological practice of anonymizing sources of evidence) who he describes as having set out to persuade the community to reject Weber's results: "Q acted as though he did not think that the simple presentation of results with only a low key comment would be sufficient to destroy the credibility of Weber's results. In other words, he acted as one might expect a scientist to act who realized that evidence and arguments alone are insufficient to settle unambiguously the existential status of a phenomenon" (Collins, 1992, 95).

In Collins's account, these efforts on the part of Q succeeded sufficiently well that the community of gravitational wave researchers largely came to agree that Weber's results did not support the existence of large fluxes of gravitational waves. (When the LIGO group later claimed the observation of gravity waves, they were using a detector with much greater sensitivity and found a much smaller flux.) But the content of this consensus made no reference to its origins in the socially efficacious efforts of Q, or any other aspect of that social process. The scientific fact became simply: No gravitational waves had been observed. In Collins's words, "*[T]he objects of science are made by hiding their social origins*" (Collins, 1992, 188, emphasis in original).

6.4 From Sociology to Anthropology of Knowledge

We have been considering sociological explanations of scientific knowledge, but sociology is not the only social science. Other social scientists have pursued lines of research that ultimately lead to the question of what, if anything, is distinctive about a *social* process of constructing knowledge.

The 1979 book *Laboratory Life: The Social Construction of Scientific Facts*, by Bruno Latour and Steve Woolgar, opens with an epigraph taken from Bloor's *Knowledge and Social Imagery*: "If sociology could not be applied in a thorough going way to scientific knowledge, it would mean that science could not scientifically know itself." Latour and Woolgar set forth their project in the context of discussions about the sociology of science and proceed to cast doubt upon the distinction – sometimes drawn by sociologists, as well as by scientists who are the subjects of sociological studies – between "social" and "technical" aspects of scientific work (the use of quotation marks around the two terms is maintained throughout their discussion of this issue). They describe their own approach as "for want of a better term, an anthropology of science" (Latour & Woolgar, 1986, 27). Latour observed the activities of scientists working in a laboratory[5] as an "ethnographic investigation of one specific group of scientists," obtained in "a particular setting," while "bracketing our familiarity with the object of our study" (Latour & Woolgar, 1986, 28–29). They describe the scientists they study as a "tribe whose daily manipulation and production of objects is in danger of being misunderstood, if accorded the high status with which its outputs are sometimes greeted by the outside world" (Latour & Woolgar, 1986, 29). They adopt as a "working definition" that they are concerned with "the *social*

[5] The laboratory was the Salk Institute for Biological Studies in La Jolla, California, although the only explicit reference in *Laboratory Life* to this fact appears in a brief introduction written by Jonas Salk himself.

construction of scientific knowledge in so far as this draws attention to the *process* by which scientists make sense of their observations" (Latour & Woolgar, 1986, 32, original emphasis).

Of course, processes can be described in different ways. Latour and Woolgar's anthropological methodology yields a social constructionism that emphasizes the avoidance of 'going native' in the study of scientists working at the lab. Their account of the laboratory as a "system of fact construction" (Latour & Woolgar, 1986, 41) avoids using the scientific and technical concepts employed by scientists in providing explanations, but instead regards the scientists' use of such concepts as something to be explained by reference to processes described in terms appropriate to the social sciences. At the core of their description of that process is the activity of *literary inscription*, the means by which workers at the lab construct facts. Latour and Woolgar distinguish *facts* using a classification of statement types, ranging from the taken-for-granted-and-not-worth-stating (type 5) to the highly conjectural (type 1). It is the type 4 and type 5 statements – those that are taken for granted, and those asserted without modal qualification,[6] especially in textbooks – that have achieved the status of facts.

The process of inscription[7] that transforms statements into facts begins with the materials and apparatus found in the laboratory. The latter include *inscription devices* that are apparatuses or configurations of apparatuses that "transform a material substance into a figure or diagram which is directly usable by" workers producing "a variety of documents, which are used to effect the transformation of statement types and so enhance or detract from their fact-like status" (Latour & Woolgar, 1986, 51, 151). The notion of an inscription device is itself "sociological by nature" because it is tied to the way in which an apparatus is used by a worker in the laboratory to perform the transformations in question (Latour & Woolgar, 1986, 89n).

From the standpoint of the scientists working in the laboratory, their work is aimed at the study of endocrinology, and in particular the regulation of the function of the thyroid gland. In pursuit of this broad aim, they

[6] A modal qualification, in Latour and Woolgar's usage, consists in a statement about a statement, most especially those that characterize a statement as being probable, uncertain, conjectural, hypothetical, and so on (Latour & Woolgar, 1986, 90).

[7] Latour and Woolgar attribute their notion of inscription ("an operation more basic than writing" that "summarizes all traces, spots, points, histograms, recorded numbers, spectra, peaks, and so on") to Jacques Derrida (1977) (Latour & Woolgar, 1986, 88n).

succeeded in synthesizing the hormone TRH (thyrotropin-releasing hormone, referred to in the La Jolla lab at the time of Latour and Woolgar's study as TRF – thyrotropin-releasing factor). TRH triggers the production of thyrotropin in the pituitary gland, and thyrotropin in turn runs the thyroid gland. For this work, Roger Guillemin of the Salk Institute shared the 1977 Nobel Prize in medicine with Andrew Schally.

Latour and Woolgar consider it a mistake, however, for an anthropologist to study the scientific process centered around the La Jolla lab from the standpoint of the scientists themselves. They do not describe the scientists' work as piecemeal experimental progress toward the eventual synthesizing of TRH, but rather as producing inscriptions that transform statements about TRH and related matters into facts. Such an account departs both from a scientist's account that explains the experimental evidence in its own terms and from a philosophical account in terms of norms of rational appraisal of that evidence. It is, in some sense, a social scientific account, but in what sense? In what way is this description social?

The second edition of *Laboratory Life* attempts to answer this question explicitly, by denying that they ever meant the term 'social' to be taken seriously. The term even disappears from the book's subtitle, now simply *The Construction of Scientific Facts*. In a postscript, they assert, "Given our explicit disavowal of 'social factors' in the first chapter, it is clear that our continued use of the term was ironic" and that by "demonstrating its pervasive applicability, the social study of science has rendered 'social' devoid of any meaning" (Latour & Woolgar, 1986, 281). The idea seems to be this: For the term 'social' to be meaningful, it must mark some kind of contrast with things that are *not* social. But by showing how 'social' scientific explanations apply to knowledge in every field of science, we dissolve that contrast: If everything is 'social,' then the term marks no real distinction and we can simply dispense with its (nonironic) use.

6.5 Criticisms: Reflexivity, Truth, and History

A persistent criticism holds that social constructionist analyses of scientific knowledge are essentially self-refuting. This criticism treats social constructionist analysis as a kind of debunking of scientific knowledge, reflected in some of the social constructionists' own language, such as Collins's claim that the "objects of science are made by hiding their social origins."

Producing scientific knowledge by hiding facts about how it is produced seems inherently deceptive. It would seem to follow that we should not give such knowledge claims the authority that scientists would like for us to grant them.

But social constructivists *also* claim that their work is simply the result of applying the methods of science to the scientific process itself. (Recall Latour and Woolgar's quotation from Bloor: "If sociology could not be applied in a thorough going way to scientific knowledge, it would mean that science could not scientifically know itself.")

Combining these two points yields the following argument: If the social constructionist account of scientific knowledge is correct, then scientific knowledge claims rest on deception regarding their justification. If a claim rests on deception regarding its justification, then it should not be trusted. The social constructionist account of any given scientific knowledge claim is arrived at by the application of scientific methods to the scientific process itself, and thus constitutes a claim of scientific knowledge like any other. Like other scientific claims, any social constructionist claim about scientific knowledge rests on deception regarding its own justification. Therefore, social constructionist accounts of science should not be trusted.

Something like this argument can be found, for example, in Mary Hesse's review of the first edition of Collins's book (Hesse, 1986), in which she places Collins's endorsement of science as "the best institution for generating knowledge about the natural world that we have" (Collins, 1992, 165) alongside his assertion that "[i]t is not the regularity of the world that imposes itself on our senses but the regularity of our institutionalized beliefs that imposes itself on the world" (Collins, 1992, 148). The first claim underwrites Collins's confidence in the very account he is providing, while the second seems to expose that account (along with the rest of the products of the scientific method) as a mere artifact of the social conditions of its production.

One response would be to attribute a special status to social scientific explanations in contrast to others. Reading social constructionist treatments of episodes of scientific inquiry often creates the impression that this attitude is being adopted.

But when the same authors confront the question explicitly, they typically embrace reflexivity and insist the kind of analysis they use in explaining scientific claims about gravitational waves, or air pressure, or quarks also applies to those very same explanations themselves. Just as the "objectivity" of scientific facts is a product of the social order in which consensus regarding those facts is achieved, so the 'objective' descriptions of how social order underwrites the formation of such consensus is likewise socially produced through persuasive actions of the social scientists who provide us with these narratives. How can the social scientist pursue such explanations without undermining their own efforts? As Collins sees it, the key is to put the activity of examining science in a different 'compartment' than the activity of doing science. The sociologist of science, like the scientists they study, seeks to make new scientific objects (perhaps facts about social order like institutions, networks, etc.). To do this, the social scientist "must ignore the social origins of these objects" just as the physicists do when they make their objects. One who is unable to engage in such compartmentalization "must either abandon sociology of scientific knowledge" or disavow any claims to "find" anything based on their studies (Collins, 1992, 188).

Another line of criticism brought against social constructionist approaches to scientific knowledge objects to their treatment of truth as consensus. The social constructionist, in seeking to give an entirely social explanation of how a scientific fact comes into being, treats that fact as consisting in something social: the agreement of the relevant scientific community that it is a fact. The proposition 'high fluxes of gravity waves reaching the surface of the earth do not exist' becomes true precisely when the community of gravity wave researchers agrees that Weber's results are not credible and that the findings of others ruling out high fluxes of gravity waves are credible. If air pressure, or quarks, or illness, or emotions are constructed by scientists, then they do not exist until that construction process is complete, and propositions that state facts about them are not true until they become accepted by the relevant scientific community. A proposition becoming true, thus, becomes dependent upon a seemingly very contingent matter of a certain opinion taking hold in some group of people.

Although such a view about truth might seem strange, this does not prove it wrong. Making sense of the notion of truth is a long-standing philosophical challenge. One influential formulation about truth is that a proposition is true just in case it 'corresponds to the facts' or 'to the way the world is.' This idea has a history going back at least to ideas expressed by Plato, but its canonical formulation for contemporary debates arrives in the early twentieth century in the work of authors such as Bertrand Russell (e.g., Russell,

1906–1907, 1918) and Ludwig Wittgenstein (Wittgenstein, 1921). A correspondence theory does not lead to the sense of strangeness that accompanies the consensus theory. But critics charge that it does not help us to understand truth better than we did before. The notion of 'correspondence' has proved difficult to render as something more than a platitude or obscure metaphor (though not for want of trying). In philosophers' efforts to make sense of truth, theories that avoid being merely platitudinous often wind up seeming counterintuitive.

However, the combination of reflexivity and a consensus theory of truth burdens the constructionist approach to scientific knowledge with a worse problem than merely seeming strange, according to an argument from Arthur Fine. An apparent advantage of the consensus theory is that it offers a 'definite' criterion for what to count as true: the consensus of the relevant community. If we want to judge whether a claim about the world is true, we look to the relevant community and ask whether it has reached a consensus. But this means that we have to "judge whether the relevant community has settled on the opinion that they have settled on an opinion about the original item. But to be able to judge whether that is true, given the consensus theory, we have to add yet another level of community judgment, and to be able to judge that." This adding of levels of community judgment has no end, and we wind up needing an "infinite tower of judgments" to give a definite answer to a question about the truth of any proposition whatsoever (Fine, 1996, 244).

As Fine goes on to note, the argument does not quite end there, for the constructionist is a "virtuoso . . . at dodging refuting arguments" (Fine, 1996). Nonetheless, this point about consensus theories of truth opens a window into some serious tensions within the social constructionist project. At the very least, we can say that this argument should disabuse one of thinking that a social constructionist approach will allow us an easy escape from the kind of problem posed by the experimenters' regress. A regress just as troubling lies hidden within the social constructionist account proposed to solve that problem.

Another line of criticism brought against social constructionist arguments disputes the accounts they offer of the resolution of scientific disagreements. In some instances, these criticisms dispute the accuracy of social constructionists' historical accounts. For example, Cassandra Pinnick (1998) charges Shapin and Schaffer with having misrepresented the methodological views of both Boyle and Hobbes in *Leviathan and the Air-Pump*. Other criticisms dispute the interpretations of how controversies are resolved.

An illuminating example of the latter concerns Collins's account of the 1970s' dispute over Weber's claim to have detected gravity waves. Allan Franklin undertook a careful review of the published papers and comments of participants in this dispute. One virtue of Franklin's work (in this and in other studies he has published of episodes in experimental physics) is that it avoids the kind of simplistic portrayal of scientists simply coming to form a belief because the evidence just *showed* it to be correct. (This seemed to be the potential straw man target of Bloor's critique of 'teleological' views.) According to Franklin, "the scientific community made a reasoned judgment" to reject Weber's claims based on "epistemological criteria" (although not based on "formal rules") (Franklin, 1994, 484, 487). Although there was a period of considerable uncertainty over whether Weber's results were credible, physicists used a number of reasonable epistemological strategies for resolving that uncertainty. For example, several different groups had performed experiments attempting to replicate Weber's findings, although none were performed in exactly the same way as Weber's, and each of these groups had failed to find any significant excess in coincidences in their data that would constitute evidence of gravity waves. Franklin quotes one physicist, commenting in 1974, on how each of these experiments yielding a negative result has some 'loophole' in the sense that there is some form that the gravity waves *might* take to which the experiment would not be sensitive. Thus, no single experiment yielding a negative result decides definitively against the claim made by Weber. However, sensitivities vary from one experiment to another, so the loopholes are different for different experiments. A loophole left open by one experiment is closed by another. "I think that when you put all these different experiments together, because they are different, most loopholes are closed" (Franklin, 1994, 478).

Franklin's account also emphasizes the numerous problems with Weber's results. In Franklin's summary these include "an admitted programming error that generated spurious coincidences between Weber's two detectors, possible selection bias by Weber, Weber's report of coincidences between two detectors when the data had been taken 4 hours apart, and whether or not Weber's experimental apparatus could produce the narrow coincidences claimed." By contrast, Weber's critics "had checked their results by independent confirmation, which included the sharing of data and analysis programs," had "eliminated a plausible source of error [in their negative results], that of the pulses being longer than expected," and had "calibrated their apparata by injecting known pulses of energy and observing the output" (Franklin, 1994, 484).

Franklin's final point about calibration deserves emphasis. To calibrate an instrument, investigators deliver a 'surrogate signal' to it and observe its outputs. They want to know how the instrument responds to inputs relevant to the phenomenon being investigated. To understand the responsiveness of various Weber bars to gravitational waves, investigators studied their response to these surrogate signals. Calibration by itself was not sufficient to resolve the dispute between Weber and his critics, because it was not certain whether the input used for calibration – an electrostatic energy pulse – was an adequate surrogate for a gravity wave. Experimenters had to resort to other kinds of arguments. But Franklin argues that calibration often serves as "a legitimate and important" and possibly "decisive" factor in determining that an experimental result is valid (Franklin, 1994, 487–488; see also Franklin, 1997 for more on the importance of calibration in the validation of experimental results).

Collins disputes numerous claims that Franklin makes and discusses important differences of methodology, but the most significant aspect of his response for our purposes is his insistence that Franklin has mischaracterized Collins's own position. Franklin's argument that the potential regress in the debate over Weber's claims about gravitational waves was broken by "reasoned argument" constitutes, according to Collins, "pushing at an open door" because it defends a thesis that Collins has not disputed (Collins, 1994, 502). However, in his formulation of the experimenters' regress quoted above, he does deny that "scientific" considerations are sufficient to break the regress, and in his response to Franklin, he says that "something in addition to experiments" is needed to settle scientific controversies, which can include "a range of things from high theoretical argument to low dirty tricks" (Collins, 1994, 502). It is, therefore, not clear whether we should accept that Collins does not dispute the claim that Franklin defends, because they seem to mean quite different things by 'reasoned argument.'

Collins claimed from the outset that the evidence cited against Weber's claim "did not *have* to add up so decisively" (Collins, 1992, 91, emphasis in original) and "rejection of the high flux claim was not the *necessary* inference

(Collins, 1992, 91, emphasis added). In responding to Franklin's citation of this claim, Collins now adds that "[i]t is quite reasonable that they were made to add up the way they did, and it would have been quite reasonable had they been made to add up another way. Franklin seems to think that reasonableness can only lead to a single conclusion" (Collins, 1994, 502). Ian Hacking identifies *contingency* as a central point of disagreement regarding social constructionism. A social constructionist will maintain that "in a thoroughly nontrivial sense, a successful science did not have to develop in the way it did, but could have had different successes evolving in other ways that do not converge on the route that was in fact taken" (Hacking, 1999, 33). We might also read Collins's comments about the reasonableness of two different conclusions in light of our discussion of Duhem's underdetermination problem discussed in Chapter 3. Collins's formulation of the problem resembles Duhem's: that the resources of deductive logic (which determine the scope of what counts as a necessary inference in the logical sense) are insufficient to decide what should be regarded as the outcome of an experiment.[8] However, Collins seems to reject Duhem's additional claim (and related claims by others) that there is a domain or mode of reasoning beyond deductive logic: "reasons that reason does not know," that constitute *good sense* and which *do* suffice to resolve univocally disputes about the inference that should be drawn from an experiment.

A more recent development opens a perspective on the debates over constructionism that suggests a transformation of some kind. Allan Franklin and Harry Collins coauthored an article in the form of an exchange, revisiting their own past disputes, noting ways in which their views, while not coming into complete agreement, had harmonized, while also commenting on the ways in which the context of their original disagreement had encouraged them to put their arguments in a vitriolic form (Franklin & Collins, 2016).[9] Without detailing the precise contours of their agreement and disagreement, their rapprochement may be summarized as such: The

[8] By his own account, Collins "did not have Duhem-Quine in mind" in formulating the experimenters' regress. The parallel being drawn here is to the 'Duhem problem of underdetermination' and not to the so-called Duhem-Quine thesis.

[9] Collins notes in their earlier dispute, "I wrote the nastiest things about Franklin that I have ever put in print and he wrote some very unpleasant things about me" (Franklin & Collins, 2016, 96). If nothing else, their subsequent work together provides a model for how respectful disagreement can yield productive cooperation.

claim of Collins and other social scientists studying knowledge production that social change contributes, in some sense and in some ways, to scientific change has been accepted by Franklin (and many other philosophers of science). Collins, in turn, 'applauds' Franklin's enumeration of strategies for legitimate argumentation for the correctness of experimental results, constituting a nonexhaustive list of strategies, some use of which is "necessary to establish the credibility of a result" (Franklin & Collins, 2016, 99–100). Franklin and Collins remain divided, however, regarding the significance of the use of such strategies. In the controversy over Weber's gravity wave claims, Franklin concludes that these strategies showed that "the scientific community was both reasonable and justified in their rejection of Weber's results," while Collins finds the expression "reasonable and justified" "entirely superfluous" (Franklin & Collins, 2016, 106–107). What remains of substance in their dispute over this episode is the precise nature and extent of the social influences on the dispute and the manner of its resolution, as well as questions of historiographical method and what sources of evidence ought to be relied upon most heavily in giving an account of the episode. These disputes can, Collins notes, be pursued in "a nicely un-insulting way" (Franklin & Collins, 2016, 117).

Part II

Ongoing Investigations

7 Scientific Models and Representation

7.1 Introduction

School groups visiting the Museum of the Reserve Bank of New Zealand (Te Pūtea Matua) can visit the MONIAC, a collection of variously sized and colored chutes and tanks mounted to a board. Created by New Zealand economist William Phillips in 1949, the MONIAC serves as a hydromechanical macroeconomic computer. Water flows through chutes labeled 'taxes,' or 'savings,' or 'consumer expenditure,' or is diverted into a tank labeled 'investement fund,' and so on. The relative sizes of the flows can be adjusted. Mechanically driven pens then chart changes in gross domestic product or interest rates.

Phillips devised the MONIAC to enable one to visualize economic changes as represented in macroeconomic theory. The basic structure of the machine was meant to serve as a physical representation of a mathematical theory of the dynamics of an economy, expressed in terms of first-order differential equations. That structure could also be modified to correspond to certain more complex dynamics incorporating higher-order effects. According to economist Walter Newlyn, who helped Phillips build his original version of the device, it was particularly useful "at that stage of perplexity and confusion which occurs when the student has encountered a considerable number of apparently conflicting ideas and is trying to reconcile them. It was with this state of confusion in mind that the model was originally conceived and it is at this stage that it has been found to be of most value" (quoted in Bissell, 2007, 72). The accuracy of the machine was claimed to be ±4% (in Bissell, 2007, 72).

The pedagogical usefulness of MONIAC seems to be related to two of its most obvious features: First, it is based on an analogy between money, which can be a challenging concept to grasp, and water, with which every human has everyday intimate contact. Second, the changing flow of water – as an analog to changes in the economy – is something directly and easily observable. (In Phillips's original version, different input flows consisted of differently colored water; the version on display in the New Zealand museum uses uncolored water flowing through transparent chutes of different colors.) In the words of one commentator speaking about Phillips's original machine, "There was something direct about the money/water analogy that an electronic analog computer, say, would have been unable to reproduce, even though the computational accuracy of the latter might have been higher" (Bissell, 2007, 73).

In this chapter, we will take a closer look at the ways in which scientists use models and the different kinds of models they use. The example of MONIAC illustrates nicely some important themes of our discussion: *Scientists develop models for a variety of purposes beyond simply the accurate representation of a scientific phenomenon.* MONIAC was not that accurate, but was well suited for achieving a certain kind of macroeconomic understanding. *Choices about the model often reflect choices about which purpose is to be prioritized over others.* Phillips set out from the beginning to produce a model that would be useful as a teaching device, relegating accuracy to a secondary virtue. *Models consist of different kinds of things.* MONIAC is an analogue, hydromechanical model based on a distinct *mathematical* model consisting of equations relating variables for price, production, consumption, and investment. The mathematical model is abstract, while the hydromechanical model is concrete and observable. *Different kinds of models can be coordinated to work together to achieve what no single model can do on its own.* By combining MONIAC with a well-constructed mathematical model, one can promote the understanding of macroeconomic dynamics while also having a chance of achieving better accuracy in economic predictions than those of MONIAC. *Models can serve as devices to promote new discoveries.* When two MONIAC devices were built and connected to one another (one built as a mirror image of the other), it allowed for the simulation of two interconnected economies, "something that appears to have led to important new insights into the dynamics of international economies" (Bissell, 2007, 73). *The usefulness of a model may rest upon an analogy between the model and what it is a model of.* As we will see, there is a difference between an analogy and a model, but they are closely related. Part of what makes MONIAC a useful model for thinking about macroeconomic phenomena is an analogy between the flow of water through a hydromechanical system and the flow of money through an economic system. For all of the obvious differences between water and money, the ways in which each flows through its relevant system resemble one another sufficiently to enable MONIAC to serve its primarily pedagogical purpose by relating a phenomenon that is less familiar and more difficult to visualize to one that is both very familiar and not only visualizable but *put on display* (including correlated visual and auditory information[1]) to the observer. To draw the analogy is to note the relevant similarity. Building the model exploits that analogy but requires additional efforts. And not every model is based on an analogy.

7.2 Types of Models

Scientists employ such a wide variety of models for such a wide variety of purposes that classification and systematization of models in science presents significant challenges. Perhaps there is no uniquely best way to categorize the kinds of models used in scientific contexts, but we can begin with an influential classification, introduced by Peter Achinstein, that distinguishes scientific models into *representational models, theoretical models, and imaginary models* (Achinstein, 1968).

Each of these types of model attempts to represent or describe something: an object or system, or a kind of object or system, or a phenomenon. (Although we will see that the way in which this applies to imaginary models poses some difficulties.) The discussion that follows will employ the term 'target' to refer generically to these 'somethings' at which models aim. We can ask of each type of model both what the model consists of and what the model is for.

Representational Models. Achinstein describes these as "a three-dimensional physical representation of an object which is such that by examining it one can ascertain facts about the object it represents" (Achinstein, 1968, 209). That is to say that a representational model consists of a physical thing or a type of physical thing. Achinstein presents a further division of this category into subtypes, some of which correspond to kinds of scale models that

[1] "Spectators could not only see the red water streaming through the pipes, but also hear the bubbling and splashing as it ran through the machine" (Morgan & Boumans, 2004, 390).

reproduce features of the target with varying degrees of fidelity. Another subtype includes *analog models* that do not reproduce features of the target but that are built based on an analogy between two "essentially unlike objects or systems" (Achinstein, 1968, 210). MONIAC is such an analog model. Representational models are commonly used in engineering (in which scale models can be usefully subjected to tests of various kinds, for example) and in the science classroom and other contexts where pedagogical aims like promoting understanding take priority. Achinstein notes the example of James Clerk Maxwell (1831–1879) drawing an analogy between the self-diffusion of molecules in a gas and the motion of a swarm of bees. Maxwell's analogy does not provide any help with calculations or testing of hypotheses, but it does provide a vivid and familiar image "the mere mention of which would provide illumination" (Achinstein, 1968, 210–211).

Theoretical Models. A theoretical model of a target system or object consists – in Achinstein's classification – not of some other object or system but "a set of assumptions" about the target (Achinstein, 1968, 212). The model of the hydrogen atom developed by Niels Bohr (discussed in Chapter 5) can illustrate Achinstein's idea of a theoretical model. The assumptions that define the model include that the electron in the hydrogen atom follows orbital paths around the proton (nucleus) that correspond to discrete, quantized amounts of angular momentum, and that energy is radiated or absorbed in amounts corresponding to the differences between the energy levels of these discrete orbits. A theoretical model in Achinstein's approach *consists* of assumptions. What a theoretical model *does* is to describe "a type of object or system by attributing to it what might be called an inner structure, composition, or mechanism, reference to which is intended to explain various properties exhibited by that object or system" (Achinstein, 1968, 213). Because the model serves a descriptive and explanatory purpose, the content of the assumptions that constitute the model must provide resources for both description and explanation. For example, the Bohr model is intended to explain this: When hydrogen gas is excited by heating or electromagnetic radiation such as X-rays, it gives off light that can be refracted into a series of discrete spectral lines, corresponding to different wavelengths of light. The model provides this explanation by describing the atom as having a structure that results in the emission of discrete wavelengths of light.

Achinstein notes that a theoretical model "is treated as a simplified approximation useful for certain purposes" (Achinstein, 1968, 214). Certain features of the target system will not be included in the model if they do not contribute to – or if they interfere with – the usefulness of the model for its intended purpose. The decision to exclude a feature of the target system, thus, rests on the judgment that any gain in accurate representation of the target would not contribute enough to the aims of the model to be worth the cost of its inclusion. Tradeoffs between descriptive detail or accuracy and other purposes served by a model are typical of the kind of practical considerations that matter for even the most highly theoretical of modeling enterprises.

Imaginary Models. In Achinstein's usage the term 'imaginary model' can refer either to a set of assumptions that serve to describe an object or system, or to the object or system being described. We can distinguish imaginary models by the fact that in proposing one, a scientist makes no commitment to the assumptions being true, or even plausible, and does not intend them to serve as "approximations to what is actually the case ... In some cases, indeed, the assumptions may be entirely fanciful and contrary to what their proponent believes to hold in the actual world" (Achinstein, 1968, 220). Achinstein provides another example from James Clerk Maxwell, who introduced a mechanical model of the electromagnetic field consisting of rows of vortices that are in rolling contact with particles, each of which is situated between two vortices spinning in opposite directions. The particles "play the part of electricity" and their "motion of translation constitutes an electric current" (Maxwell, 1861, 346).

So, an imaginary model consists in either assumptions imagined to be true, or an imagined object described by those assumptions. What an imaginary model does is guide thinking about what an object or system could be like, "if it were to satisfy certain conditions initially specified." Maxwell's imaginary model of the electromagnetic field describes what it could be like, if it were a strictly mechanical system fully governed by the laws of Newtonian mechanics. Maxwell does not propose that this is what the electromagnetic field is actually like, or even that it would be as his model describes if it were a strictly mechanical system. He only proposes that, under that condition, it *could* be as described by the model (Achinstein, 1968, 220–221). By guiding thinking about this question of what a system could be like under a certain condition, one can promote a number of other scientific goals, such as understanding constraints on how one might accurately describe an actual system, or developing ideas for what a system is actually like (Achinstein attributes this aim to Maxwell in his imaginary model of the electromagnetic field).

7.3 Models and Idealization

By their nature, models differ from their targets in some respects. Importantly, models involve idealizations of their targets that are crucial for many of their uses. Models involving idealizations are often employed when trying to solve a problem or determine a quantity in the face of complexities beyond what a scientist can – or has sufficient reason to – include in their representation. In Boyle's experiment using the J-tube that led him to the reciprocal relationship between volume and pressure of a gas at a constant temperature, he did not measure volume directly, but instead measured the distance between the top of the shorter, sealed leg of the tube and the top of the mercury column in that leg. This would provide an exact measure of volume only if the diameter of the tube were constant over the distance of that portion of the tube, which Boyle knew quite well it was not (both because the end of the tube was not perfectly flat and because glass-blowing techniques could not ensure a perfectly straight length of tube with walls of perfectly uniform thickness). By treating the tube as a device for measuring the volume of gas in the way he did, Boyle effectively relied on a fictional device with a geometry that approximated the actual apparatus in his experiment.[2]

Other idealizations introduce more radical departures from reality. To determine the orbit a planet would follow if the force of gravitational attraction between it and the sun were proportional to the inverse square of the distance between the two, Isaac Newton initially assumes that the sun is at rest, although, according to his own laws of motion, the sun is subject to

[2] John Norton proposes that the term "approximation" be used to refer to *propositions* that describe a target system inexactly, while "idealization" be used to refer to a *system (real or fictitious) distinct from the target system*, "some of whose properties provide an inexact description of some aspects of a target system" (Norton, 2012, 209). I have interpreted Boyle's procedure as introducing a fictitious system, making it a case of idealization, rather than as relying on a series of descriptions of that system that approximate the actual volumes of gas in the experiment, which would make it a case of approximation. Norton argues that the distinction becomes more consequential in certain contexts than it is here, particularly in the use of limits in statistical mechanics.

an accelerating force in the direction of the planet in question. A biologist seeking to develop a model to represent changes in a species population over time might assume that every member of that population has an equal probability of mating with every other member of that population. Such model assumptions exemplify a characteristic depiction of idealizations: They are "assumptions made without regard for whether they are true and often with full knowledge that they are false" (Potochnik, 2017, 18). An alternative characterization highlights idealization as "a deliberate simplifying of something complicated (a situation, a concept, etc.) with a view to achieving at least a partial understanding of that thing" (McMullin, 1985, 248).

7.3.1 Galilean Idealization

At first glance, it may seem that idealization is a strategy adopted out of necessity that one would prefer to avoid. To idealize is to accept that one's ability to represent the order of nature is outstripped by the complexities and messiness of the world as it actually is. But idealization is much more than this. Idealization can be part of a strategy for demonstrating the fruitfulness and explanatory power of a theoretical model. Ernan McMullin introduced the term 'Galilean idealization' to refer to a group of techniques of idealization that enable, in the right circumstances, such successful theoretical engagement with physical phenomena.[3] Of particular importance for our purposes is the Galilean technique of *construct idealization*, in which the conceptual representation of the object of inquiry is simplified. McMullin distinguishes *formal* and *material* conceptual idealizations.

In its formal mode, idealization omits or simplifies features "known (or suspected) to be relevant to the kind of explanation being offered" so that one may obtain a result of interest (McMullin, 1985, 258). Newton was aware that the gravitational forces exerted by the planets would cause the sun to move a little. Newton realized that he could easily derive laws describing approximate regularities in the motions of the planetary orbits (i.e., Kepler's

[3] McMullin's use of the term "Galilean" is not meant to express that all these techniques were originated or perfected by Galileo, but that each of them "played a distinctive part in shaping the 'new science'" of mechanics that Galileo sought to inaugurate (McMullin, 1985, 248).

laws) by omitting this feature. Arriving at more exact descriptions of orbits taking into account the sun's motion would be much more difficult. Deriving Kepler's laws made for an effective argumentative strategy because they expressed regularities already enjoying strong empirical support from the astronomical data available at the time. But stronger validation of a model is produced when an investigator adds back details previously omitted or simplified in an initial formal idealization and is thereby able to derive regularities that accord better with experimental data, or predict new findings subsequently supported by experimental results. This process is often referred to as 'deidealization.'

The deidealization of Newton's gravitational model of the solar system required the development of *perturbation theory*. Newton himself was able to make progress only on the application of this approach to the problem of the moon's orbit around the earth. Others would extend this technique to many other phenomena.

In an instance of material idealization, the investigator omits features of a system under study that they judge as irrelevant to their inquiry. When Newton derives Kepler's laws from his gravitational model of the solar system, he omits the motion of the sun not because it is irrelevant but because he judges that the difference it makes will be small enough to allow the derivation of laws for planetary orbits that approximate closely the actual orbits. By contrast, he omits the chemical composition of the sun and planets because it is not – in the theoretical framework of his laws of motion and of gravity – relevant to the question of how the planets move.[4] Treating the sun as stationary is a formal idealization; omitting any reference to its chemical composition is a material idealization. In other cases, a feature may be omitted from a model because it is not yet conceived of. McMullin gives the example of the property of spin as an omission from Bohr's model of the hydrogen atom. Spin is a kind of intrinsic angular momentum – attributed to particles such as electrons – that was introduced in response to phenomena the Bohr model could not account for.

[4] A little more carefully: Whatever difference the chemical composition of, say, the planet Mars might make to its orbit will only be, in Newton's framework, through its effect on those features of Mars that determine its mass distribution (i.e., density, volume, shape, etc.). Once those features are specified, it will be irrelevant to the gravitational forces exerted by and on Mars that a certain percentage of the mass consists of, say, hydrogen.

As McMullin notes, whether electrons have spin "was a question that simply had not been asked, and that could not be answered within the original model. But it was a question that at some point could easily enough suggest a fruitful line to follow . . . The model operated here not as a starting-point for strict inference but as a source of suggestion" (McMullin, 1985, 263–264).

Galilean idealization aims at understanding phenomena by constructing a series of models that begin with simplifications and omissions, and then showing the validity of the ideas that motivate those models by strategically deidealizing, or removing simplifications and omissions, with the aim of achieving greater empirical success (more accurate predictions, more comprehensive explanations). As Michael Weisberg has noted, Galilean idealization is "important in research traditions dealing with computationally complex systems" and is pragmatically justified in those contexts: To try, at the outset, to include every potentially relevant detail in one's model would leave one stuck with a computationally intractable model. As computational power increases, and with the development of new computational methods, investigators become better equipped to deidealize their models (Weisberg, 2007, 641–642).

7.3.2 Minimalist Idealization

Explanatory aims can justify omitting details from models, even though they may be relevant to a complete characterization of the behavior and characteristics of a system. Weisberg proposes that *minimalist models* are idealized to include "only those factors that *make a difference* to the occurrence and essential character of the phenomenon in question" (Weisberg, 2007, 642). Minimalist and Galilean idealized models do not differ intrinsically but in the rationales for their idealizations. The minimalist idealizer justifies their procedure by arguing that the idealized model captures 'what really matters' for the phenomenon in question, whereas the Galilean idealizer justifies the omission of details as a practical necessity, to be vindicated by subsequent deidealization.

Weisberg's idea of a minimalist model seems to assume that the included factors will 'make a difference' in the sense of being causally relevant to the phenomenon, that is, they will be causal factors shared between the model and the target. Using examples from fluid mechanics and models of populations in evolutionary biology, Robert Batterman and Collin Rice argue that the explanatory work done by minimal models need not make use of causal or any other features that are common to both the model and the target system. Some explanations based on what they call 'minimal models,' instead, show how members of a class of target systems share a behavior because of the ways in which the distinguishing features of those systems are *not* relevant to their behavior (Batterman & Rice, 2014).

7.3.3 Multiple-Models Idealization

Faced with the challenge of modeling a very complex phenomenon, scientists commonly find that no single model suffices for all of their aims, whether explanatory, representational, or predictive. Multiple-model idealization (MMI) occurs when scientists resort to "multiple related but incompatible models, each of which makes distinct claims about the nature and causal structure giving rise to a phenomenon" (Weisberg, 2007, 645). Weisberg mentions the National Weather Service's reliance on multiple models of global circulation patterns for the purposes of modeling and predicting the weather. The aims and justifications for MMI are sufficiently diverse to challenge efforts at systematizing the practices of MMI. Like minimalist modeling (and unlike Galilean idealization), MMI is typically not justified by the potential for successful deidealization. For example, it may be that – as in the case of the models used for predicting weather – the justification aim is to make reliable predictions, and the use of multiple idealized models is based on their ability to yield such reliability.

A particularly interesting kind of rationale is that idealizations of the sort employed in MMI may provide a means toward developing "truer theories," that is, such models allow investigators to "get new or better information about the processes being modeled" (Wimsatt, 2007, 100). William Wimsatt lists twelve ways in which scientists can use "false models" to learn more about phenomena of interest. (Wimsatt's concern is not exclusively with MMI in Weisberg's sense; his "false models" include not only idealized models but also models that are false in other ways, such as being incomplete or having only local applicability.) For example, an "incorrect simpler model can be used as a reference standard to evaluate causal claims about the effects of variables left out of it but included in more complete models, or in different competing models to determine how these models fare if these variables are left out" (Wimsatt, 2007, 104). Wimsatt's list emphasizes the ways in which an idealized or otherwise "false" model may, nonetheless, support a strategy of localizing the errors the false model introduces to some "parts, aspects, assumptions, or subcomponents of the model" and thus work in a piecemeal way toward a better understanding of the phenomenon under investigation (Wimsatt, 2007, 103).[5]

To help systematize our understanding of the various idealization strategies in scientific modeling, Weisberg proposes that we consider *representational ideals* that constitute the goals of these practices. *Completeness*, for example, specifies that "each property of the target phenomenon must be included" and "the best model is one that represents every aspect of the target system and its exogenous causes with an arbitrarily high degree of precision and accuracy" (Weisberg, 2007, 649). Completeness might exceed what scientists can accomplish, but, Weisberg claims, it constitutes the representational ideal of Galilean idealization. Minimalist idealization, by contrast, aims at an ideal, called *1-Causal* by Weisberg, that "says the best model is the one that includes the primary causal factors that account for the phenomenon of interest, up to a suitable level of fidelity chosen by the theorist" (Weisberg, 2007, 655). MMI does not aim at any one representational ideal but may be guided by "[p]retty much any representational ideal" (Weisberg, 2007, 655).

Angela Potochnik has criticized Weisberg's categorization of kinds of idealization demarcated by their representational ideals on the grounds that it fails to reflect the "many intertwined reasons to idealize" that motivate idealizations in scientific models. Potochnik notes how the reasons to idealize arise from pervasive features of the objects of scientific inquiry ('the world') and the subjects carrying out that inquiry (scientists). The world exhibits complex causal structures, while scientists are constrained by cognitive limits. From these two basic features of the scientific enterprise arise a variety of justifications, often overlapping, for idealizing. Accordingly, idealizations are, Potochnik argues, *rampant* in that they "exist throughout our best scientific representations, and they stand in even for important causal influences." They are also *unchecked* insofar as "there is little focus on eliminating idealizations or even on controlling their influence" (Potochnik, 2017, 57–58).

[5] Wimsatt notes that the possibility of this kind of localization of error is denied by "the so-called Quine-Duhem thesis" and comments that if this thesis were true "[n]ot only science, but also technology and evolution would be impossible" (Wimsatt, 2007, 103).

7.4 Models and Theory Structure

A further use of the term 'model' that has importance for philosophical thinking about science and scientific theories comes from the study of logic and semantic relations. Before explaining this usage of the term 'model,' let's consider a philosophical question for which it becomes relevant: What is a theory?

Previously we discussed how logical empiricism relies on a certain answer to this question (and their falsificationist foes agree): A scientific theory consists of a structured set of *sentences*. Consequently, we can apply the apparatus of formal logic *directly* to the theory.

The logical empiricists thus hold a *syntactic* conception of scientific theories ('syntactic' because of its emphasis on the construction and ordering of sentences out of their elements, i.e., the *syntax* of sentences). More precisely, they consider a theory to consist of a set of sentences with a certain logical structure. First, there are sentences (*axioms*) that use a theoretical vocabulary to articulate the basic principles of the theory. The *correspondence rules* relate the theoretical terms in the axioms to terms in the observational language. By means of the correspondence rules one can then deduce from the axioms the observational implications of the theory. The theory then consists of the axioms, the correspondence rules, and all of the other sentences that can be deduced from them and their deductive consequences.

Part of the decline of the logical empiricist approach has involved a partial turn away from the syntactic view as a number of philosophers of science have embraced some version of a *semantic view* of scientific theories (Sneed, 1971; Stegmuller, 1976; Suppes, 2002; van Fraassen, 1972). Whereas syntax is concerned with the structure of sentences, semantics is concerned with relations between language (such as sentences) and whatever it is that we use language to talk about; semantics, thus, is concerned with such relations as meaning, reference, and representation. Accordingly, the turn away from the syntactic view is associated with a broader rejection of the tendency exemplified in logical empiricism of treating philosophical problems as primarily problems of *language*. This rejection is neatly expressed by Bas van Fraassen: "The main lesson of twentieth-century philosophy of science may well be this: no concept which is essentially language-dependent has any philosophical importance at all" (van Fraassen, 1980, 56).

The semantic approach to theories holds that the structure of a theory should not be thought of in terms of the logical relations between sentences.

Instead, the structure of a theory consists of the class of its *mathematical models*. When scientists engage in theoretical discourse, the sentences they use do not *constitute* the theory, but instead *describe* the theory in the sense that a sentence describes the structures that are its models. If we wish to understand what a scientific theory is, we should therefore look to the models rather than the sentences. As stated by Suppe, "Theories are extra-linguistic entities which can be described by their linguistic formulations. The propositions in a formulation of a theory thus provide true descriptions of the theory, and so the theory qualifies as a model for each of its formulations" (Suppe, 1977, 222).

Understanding the semantic approach requires considering what constitutes a mathematical model of a theory. Different approaches involve different conceptions of the kind of mathematical models that constitute a theory, but they agree that the models that constitute a theory are a class of *mathematical things and their relations*, not sentences that are made true by those things. For example, in one such approach, a model of a theory is specified through an abstract mathematical space (somewhat like that represented by a Cartesian coordinate system), with its dimensions representing variables that characterize a physical system. Specifying the theory using such a space involves specifying models as sets of points in the space that represent *possible states of a system* to which that theory applies, or sets of *possible trajectories through the space as described by laws* of the theory (Beatty, 1981; Giere, 1988, 2004; Lloyd, 1994; van Fraassen, 1989). The class of models thus characterized directly specifies what the theory is. A variant on this view regards the class of models as something to be itself defined by a set of axioms made true by those models. Although the axioms do not constitute the theory, they define the class of models that do (Sneed, 1979; Stegmüller, 1979).

On the semantic view, the way in which truth relates to theories is indirect. The models that constitute the theories are neither true nor false, but sentences describing those models are. These features of the semantic view of theories have significance for philosophical debates regarding scientific realism, to which we will turn in Chapter 10.

7.5 Models Connecting ... to What?

One of the most important ways scientists use models is as a means of making connections between two ways of representing things. For example, a theory and a body of data are different in numerous ways. How do data become relevant for deciding the acceptability of a theory? A theoretical description typically involves one or more parameters that are defined at a high level of generality and as continuous quantities. A population mean for some characteristic is a good and somewhat simple[6] example. Maybe you are a fox enthusiast and are curious to know the mean number of kits (also called cubs) in a red fox litter. Let us represent this quantity with a parameter θ. The value of θ could in principle be represented by any positive real number (setting aside limits imposed by biological plausibility). Hence θ is a *continuous* quantity. Moreover, as a theoretical statement about a *population*, we can think of the statement about the mean number of kits in a red fox litter not just as assigning a value to θ but as describing a potentially infinite population of vixens (female foxes) and their litters. We take this to be the import of the statement about θ not because we expect there to be an actual infinity of such litters but because this preserves a generality to our theoretical statement that would be lost if we took it only to describe some particular finite number of litters.

Any conclusion we draw about θ, however, will be based on data that constitute values of *discrete* quantities (such as counting the number of scars left on the placental lining of a vixen's uterus, indicating where a fox fetus has attached; Reynolds & Tapper, 1995) drawn from a finite number of observations, that is, from a sample of the population.

To breach this kind of discrepancy between theory and data, scientists must somehow *create* a connection between theory and data. Patrick Suppes proposes that the means by which this connection is created is through "a hierarchy of models of different logical type" (Suppes, 1962, 253). At the top of this hierarchy is a *model of the theory*. In Suppes's conception, a model of a theory is "a possible experimental outcome in the sense of the theory," that is, a "possible realization" of the theory (Suppes, 1962, 253). If our theory consists of the statement 'the mean number of kits in litters of red fox vixen is θ,' then this kind of model would include a representation of a possible population of pairs of vixens with their litters, as well as a representation of a value for θ – some real number greater than or equal to zero. One might think of this model of the theory as a representation of an answer to the scientific question

[6] Not as simple as one might suppose, though, as a dive into the subtleties of determining litter sizes for the red fox will quickly reveal.

addressed by the experiment being performed – in this case 'What is the average litter size for a red fox?' There is a class of models constituting answers to this question, all having the structure just described but with different possible populations of vixen–litter pairs and different values of θ. But which is the right answer, and how can we use the data to figure that out?

One problem is that "a possible realization of the theory cannot be a possible realization of experimental data," both because any actual experimental data will be finite and not include the entire population of all possible vixen–litter pairs, and because the parameter θ "is not directly observable and is not part of the recorded data" (Suppes, 1962, 253). Relating the experimental data to the theory requires, in addition to the model of the theory, at least two additional models: a model of the experiment and a model of the data.

Eschewing technicalities, we can say for our present purposes that the *model of the experiment* represents the means by which the investigator will relate the information contained in the experimental data to the possible models of the theory. This will typically include a representation of all possible datasets the experiment might yield and a plan or *testing rule* for both choosing a theoretical possibility (typically a class of related possibilities) as that *favored* by the experimental data and identifying which theoretical possibilities are to be considered as *excluded* by or incompatible with the data. It will also include a probability function that describes the statistical relationship between different theoretical possibilities that might be chosen by the testing rule and the different possible datasets (more about this in Chapter 9). To produce such a model requires that numerous features of the experiment be specified, such as the kind of data to be collected, the structure of the dataset, the procedure by which the end of data collection will be determined, and more.

The model of the experiment describes the way in which the experimenter uses information contained in the data to arrive at a conclusion at the level of theory, but it does not include a reference to the actual data in their entirety, which will be highly specific and idiosyncratic, including numerous details that could not be anticipated in the design of the experiment. (Who could have predicted that the thirteenth vixen observed would be found to have given birth to thirteen kits in a single litter?)

Drawing a conclusion from the experimental data requires not just the data themselves but a *model of the data*. This is true even for our very simple example of inferring a population mean from direct observations of a sample drawn from that population. The model of the data in this case might involve little more than the sequence of vixen–litter observations recorded as numbers, such as $\{X_i\}$ with $i = 1, \ldots, n$, where X_i is the number of kits in the litter for the ith vixen out of n observed, from which a sample mean $\overline{X}$ is then calculated.

The idea of a model of data – or data model – deserves special emphasis here. Data are sometimes treated as unproblematically providing scientists with "an unmediated window onto the world" (Bokulich & Parker, 2021, 3). The etymology of the word itself contributes to the problem. Datum – the singular form of data – derives from the past participle *datus* (something given) of the Latin verb *dare* (to give). Data are, etymologically speaking, 'given,' and it is natural to think that they are given by 'the world' in some sense. But our fox–litter example itself shows how this is not the case. One can count the number of kits in a fox den (also called an *earth*), or one can count placental scars. "Because some cub mortality takes place between birth and weaning, observation at earths must underestimate the number of cubs born, while placental scar counts overestimate the number of cubs weaned" (Reynolds & Tapper, 1995, 109). Additional complications arise when one starts to consider the difficulties of obtaining these different kinds of data, and doing so in a way that does not bias the result. Public discourse around science frequently emphasizes the importance of data in a way that suggests that data themselves are unproblematically given and trustworthy: Policy decisions are presented as 'data driven,' or we are told that policy-makers are 'following the data.' Careful examination of what is involved in collecting good, usable data and drawing relevant and reliable conclusions from it will quickly cure one from taking facile use of these kinds of catchphrases very seriously.

Attending to the ways in which data are collected and used across the sciences reveals that conclusions are not, in fact, drawn *directly* from data. Indeed, Suppes's crucial insight is that we *cannot* draw conclusions of a general sort directly from data. To learn from data requires that they be transformed into a usable form, and models are crucial to such transformations.

The hierarchy of the levels of experimental 'theories, models, and problems' that Suppes has proposed features two more levels below models of data: *experimental design*, which is concerned with formal matters of experimental design not specific to the theory under test (e.g., randomization) and, at the bottom, "*ceteris paribus* conditions," which include "every intuitive consideration of experimental design that involves no formal statistics" (Suppes, 1962, 258). Above the level of *ceteris paribus* conditions, Suppes proposes that the apparatus of a hierarchy of models allows systematic evaluation of theory by means of experimental data to proceed in a "purely formal" manner (Suppes, 1962, 260).

Subsequent work has drawn fruitfully on Suppes's idea of models connecting theories to data (e.g., Mayo, 1996), but has also revealed limitations of his strictly formal and logical approach to models. The collection of data (with its attendant 'intuitive' consideration of *ceteris paribus* conditions) and its subsequent modeling are not always clearly delineated, and the very distinction between data and a data model, which Suppes takes for granted, requires investigation (Harris, 2003). Sabina Leonelli argues that Suppes's approach "fails to tackle critical questions around the source of the epistemic value of data, and the relation between data and models" because it does not adequately address this distinction, nor can it be applied to cases that involve qualitative data (Leonelli, 2019, 7). Based on a case study drawn from plant phenotyping involving the use of images as data, with multiple stages of data transformation and modeling, Leonelli argues that the very distinction between data and models depends not on any intrinsic characteristics but on "the interests and evaluative criteria of the researchers using them, which in turn determine their epistemic role" (Leonelli, 2019, 23). Work such as Leonelli's (see also Bokulich & Parker, 2021) makes clear that the models invoked by Suppes and the strict semantic view of theory structure are only part of the story of how models figure in science.

7.6 The Autonomy of Scientific Models

The models in Suppes's scheme enable one to perform an experiment to test a theory, making the models subordinate to the theory. Similarly, the strict semantic view of theory structure centers on an understanding of models that defines their role in relation to theory. Studying scientific practices of modeling reveals a more complicated relationship between models (and modeling) and theories (and theorizing). Margaret Morrison and Mary Morgan argue on the basis of the construction, function, and use of models that they are "partially independent of both theories and the world" and thus "can be used as instruments of exploration in both domains" (Morrison & Morgan, 1999, 10). They note that models often are not strictly theoretical nor strictly drawn from data but involve elements of both. For example, models of data often depend upon theories in significant ways (Antoniou, 2021; Harris, 2003). Models can be thought of as tools or instruments that scientists use; but to the extent that scientists use models for *learning*, either about the world or about scientists' theories about the world, they have a representational function that is missing from tools like hammers or irrigation systems.

We have seen, in Achinstein's classification, how models can serve a variety of scientific aims, so it should not come as a complete surprise that, as Morrison and Morgan argue, the construction of models, their functions, the ways in which they represent, and the ways in which they enable scientific learning cannot be reduced to rules, nor subsumed within the formulation or testing of theories. The autonomy of models includes their ability to explain phenomena in ways that do not derive from theoretical principles. One of the themes of the autonomous view of models is that models serve a 'mediating' function in forging a relation between the world and scientific theories. From this point of view, *theories* in physics, for example, do not in any straightforward sense (if at all) represent what happens in the world (Cartwright, 1983, 1999a, 1999b). That representation, when it is achieved, comes from the building of models, perhaps with input from theoretical principles, but also drawing upon many other resources. Marcel Boumans describes this process as an "integration" of heterogenous ingredients including "theoretical notions, metaphors, analogies, mathematical concepts and techniques, policy views, stylized facts and empirical data" (Boumans, 1999, 94).

Disagreements over how representation is achieved in science, and over the relationship between models and theories, relate strongly to disagreements over notions like truth, resemblance, and usefulness. These quickly spill over into debates between scientific realists and antirealists, to be treated at greater length in Chapter 10. For now, take note of the discrepancy between McMullin's treatment of Galilean idealization and Cartwright's view of the autonomy of model representation. The ultimate goal of a Galilean idealizer, according to McMullin, is to vindicate the truth of a theory by showing how an idealization based on that theory can be used to yield successively better empirical results by the process of deidealization.

He presents this explicitly as a response to Cartwright's claim (in her 1983 *How the Laws of Physics Lie*) that "the fundamental laws of physics do not describe true facts about reality." *Representative models* can do this, but their construction is achieved not by simply deriving them from a theory but by the kind integrative procedure discussed by Boumans, drawing on heterogenous resources (Cartwright, 1999a, 1999b). Cartwright rejects the idea that physical theory serves as a source from which we can derive representations. She proposes to replace this 'theory as a vending machine' idea with the idea of 'theory as a toolbox': Theories provide means of assembling, from various materials, representative models, as well as *interpretative* models that transform the abstract concepts of physical theory in more concrete ways (Cartwright, 1999a).

7.7 Simulation Models

In the mid twentieth century, physicists and mathematicians developed techniques for using digital computers to perform simulations as a strategy to deal with the difficulties of determining analytically correct solutions to theoretical equations in nuclear physics. This development has transformed scientific practice across a wide range of fields that routinely use computer simulation as an investigative tool. Computer simulations constitute a practice of modeling in science insofar as simulations require models as a *basis* and produce models as *output*.

Not all simulations involve the use of computers.[7] The MONIAC described above may be thought of as performing a simulation of a certain kind of economic system, with the charts of simulated quantities like gross domestic product and interest rate serving as the output of simulation. As with *model* or *modeling*, *simulation* and *simulating* are concepts that evade easy capture in a definition. Stephan Hartmann's claim that "*a simulation imitates one process by another process*" certainly identifies a salient point (Hartmann, 1996, 82, original emphases). My aim in this section is not to pursue these quests for definition, nor to aspire for the heights of a general philosophical account of computer simulation. I will content myself to sketch some of the main

[7] And not all computers are digital. See, for example, Maley (2023) for a discussion of analog computation and its relationship to digital computation. See Humphreys (2004, 125–128) on analog computer simulations.

questions that have preoccupied philosophical discussions of simulation in science and some of the many reasons to think that these questions are important.

The historical origins of computer simulation lie in a difficulty a scientist faces when they have a mathematical theory relevant to a problem they are trying to solve but they know no direct way to apply that theory. In 1945, physicists and mathematicians at the Los Alamos laboratory realized that the ENIAC, the first electronic computer (and inspiration for the MONIAC) developed at the University of Pennsylvania, could be used to develop a "preliminary computational model of a thermonuclear reaction" (Metropolis, 1987, 126). Crucial to the achievement of this goal was the development of the Monte Carlo method, "a statistical approach to the study of … integro-differential equations that occur in various branches of the natural sciences" (Metropolis & Ulam, 1949, 335).

We will forego mathematical details of the method here, but an appreciation for the Monte Carlo method might be gained from a very simple illustration of the method's central idea, which is to use a kind of random sampling to arrive at approximate solutions to mathematical problems. We know that the area of a circle with radius r is equal to πr^2 (see Figure 7.1). We also know that a square within which such a circle is inscribed, such that the circumference of the circle just touches each side of the square, will have an area of $4r^2$. We can deduce that the ratio of the

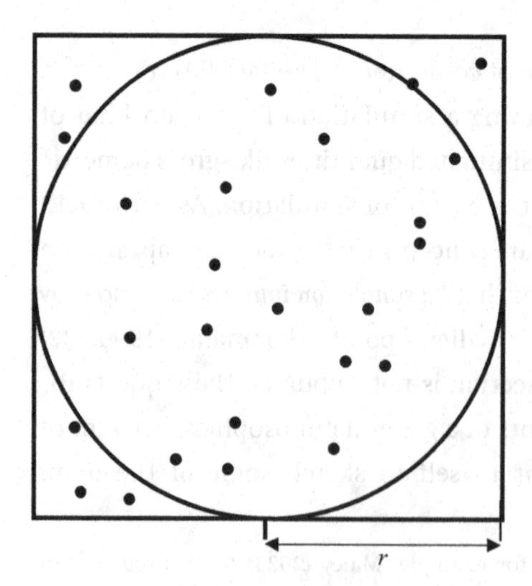

Figure 7.1 Using a sampling technique to estimate $\pi/4$.

circle's area to the square's is equal to $\pi/4$. But what is the value of π? By selecting points randomly inside the square and then calculating the proportion of those inside the circle, one can estimate $\pi/4$, and hence π. In other words, as the sample of points grows, the value of the ratio approaches the value of $\pi/4$. To underscore the potential for a technique like this to be helpful in solving problems involving integral and differential equations that resist exact solution, it will perhaps suffice here to note how learning to compute integrals is often presented as a means for calculating the 'area under the curve' that represents the mathematical function to which the integral is being applied.

Monte Carlo techniques are widespread in computer simulations. Another common approach is the use of *cellular automata* (Keller, 2003). A cellular automata model consists of a set of cells, with a defined set of possible states that each cell can be in. Rules specify how the state of a given cell changes, depending upon the state of neighboring cells. A simple yet interesting example is 'the game of life,' developed by John Conway in 1970 (and now available as an app that you can play with on your phone). Each cell (represented as squares in a grid) is either 'live' or 'dead,' and transitions from one state to another depend on the states of the cells that are its direct neighbors. For example, if a live cell has fewer than two live neighbors, it would die. The interest of the game is to see what happens to various initial configurations of live cells: Do they all die off quickly? Does the pattern of live cells change in interesting ways over time? Configurations called 'gliders' will yield the same pattern of live cells indefinitely, but moving across the array of cells. If you think of a cluster of live cells as an organism, you will find patterns that yield movements that call to mind biological processes like reproduction, predation, and so on (Gardner, 1970; Weisberg, 2013, 129–131).

Michael Weisberg cites the game of life as an example of a simulation model that has no target. The appearance of behaviors resembling those of living organisms – resulting from applying rules that make no reference to organisms but only to whether a given cell is live or dead – suggests that the game of life could be used to study life itself.[8] Nonetheless, the game of life does not represent any actual biological system or phenomena. It is an

[8] Daniel Dennett has also invoked the game of life to discuss philosophical questions about the relationship between phenomena described at different levels (Dennett, 1991).

example of a model studied primarily for its "intrinsic interest," yet has proved sufficiently rich as an object of mathematical and computational interest that it has been exploited to yield insights into other sciences (Weisberg, 2013, 130–131).

Most uses of simulation in contemporary science, however, are intended to relate more directly what actually happens, or what is expected to happen, and some of the most challenging philosophical questions about simulation have to do with our grounds for trusting what a simulation says about these things. Eric Winsberg's discussion of these issues begins with an 'ideal model' of the simulation process, which he then proceeds to demolish. In this way we can learn about the methodology of computer simulation by seeing the ways in which the ideal 'textbook' model of it departs from practices.

7.7.1 The Autonomy of Simulation Models

Winsberg's 'ideal model' begins with a *theory* and uses that theory to develop a *model* of the system to be studied. The model consists of "an abstract description of the physical system ... and some differential equations, provided by theory, that describe the evolution of the values of the variables in the abstract model" (Winsberg, 2010, 10).[9] The *treatment* assigns values to parameters and initial values to variables in the model. The model and its treatment are then "combined to create a *solver* – a step-by-step computable algorithm that is designed to be an approximate, discrete substitute for the continuous differential equations of the model" (Winsberg, 2010, 10–11, original emphasis). The solver is then run on a computer yielding *results*. The results will consist of simulated data.

Winsberg acknowledges that the sequence of "theory → model → treatment → solver → results" "captures a great deal about many computer simulations of physical systems" (Winsberg, 2010, 11). Yet it obscures important features of the methodology of computer simulation. For example, the transition from the differential equations of theory (which are defined over continuous quantities) to discrete difference equations is

[9] Note how this idealized description of the method suggests the 'vending machine' view of the relation between theories and models criticized by Cartwright. The needed differential equations are simply provided by the theory.

anything but straightforward; the most obvious approaches may be too computationally expensive, or they may produce unreliable results due to rounding errors or other divergences. The solution may involve introducing elements to the solver "that have no direct connection to the original differential equations of the model" but are "rough-and-ready, theoretically unprincipled model-building tools [involving] relatively simple mathematical relationships that are designed to approximately capture some physical effect in nature that may have been left out of the simulation for the sake of computational tractability" (Winsberg, 2010, 12). The computational cost of a simulation can grow rapidly as the scale at which physical processes are represented grows smaller, or more fine-grained. Processes occurring at scales smaller than a model represents may, nonetheless, be relevant to the reliable characterization of a system's behavior. The solution many involve inventing a mathematical scheme to capture aggregate effects of processes that "slip between the cracks of the discretization grid" (Winsberg, 2010, 14). The mathematical scheme comes not from the theory but from a careful study of the divergences between simulation results and observations, providing an experimental input into the simulation model that complements its theoretical basis.

An example from the history of Global Climate Models (GCMs) illustrates how the development of simulation models can exemplify Morrison and Morgan's thesis of model autonomy (Lenhard, 2007). Developing a theoretical approach to understand and predict atmospheric dynamics was a significant challenge of the mid twentieth century. Norman Phillips, working at the Institute of Advanced Studies, achieved a breakthrough, reproducing atmospheric wind and pressure relations. He introduced a discrete spatial grid of the atmosphere. He reformulated the equations of hydrodynamics, which used continuous variables for space and time, in ways applicable to his discrete representation. This reformulation – "a conceptually new type of task" (Lenhard, 2007, 179) – replaced the continuous-variable partial *differential equations* of hydrodynamic theory with distinct *difference equations* in terms of discrete quantities. Unlike in our simple geometric example of using Monte Carlo techniques to approximate the solution to a mathematical problem, the simulation model did not produce numerical solutions to the equations of the original theory. "Instead, a simulation model was added, and its difference equations were integrated numerically. It was the generative mechanism of the difference equations that then imitated the atmospheric phenomena so astonishingly well" (Lenhard, 2007, 181). The theoretical equations were important for the development of the simulation model, but not sufficient.

Phillips's model encountered a limit in its ability to imitate atmospheric phenomena "astonishingly well." The results produced by the model would "explode" after about four weeks: "the stable flow patterns dissolved into chaos" (Lenhard, 2007, 183). The problem was suspected to lie in *truncation*: numbers are stored in computer memory only up to a fixed number of decimal places, making small errors intrinsic to the representation of natural quantities in computers. Running a computer simulation model requires very large numbers of iterations of recursive procedures, meaning that an initial input produces an output that is then the input to a new round of computation, which then becomes the input to the next round, and so on. The result is that "even the smallest error can lead to major errors over the course of millions of iterations" (Lenhard, 2007, 183).

Investigators tried different solutions to this problem, but the most successful one came from meteorologist Akio Arakawa at UCLA. Arakawa's key insight, according to Lenhard, was "that one could, and, indeed, that one should, dispense completely with any search for a true solution to the primitive equations" (Lenhard, 2007, 184). This did not mean ignoring the basic theoretical equations entirely, but it did mean that Arakawa was free to introduce assumptions in the development of his simulation model "which ran directly counter to experience and to the laws underlying the theoretical model," such as assuming that kinetic energy is conserved in the atmosphere when it is well known that it is not (friction transforms kinetic energy into heat in the atmosphere just as it does elsewhere) (Lenhard, 2007, 184–185).

Lenhard interprets Arakawa's solution to the stability problem in GCMs as an exploitation of "the autonomy of simulation modeling" that was justified by its success at producing results with behavior typical of the "real atmosphere." The point is not that the simulation constituted a black box providing a reliable output for completely arbitrary or mysterious reasons. On the contrary, the Arakawa approach relied on, in Norman Phillips's words, "a *physically meaningful* finite-difference scheme that will simulate the nonlinear process in the equations of motion" (quoted in Lenhard, 2007, 187, emphasis in original). Rather, the point is that "something 'physically meaningful' need not be constructed 'from below' out of the analysis of basic physical laws but can be obtained 'from above' by adjusting it to fit the phenomena" (Lenhard, 2007, 187).[10]

7.7.2 Simulation and Experimentation

Like simulation, experimentation is related to but distinct from theorizing. In many respects, the methodology of computer simulation resembles or is even indistinguishable from that of scientific experimentation. A debate has ensued over the relationship between experimentation and simulation. The question is sometimes posed as: Are simulations a kind of experimentation? If they are not, are they just a computational method for making predictions? Or are they some kind of hybrid (Morgan, 2003)? A related question asks whether data produced by simulation are 'equivalent' or 'epistemically equivalent' to data produced by experimentation.

A conclusive answer to whether simulations are a kind of experiment would require clear definitions of the two categories: What counts as a simulation, and what counts as an experiment? The question is further complicated by the wide variety of ends pursued and means employed in both activities. In the absence of clear and precise definitions, the debate has been about the relative significance of similarities and differences.

The debate starts from a consensus: *If* we are to regard computer simulation as, in some sense, experimental, the system upon which an experimental intervention is performed is the computer on which the simulation is run. One argument against identifying computer simulations with experiments emphasizes that this is not true of experiments performed on fruit flies, or the nuclei of atoms: "[I]n an experiment, one is controlling the actual object of interest [but] in a simulation one is experimenting with a model rather than the phenomenon itself" (Gilbert & Troitzsch, 2005, 14). The 'object of interest' in many experiments, however, is distinct from the direct target of intervention. For example, in an animal study to determine the

[10] There is no governing body in philosophy that sets rules for metaphors, including those of altitude invoked here. Lenhard's metaphor puts theory below phenomena, presumably to invoke the idea of theory as *basis* or *foundation*. Often the poles are reversed, with theory residing above (in the realm of Platonic forms?) while data and observations occupy the humble earthly realm below.

effectiveness of a drug intended for human use, the object of interest might be taken to be the human response to the drug, while the target of intervention is the physiological response in a model organism such as a rat (Parker, 2009, 485). Isabelle Peschard refines Gilbert and Troitzsch's argument by introducing a distinction between the epistemic *motivation* for an experiment and its epistemic *target* (Peschard, 2010). The motivation for doing the animal study may be to learn about its potential usefulness in humans, but the direct inferences drawn from the experimental data concern the target that is experimented upon, which is the rat. By contrast, the epistemic target of a computer simulation is not typically thought of as the computer on which the intervention is performed but rather the system (atmospheric circulation, subatomic particle interaction, etc.) that is represented in the simulation.

Wendy Parker has argued that computer simulation *studies* count as "material experiments in a straightforward sense" (Parker, 2009, 495). Parker defines a simulation (with a computer or otherwise) as a "time-ordered sequence of states that serves as a representation of some other time-ordered sequence of states" (Parker, 2009, 486; see also Hartmann, 1996). An experiment, by contrast, is not a sequence of states but 'an investigative activity that involves intervening on a system' to learn about how the system changes consequent to that intervention. Simulations, including computer simulations, in which the sequence of states doing the representing are states of a programmed digital computer, are not experiments as such. But a *computer simulation study* – which involves "setting the state of the digital computer from which a simulation will evolve, triggering that evolution by starting the computer program that generates the simulation, and then collecting information regarding how various properties of the computer system, such as the values stored in various locations in its memory or the colors displayed on its monitor, evolve in light of the earlier intervention" – does constitute an experiment. The experiment is performed on the computer, but with the intent to provide evidence that supports inferences regarding a target system that is represented by the sequence of states of the computer (Parker, 2009, 488). That computer simulations can be considered a kind of 'computer experiment' is also advocated by R. I. G. Hughes, who emphasizes that, lacking other data, we can never evaluate the information these experiments provide, since they inform us about "hypothetical worlds" that may or may not be our own (Hughes, 1999, 142).

To the extent that arguments like Peschard's against assimilating simulation to experimentation depend upon emphasizing material contact with an epistemic target in experimentation, they face a problem emphasized by Margaret Morrison: Models, often nonmaterial, play extensive roles in the experimental process, a fact that "calls into question appeals to materiality that are allegedly responsible for grounding ontological claims" about what can be concluded from experimentation as opposed to simulation (Morrison, 2009, 40).

The dispute, as Eric Winsberg notes, seems to rest largely on emphasizing similarities or differences. Could we dissolve Peschard's distinction between experimentation and simulation by insisting that the epistemic object of a computer simulation really is the computer and, in this way, take seriously the idea of the computer simulation as a 'computer experiment'? The answer seems to turn on "nothing but a debate about nomenclature" (Winsberg, 2022). In addition, note that insofar as the debate turns on a question about how to identify the epistemic object of an inquiry, it rests on an assumption that experimental inquiries have uniquely defined epistemic objects.

A detailed examination of any but the simplest experimental inquiry, whether in high-energy physics or in plant science, reveals that such an inquiry has a fine structure with multiple related interventions and inferences regarding different objects or targets (see Mayo, 1996). Sabina Leonelli's discussion of a research project in plant phenotyping provides a nice illustration of the problems with the assumption that experiments have single epistemic objects. Researchers in Leonelli's case study sought to use images of roots to investigate the relationship between root structures and soil conditions. Leonelli's analysis identifies seven stages of data analysis, noting shifts in what counts as data and even changes in the relevant scientific background needed for the execution of each stage. For example, stage 4 involved "evaluating which measurements could be effectively extracted from the imaging data through computational means, so as to make it possible to accurately and consistently compare root systems," a task overseen by a "highly skilled computer scientist with decades of experience," whose approach focused on determining "which properties of the images at hand would be most easily and reliably amenable to analysis through existing computational tools" (Leonelli, 2019, 12). In this example, one part of an experiment concerning plant morphology and the fit between root structures and soil conditions consisted of a computer science investigation.

The latter examined the fit between images of root structures and the computational tools available for extracting and analyzing information from those images.

The identification of a single epistemic object of inquiry distinct from what is investigated in a computer simulation becomes even more dubious in experimental fields such as high-energy physics, in which simulation plays an important role in the generation of experimental data. In these settings, the investigative activities required for the experiment include inquiring whether the simulations employed are adequate for the purpose of producing a reliable and secure result (Beauchemin, 2017; Boge, forthcoming; Karaca, 2023; Massimi & Bhimji, 2015; Morrison, 2009, 2015; Ritson & Staley, 2021).

The complications we confront in answering the questions 'Are computer simulation studies a kind of experiment?' or 'Do the results of computer simulations have an epistemic value comparable or equivalent to the results of experiments?' prompt us to ask about the importance of these questions. What is it that depends on the answers to these questions?

Perhaps it is a question of trustworthiness. People seeking to influence your behavior will often appeal to what 'studies show,' usually meaning the results of some kind of experiment. The assumption is that at least some experiments produce evidence that can be treated as a trustworthy basis for deciding how to behave. If computer simulation studies are a type of experiment, then perhaps they, too, provide trustworthy results that we can rely on when making decisions.

Note the attention these issues have received in relationship to climate projections relying on computer simulations. In the public discourse around climate change, some 'climate skeptics' have focused in particular on the use of computer simulations as a reason for doubting such projections. Some philosophical contributions have sought (1) to provide insights into the methodology of computer simulations in a way that can provide a better informed basis for understanding what such simulation studies can and cannot accomplish, (2) to give a nuanced account of how to evaluate their reliability, and (3) to provide guidance in how the results of such studies might inform policymaking (Lloyd, 2010; Oreskes, Stainforth, & Smith, 2010; Parker, 2008, 2009, 2010, 2011; Smith & Petersen, 2014).

The question of which computer simulation results to trust is not easily answered. Having a well-established theoretical background for a simulation helps, but is not sufficient. Some form of *robustness* is frequently invoked, sometimes meaning that different, independently developed simulation models agree in their results. But robustness is no guarantee of reliability and can be difficult to assess. The requirement of independence is supposed to guard against agreement among simulation results that is due to shared but potentially error-inducing features. But simulation models can involve massive amounts of computer code. They are developed over time and are sometimes related to one another in complicated ways, making independence difficult to discern.

As our discussion of climate model simulations illustrates, the trustworthiness of computer simulation results cannot be settled with a simple, blanket judgment. But the same point applies to experimentation. Experimental results can be misleading or unreliable. Perhaps the experiment was badly designed, poorly executed, or simply not capable of answering the question at which it was directed. As a consequence, declaring computer simulation to be – or not to be – a kind of experimentation will not on its own provide us with a basis for declaring the results of simulation studies to be trustworthy or not.

7.8 Conclusion

When Plato sought to answer the question "What is justice?" in his dialogue *Republic*, he developed a "theoretical model[11] of a good city" (Plato, 1992, 472e). He then developed a model of the just person through an extended analogy to his model of civic justice. Having decided that calling a person just does not require perfect resemblance to "the just itself" but only that the person "comes as close to it as possible," Plato has Socrates announce that "it was in order to have a model that we were trying to discover what justice itself is like and what the completely just man would be like, if he came into being" (Plato, 1992, 472b–c). Clearly, the use of model building in inquiry, including the pursuit of a model as the goal of inquiry, is neither a recent development nor unique to modern empirical science.

But modeling and models have changed over the years, as new techniques of modeling and new ways of using models have developed. Alongside these

[11] Plato's word here is παράδειγμα, translated by G. M. A. Grube as 'model.' Other translations give 'pattern' here.

changes, philosophers of science have gained a greater appreciation for the ways in which modeling is independent of theorizing (the 'autonomy' of models), and along with this appreciation they have contended with the idea that modeling practices contribute, sometimes directly, to our knowledge of the natural and social worlds. Models have been found to be central to understanding numerous important dimensions of science in addition to those discussed here, including exploration (Gelfert, 2016), measurement (Tal, 2017), and evaluation of uncertainty (Staley, 2020). Because of the centrality of models and modeling in science, the questions discussed in this chapter will remain important for issues to be taken up in the following chapters, particularly in relation to the use of probabilities, scientific explanation, and the pursuit of truth in scientific inquiry (the debate over scientific realism and antirealism).

8 Reasoning with Probability: Bayesianism

8.1 Introduction

Part I presented several challenges to the idea that disputes over the status of scientific theories can be resolved unambiguously by appealing to the evidence. We have encountered arguments that purport to tell us that evidence in some way or another *underdetermines* the choice of theories; that the very meanings of the words used to describe the evidence may incommensurably depend on the theory one chooses when interpreting the evidence; that different scientists will disagree about the value of various features of the evidence; that even when the evidence seems to point strongly against a theory, one may rationally hang on to that theory; and even that for any given body of evidence, opposing conclusions based on incompatible methodological rules can equally advance scientific knowledge. Moreover, we considered the proposal that when scientists do reach agreement, they do so as a result of social forces rather than achieving consensus by considering evidence independently of social forces.

Taken together, such arguments appear to throw a rather large quantity of cold water on whatever embers of hope may remain for the logical empiricists' project of providing a neutral framework that scientists could use to evaluate the degree to which any given body of data supports or confirms any given hypothesis.

Many contemporary philosophers of science, however, have sought an alternate route by which they might pursue something like this project, in ways that seek to accommodate some of the postpositivist insights discussed in earlier chapters. The key to the approach they have adopted lies in a certain way of thinking about *probability*.

Probability theory is a well-developed area of mathematics, but statements that use the language of probability lend themselves to different interpretations. In this chapter and the next we will consider two approaches to scientific reasoning that invoke two quite distinct interpretations of probability. The approach we consider here – the *Bayesian approach* – attempts to build a theory of *rational belief* on the mathematical framework of probability theory. (The name derives from the Rev. Thomas Bayes, whose posthumously published essay (1763) included the eponymous theorem that lies at the heart of the Bayesian approach.) Although we will emphasize the conceptual underpinnings and consequences of this approach, a satisfactory appreciation for the theory (as well as for the rival error-statistical approach discussed in Chapter 9) cannot be achieved without a little understanding of some basics of probability theory. So let us start with a brief tour of the probability calculus. A little algebra and elementary set theory is all that is prerequisite to the mathematics that follow.[1]

8.2 Probability Theory

The mathematics of probability theory was developed in the seventeenth century, when it was used to solve problems regarding games of chance. Here is a very simple game involving a standard six-sided die:

Game 1: A single die is tossed one time. If it lands with the 4 facing up, you win. Otherwise, you lose.

It would be helpful to you – especially if you plan on wagering on the outcome of this game – to know the probability that you will win. This probability question seems easy to answer: If the die is fair, then there is a one-in-six chance of it landing with the 4 facing up, and hence you winning. One hardly needs a mathematical theory to figure that out.

More complicated games require more sophisticated treatments. For example, suppose you are given a choice between two games.

Game 2: A single die is tossed four times, and you win if it lands with 6 facing up at least once.

[1] The requirements of a brief and mathematically shallow treatment entail that we will have to gloss over some important conceptual issues that would emerge from a more rigorous discussion (see Cox & Hinkley, 1974; De Finetti, 1972; Kolmogorov, 1950; Savage, 1972). For a good philosophical introduction to probability and induction, see Hacking (2001), which influenced my own presentation. Howson and Urbach (2005) introduce the topic in the guise of a vigorous defense of personalist Bayesianism.

Game 3: A pair of dice is rolled twenty-four times, and you win if the dice come up with both dice showing 6 at least once.

Which of these games should you prefer to play if you wish to maximize your chance of winning?

The latter problem is known as 'Chevalier de Méré's problem,' named for the French nobleman with a penchant for gambling his unearned wealth who, according to legend,[2] posed the problem to Blaise Pascal (1623–1662), a pioneer of the mathematics of probability. De Méré had an intuition (correct, as it turns out) that the first game would give him a higher probability of winning than the second. But actually demonstrating that this is the case requires having a set of rules for determining probabilities.

8.2.1 Models: Representing the Problem

First we need a way of representing the problem we want to solve. For that purpose mathematicians rely on the concept of a *random variable.* For our purposes, an intuitive understanding of this concept will suffice: A random variable is one capable of taking on various values (either actual or potential), such that a probability distribution can be assigned to those values. A random variable might be such that it can take on only *discrete* values (like the six possible outcomes for the number showing on a single die), or it might be *continuous* (like a random variable for the heights of students in a given classroom). For Game 1 we might choose to represent the outcomes of the first game using a single discrete random variable X_1. (Following standard practice, we will use uppercase letters for random variables and lowercase letters for the observed values of those random variables.) The possible values of X_1 (the *outcome space* Ω) are given by the set $\{1, 2, 3, 4, 5, 6\}$. The outcome space for any process (game, experiment, etc.) is a set consisting of *all possible distinct outcomes* for that process. The question of how probable it is that you will win Game 1 then amounts to asking for the value of p in the equation $Pr_M(X_1 = 4) = p$, which we can read as 'the probability that $X_1 = 4$ is equal to p.' (What is the subscript M doing there? Good question! Hang on for a few more sentences and we will find out.)

[2] Legend does not even attribute the correct problem to the Chevalier. Ian Hacking provides a corrective to legend (Hacking, 1975, ch. 7).

The mathematical theory of probability, thus, gives us a means of representing the question we want to ask, and you might think it would be nice for it also to provide us with a means of answering that question. However, the theory of probability is of no help in answering this question. It does not give us a set of principles from which we can deduce the answer. Fortunately we do not need the theory of probability for the answer. If the die is really fair, the answer is $p = 1/6$, and we are just going to assume that the die *is* fair. More precisely we will suppose that

$$p_i = 1/6 \text{ for } i = 1, 2, 3, 4, 5, 6 \tag{8.1}$$

where $p_i = Pr_M(X_1 = i)$. The probability is thus *uniformly distributed* across the outcome space.

It will at this point seem that the theory of probability has done precious little for us beyond introducing some distracting notation, since you did not need any of what I said in the few paragraphs above to know that the probability of getting a '4' on a single throw of a fair, six-sided die is 1/6. The practical usefulness of the theory will become more apparent when we tackle less obvious questions such as that posed by the Chevalier.

Philosophically, however, we have taken a significant though easily overlooked step, which has to do with that subscript M. In our attempt to answer the question of the probability of Game 1, we have had to make some assumptions. Some of these have already been mentioned explicitly, such as the assumption that the die is fair, and that the die is tossed just once. Others may be less obvious, such as the assumption that no player is allowed to veto the outcome of a toss. Others may seem too obvious to state explicitly, being essential to the very possibility of playing Game 1, such as the assumption that the game is played where there is sufficient gravity to make the die land with a determinate side facing up (or some other mechanism to ensure definite outcomes). As diverse as these assumptions may be, they all show up in one place in our probability calculations: a *model* – in this case a mathematical model – of the problem we are trying to solve. We derive the answers to questions about probability from mathematical models that encode all the assumptions about the scenario that are relevant to the probabilities that interest us. For this reason, particular probability assignments here will often be given with a subscript denoting the name of the model from which that assignment is derived. We can think of the model M we just used as consisting of the random variable outcome space $\Omega = \{1, 2, 3, 4, 5, 6\}$ and the uniform distribution assumption $Pr_M(X_1 = i) = Pr_M(X_1 = j)$ for all $i, j \in \Omega$. (Having made the philosophical point, we will sometimes omit the subscript when the model being used is obvious.)

We may now distinguish two importantly different questions: (1) What is the value of the probability I am trying to determine, *as given by the model I am using*? (2) Is the model I am using *adequate* for determining the probability that interests me? One may arrive at a perfectly correct answer to a question of type 1, but find that the answer is useless because the answer to the corresponding type 2 question is 'no.'

8.2.2 Axioms of Probability Theory

For the purposes of calculating odds in games of chance, it makes sense to think of probabilities as mathematical functions whose domains consist of sets of *events*, which we can represent in our model in terms of the values taken on by random variables. In Game 1 an event might be 'the die lands with 3 facing up' and we would represent that with $X_1 = 3$. (We will shortly extend our conception so that we can think of probabilities as applying to sentences.) We will also count as events outcomes with more complicated descriptions, such as 'the die lands with either 3 or 5 facing up.' We want probabilities to apply to these as well.

So it will be helpful to think of the domain on which the probability function is defined as constructed from set-theoretic combinations of the elements in the outcome space, such as $\{X_1 = 3\} \cup \{X_1 = 5\}$, that is, the *union* of the sets $\{X_1 = 3\}$ and $\{X_1 = 5\}$. Let's call the set S of *all* such possible combinations, for a given problem, the *outcomes set* for that problem.

Now we can characterize a probability function as a function Pr on the set of all subsets of the outcomes set S, with a range consisting of the closed interval [0, 1], such that the following *axioms* are satisfied:

Axiom 1: $Pr(S) = 1$.
Axiom 2: $Pr(A) \geq 0$, for any event $A \subseteq S$.
Axiom 3: Suppose that $A_1, A_2, \ldots, A_n \ldots$ constitute a (possibly infinite but) countable sequence of events that are mutually exclusive (i.e., for any $i, j = 1, 2, \ldots, n, \ldots$ such that $i \neq j$, $A_i \cap Aj = \emptyset$). Then:

$$Pr(A_1 \cup A_2 \cup \cdots \cup A_n \cup \cdots) = Pr(A_1) + Pr(A_2) + \cdots + Pr(A_n) + \cdots. \tag{8.2}$$

Expressed more intuitively, Axiom 1 states that a probability function assigns the number 1 to the event consisting of 'any of the possible outcomes occurs.' Axiom 2 states that a probability function assigns a number greater than or equal to zero to every possible outcome. Axiom 3, often called the axiom of *countable additivity*, tells us how to determine the probability of disjunctions of possible outcomes for the case where such outcomes themselves are mutually exclusive (there is no possibility that they can coexist), given the probabilities of the disjuncts: The probability of the disjunction is equal to the sum of the probabilities of the disjuncts considered individually. So, the probability (based on the model M) of the event $\{X_1 = 3\} \cup \{X_1 = 5\}$ (the die lands with either 3 or 5 facing up) is equal to the sum $Pr_M(X_1 = 3) + Pr_M(X_1 = 5)$.

These axioms form the foundation for probability theory (Kolmogorov, 1950) and suffice to define what a probability function is. However, they do not tell us how to handle probabilities in cases where the probability of an outcome of some process depends on the outcome of some previous iteration of that process. (For example, we could use it for calculating the probability of drawing a certain card or combination of cards from a deck if after each draw we replace the card and reshuffle; we cannot yet use it to determine the probability of drawing three consecutive spades without replacing cards.) It is useful, then, to add to our axioms a further definition (we may also consider it a fourth axiom):

Definition 1 (conditional probability):

$$\Pr(A|B) = \frac{Pr(A \cap B)}{Pr(B)},$$

where $A, B \subseteq S$ and $Pr(B) \neq 0$.

This rule does two things for us. If A is *probabilistically independent* of B (i.e., $Pr(A|B) = Pr(A)$), then we can derive a rule for calculating the probability that both A and B happen that does not directly involve conditional probabilities: $Pr(A \cap B) = Pr(A) \times Pr(B)$ (the *rule of multiplication*). But it also tells us how to handle cases in which outcomes are probabilistically dependent upon one another, which will be important for understanding Bayesian reasoning.

8.3 Probability Ideas and Science

Of course we are looking for more than just a set of rules for working with probabilities. We want to see how ideas about probability might help us make sense of the role evidence plays in scientific inquiry. In this chapter and the next, we will pursue two avenues of thought that aim to provide a reasoned link between the data scientists produce and the conclusions they draw. These two avenues exploit quite different ideas about what probabilities represent.

In the opening passages of his *The Emergence of Probability*, Ian Hacking writes: "Probability has two aspects. It is connected with the degree of belief warranted by evidence, and it is connected with the tendency, displayed by some chance devices, to produce stable relative frequencies" (Hacking, 1975, 1).

In this chapter, we consider the *Bayesian* approach to scientific reasoning, which exploits the 'degree of belief' aspect of probability. The *error-statistical approach*, discussed in Chapter 9, exploits the 'relative frequencies' idea.

Thinking of probabilities in terms of relative frequencies might seem to be the most natural way to think about the kinds of examples considered so far. If pressed to explain what it means to say that the probability of winning Game 1 is 1/6, we might give the following rough account: If you play Game 1 over and over again, then, on average, you are going to win about one out of every six games (we will refine this rough parsing in Chapter 9). But we can take that idea and use it to introduce the 'degrees of belief' idea. Once introduced, we will find that the 'degrees of belief' idea can be applied to many situations where relative frequencies are simply inapplicable.

8.4 Bayesian Probability

8.4.1 Prizes

Let us return to the choice between Games 2 and 3. Knowing the rules for calculating probabilities, one can establish that the probability of winning Game 2 is $671/1296 \approx 0.518$, while the probability of winning Game 3 is ≈ 0.491. (Can you figure out why? The solution is explained in the appendix to this chapter.) Supposing that the same desirable prize is given for winning either game (a box of delicious horseradish doughnuts, perhaps), that there is no cost (or no difference in cost) to playing either game, but that you may play only one, you should *prefer* playing Game 2 over Game 3. So, knowing the probabilities of such outcomes – in combination with information about the relative values you assign to those outcomes – can lead to conclusions about what preferences you should have.

This suggests a procedure by which – reversing things – we start with your preferences and arrive at conclusions about how probable you *believe* the various possible outcomes of a process are. For example, I might offer you a choice between receiving a valuable prize if you win Game 1 and receiving the same prize if it rains tomorrow. If you choose to receive the prize on the condition that it rains instead of winning Game 1, this indicates that you consider rain to be *more probable* than winning Game 1, that is, Pr_{you} (Rain) > Pr_{you} (Winning Game 1).[3] We can further constrain the range of values for your probability assignment by offering more choices. If you are offered the additional option of receiving the same prize for winning Game 2 and prefer that option over receiving the prize if it rains, then we can conclude that Pr_{you} (Rain) < Pr_{you} (Winning Game 2). And if you know the values of the probabilities of winning in those two games, we can conclude that $1/6 < Pr_{\text{you}}$ (Rain) $< 671/1296$.

Using that procedure requires us to attribute to you some beliefs about probabilities of other outcomes that we can use as a reference. By shifting from prizes to *wagers*, we can eliminate such assumptions and go directly from your beliefs about the odds at which a bet on a proposition is fair to the probability you must attribute to that proposition in order to satisfy what Bayesians consider a requirement of rational belief. First, a bit more terminology.

8.4.2 Wagering

We may think of a wager as an agreement between two parties, *S* and *T*, regarding a proposition *P*. If *S* 'bets on *P*' and *T* 'bets against *P*,' that means that *S* pays an amount of money *X* (*S*'s bet) for the opportunity to win a sum

[3] The subscript in $Pr_{you}(P)$ does not mean exactly the same thing as it does in $Pr_M(A)$. In the latter, it indicated that the probability value in question was that assigned to event *A* by model *M*. In the former, the subscript denotes an agent ('You') capable of forming and maintaining beliefs (an 'epistemic agent' in philosophical jargon), and the probability value is that which, in virtue of that agent's beliefs, she attributes to proposition *P*. We might relate these two cases in the following way: In cases involving personal probabilities, the subscripts on the probability functions refer to models of the beliefs of the epistemic agents in question, instead of models of chance mechanisms. Notice now that we should therefore regard the symbol 'Rain' in Pr_{you} (Rain) as denoting the *proposition* 'It will rain tomorrow' rather than the *event* of rain occurring tomorrow.

$X + Y$ (for a net gain of Y), subject to the condition that S wins that amount if and only if P is true. T pays the amount Y (T's bet) for the opportunity to win the sum $X + Y$ (a net gain of X) subject to the condition that T wins that amount if and only if P is false. The quantity $X + Y$ is known as the *stake*. S's betting rate is equal to the ratio of her bet X to the stake $X + Y$.

Suppose that Samiksha and Tatyana agree on a bet such that Samiksha will pay Tatyana \$5 if Tatyana can eat ten saltine crackers in 60 seconds without drinking any liquid. If Tatyana cannot do this, then Tatyana will pay Samiksha \$5. Let *Salt* stand for the proposition that Tatyana succeeds at this task. Then Tatyana's betting rate on *Salt* of $\frac{\$5}{\$10} = 0.5$ is equal to Samiksha's betting rate against *Salt*.

We might be tempted at this point to conclude that they both consider Tatyana's success to have a probability of one-half. But people who engage in gambling (especially casino operators) strongly prefer to take on wagers that give them an *advantage*. For example, Samiksha may have seen lots of other people attempting the saltine cracker trick and fail, and she may decide on this basis that the probability of Tatyana succeeding is in fact much less than one-half, that is, $Pr_{\text{Samiksha}}(Salt) << 0.5$.

Notice that if that were the case, it would not be reasonable for her to switch to the opposite side of the bet at the same betting rate of 0.5, because that would, according to her own assessment, put her at a disadvantage. If she considers the probability of Tatyana's success to be very low, she should only accept a bet *on* Tatyana's success if the payoff is large enough relative to her own bet to make the wager at least fair (and preferably advantageous) for her.

Whereas real-life wagering typically involves the pursuit of opportunities to bet at *unfair* rates, a prominent Bayesian approach relates probabilities to beliefs by construing personal probabilities in terms of *fair betting rates*. In our example, suppose the following holds: If offered the opportunity to bet *on Salt* at a rate of 0.5, Samiksha would prefer to bet *against Salt* rather than *on* it. If the betting rate *on* is 0.1, she would prefer to bet *on*. But suppose we find that at a betting rate *on* of 0.3, she is indifferent as to which side of the bet she would take.[4] We then call this Samiksha's *fair* betting rate: She does not consider such a bet to offer either side an advantage.

[4] Note the use of the subjunctive 'would' in the three preceding sentences. We do not wish to assume in general that fair betting rates only exist for agents engaged in actual

For *personalist* (also called *subjective*) Bayesians, a person's degrees of belief in the propositions in a set should be understood as the betting rates that they consider fair for those propositions, that is, the rates at which – were one (hypothetically) to agree to wagers regarding them – one would be indifferent between wagering on or against them. Moreover, the personalist Bayesian holds, a person's degrees of belief are collectively *rational* just in case they are probabilities, which is to say that they satisfy the axioms of the probability calculus given above. Sets of beliefs that meet this requirement are by definition *coherent*.

8.4.3 Rationality

The Bayesian approach encompasses both a view about probability and a view about rationality, but Bayesians differ about how to connect these two ideas. They agree on this much: To have beliefs that fail to satisfy the probability axioms (that fail to be coherent) is to be (in some way) irrational. The question is, what makes this so? Here we will content ourselves with a quick, somewhat rough description of two main positions, which we might label the *pragmatic* view and the *logical* view.

According to the pragmatic view, what is irrational about holding beliefs that fail to satisfy the probability axioms is that one is then subject to what is sometimes called a *Dutch Book*,[5] or a *sure-loss contract* (De Finetti, [1937] 1964; Hacking, 2001, 169).

Suppose that Tatyana asks Samiksha what she considers to be a fair betting rate on *Salt*, and Samiksha replies that she considers a rate of 0.3 to be fair. Asked what she considers a fair rate on $\neg$*Salt*, Samiksha says that 0.8 would be fair. Samiksha's fair betting rates for these two propositions clearly cannot be construed as probabilities. *Salt* and $\neg$*Salt* constitute exclusive alternatives, so any probabilities assigned to them must satisfy Axiom 3: $Pr(Salt \vee \neg Salt) = Pr(Salt) + Pr(\neg Salt)$. But by Axiom 1, $Pr(Salt \vee \neg Salt) \leq 1$.

wagering. A person may, for example, have views about what would constitute fair betting rates but refuse to enter into wagering for reasons of moral or religious principle.

[5] The origins of this term remain a mystery. Hacking, in his generally excellent text (2001, 169), claims to have found the term used 'in passing' in Ramsey (1950), but this is incorrect, as pointed out by Humphreys (2008), who discusses the history of the term and its obscure origins.

These two requirements cannot jointly be satisfied if $Pr(Salt) = 0.3$ and $Pr(\neg Salt) = 0.8$.

This suffices to show that Samiksha's fair betting rates in this case cannot be considered probabilities (they violate the rules probabilities must obey) but does not yet tell us what is irrational about holding such beliefs. According to the pragmatic view, the problem lies in the fact that, according to her own beliefs, Samiksha should be willing to take either side of both a wager on *Salt* and a wager on ¬*Salt*, provided that the wagers take place at the rates she considers fair. So, she should be willing to let Tatyana decide which side she will take. This allows Tatyana to arrange, with Samiksha's compliance, two wagers – one on *Salt*, the other on ¬*Salt* – such that Samiksha, no matter what happens, will suffer a net loss and Tatyana will enjoy a net gain.

For simplicity, we assume that in both wagers the stake is $10. In the wager concerning *Salt*, Tatyana instructs Samiksha to bet *on Salt*, while Tatyana bets *against*. The rate for a bet *on* is 0.3, so Samiksha puts up $3 against Tatyana's $7. In the wager regarding ¬*Salt*, Tatyana has Samiksha bet *on* ¬*Salt*. The rate for a bet *on* is 0.8, so Samiksha puts up $8 while Tatyana bets $2. That Samiksha will lose money whether *Salt* turns out to be true or false can be seen in Table 8.1.

Clearly Samiksha is destined to lose money if she agrees to this combination of bets, although she regards them as fair. Does that make her irrational? And even if it does, what do such examples have to do with wagering preferences relevant for beliefs about scientific questions, or, for that matter, any beliefs on which one is not contemplating a wager?

Let us take the second question first. The pragmatic Bayesian emphasizes that analyses framed in terms of wagers are simply idealizations of real-life

Table 8.1 *A sure-loss contract for Samiksha (S). Tatyana (T) wins no matter what.*

	Subject of wager					
	Salt (*S* bets on)		¬*Salt* (*S* bets on)			
	S's Gain (loss)	*T*'s Gain (loss)	*S*'s Gain (loss)	*T*'s Gain (loss)	*S*'s net	*T*'s net
Salt true	$7	($7)	($8)	$8	($1)	$1
Salt false	($3)	$3	$2	($2)	($1)	$1

decisions generally. Frank P. Ramsey, whose essay "Truth and Probability" is among the foundational works in the pragmatic Bayesian tradition (though it holds hints of the logical approach as well), states this clearly. Ramsey comments that his account of degrees of belief "is based fundamentally on betting, but this will not seem unreasonable when it is seen that all our lives we are in a sense betting. Whenever we go to the station we are betting that a train will really run, and if we had not a sufficient degree of belief in this we should decline the bet and stay at home" (Ramsey, 1950, 183). For Ramsey, measuring degrees of belief concerns beliefs "*qua* bases of action" (Ramsey, 1950, 172), and in choosing to act, we are concerned with the question of gains and losses, not, of course, in just the monetary sense but in the broad sense in which we value the possible outcomes of our choices differentially. Of course, scientific beliefs may also serve, in principle, as bases of action.

Somewhat more vexing is the question of why being subject to a sure-loss contract should by itself constitute a breach of rationality. Although few people prefer to lose money, doing so is not in itself irrational. Even if we turn away from strictly monetary wagers and think of sure-loss contracts as stand-ins for a broader class of doomed undertakings, embarking on something that is guaranteed to fail to produce results that you desire is not *obviously* a failure of rationality with regard to one's beliefs. Moreover, as Alan Hájek has pointed out, for every sure-loss contract a person is subject to on account of noncoherent beliefs, there is also a sure-gain contract (Hájek, 2008). Were Tatyana to feel charitable toward Samiksha, she could insist that Samiksha bet *against* both *Salt* and $\neg$*Salt* at the above rates and thus guarantee that Samiksha would enjoy a net profit, regardless.

Even accepting that it is irrational to agree to a sure-loss contract, it does not *immediately* follow that it is irrational to hold beliefs that violate the probability calculus, as Henry Kyburg argues (Kyburg, 1978). Recall that Samiksha's incoherent betting odds only lead her into a Dutch Book when she agrees to combine bets in a certain way. Surely she can avoid this undesirable consequence by declining to accept a bet of this kind. The connection between avoiding Dutch Books and rationality might be reestablished by insisting that a rational person must be willing to accept combinations of bets, but it is hard to see on what basis such a principle might be defended.

The *logical approach* to connecting probability and rationality tries to avoid such problems by avoiding Dutch Book arguments in favor of an appeal to the requirements of consistency. The personalist Bayesians Colin Howson and Peter Urbach have argued for this approach. They wish to divorce the foundations of Bayesian probability theory from any consideration of the preferences of human agents. On their approach, probability theory should not be thought of as a theory of rational belief directly but rather as a species of logic distinct from deductive logic. Just as deductive logic constrains the consistent assignment of truth values, "the logic of consistent assignments of *fair betting quotients* determined by appropriate constraints on them we can legitimately call *probability logic*" (Howson & Urbach, 2005, 66, emphases in original). They, moreover, propose that such a logic can be developed independently of considerations of expected gain or loss if one assumes only the following principle:

Principle 1: *If the sum of finitely many . . . fair bets uniquely determines odds O in a bet on proposition a, then O are the fair odds on a.* (Howson & Urbach, 2005, 67)

In other words: adding together bets that are individually fair yields a compound bet that is also fair.

One might wonder how well the purely logical approach can succeed in excluding preference-based considerations while also relying on the notion of *fair* betting rates. The pragmatic approach understands a fair betting rate as a rate at which a person has no preference regarding which side of a hypothetical bet she would take. Howson and Urbach, however, propose to bypass this issue by identifying a person's fair betting rates instead with that person's *chance-based odds* regarding the proposition at issue. Chance-based odds are "*judgments about the relative likelihoods* of the relevant proposition and its negation" (Howson & Urbach, 2005, 53, emphasis in original). Although these are conceptually distinct from fair betting odds, they are numerically equal to them, because the betting rate a person considers fair for a given proposition is that which the person considers to balance the risk between two sides of a bet. If Samiksha considers the chance-based odds of *Salt* being true to be 0.3, then that is also the betting rate at which neither side of a wager regarding *Salt* would enjoy an advantage over the other, according to Samiksha's estimation. What Samiksha would *prefer* is a separate question. This logical approach to Bayesianism enjoys some advantages over the pragmatic approach. Critics have charged the latter with being both too psychological and anthropocentric to provide an adequate account of rational inductive inference. The psychological component of the pragmatic approach has been accused of relying on an unrealistic theory about human preferences and beliefs. By classifying Bayesian probability theory as a purely formal *logic*, Howson and Urbach seek to avoid these complications.

That avoidance comes with some costs, however. The pragmatic approach aspires to give an account of the *meaning* of probability statements in terms of the belief states of agents. But the logical approach effectively abandons this interpretive project, telling us only that a person's probability for a given proposition reflects that person's 'chance-based odds' that are in turn 'judgments about the relative likelihoods' of the proposition and its denial. But this tells us nothing about the *meaning* of statements about 'chance-based odds,' or what in general makes such statements true or false.

Howson and Urbach's probability logic also suffers from an important *weakness* of deductive logic. Because it is only concerned with consistency, deductive logic tells us only what conclusions follow from which premises (those whose denial would be inconsistent with the given premises) but nothing about which premises we should begin with. Consequently, deductive logic can only tell us which combinations of beliefs we *might* hold consistently but not which of those we *should* believe. Philosophers have often thought that inductive reasoning should fill this gap in the import of deductive logic and supply us (fallibly, of course) with premises we can then use in the context of deductive reasoning. But, by their own admission, Howson and Urbach's probability logic can no more do this than can deductive logic. It tells us what combinations of probabilities may consistently be held in combination but not which – of the infinitely many such combinations for any given set of propositions – is the right probability assignment for that set. They regard that task as unachievable by any account of inductive reasoning.

8.5 Bayes's Theorem and Updating

We have thus far been focusing on the foundations of the Bayesian account, but not yet gotten to Bayes's theorem, the powerful result from which Bayesianism derives its name.

Before stating the theorem, let us note that in our concern with foundations, we have been dealing only with unconditional probabilities like *Pr(Salt)*. What about conditional probabilities, like 'the probability that Tatyana can eat ten saltines in 60 seconds without drinking anything, *given that* Tatyana is a world champion in competitive eating'? Without going into details, we can simply note that we can extend the wagering analysis we employed previously quite naturally with the concept of a 'conditional bet,' in which the parties agree to a wager at a certain rate, but on the condition that a certain proposition is true. If that proposition turns out not to be true, then there is no wager.

Bayes's theorem concerns conditional probabilities. Bayesians contend that the axioms of probability theory constrain what combinations of beliefs may consistently be held at a given time (called *synchronic* constraints). But to give an inductive account of how science progresses, or more generally how people learn, we need a *diachronic* perspective that can handle the situation where a person receives new information. How should beliefs *change*?

The Bayesian approach to this supposes that a rational agent starts with a certain set of degrees of belief that can be represented with a *probability function*; that is, for every proposition p that is a subject of belief for an agent S, there is a probability that represents S's degree of belief in p. We will call that initial probability function Pr. Suppose that we are concerned with S's belief regarding some particular claim H and we are interested in how that belief should change upon learning the truth of some proposition E (in this standard notation, H can be thought to suggest 'hypothesis,' and E 'evidence'). To answer that question, we need some additional information: (1) What is the probability that E is true, supposing that H is true? And (2) What is the probability that E is true, independently of the truth of H? (The latter is sometimes called the *expectedness* of E.) Given this information, we can apply Bayes's theorem, which in its simplest form can be stated as:

Theorem 1 (Bayes's theorem):

$$Pr(H|E) = \frac{Pr(E|H)Pr(H)}{Pr(E)}$$

where Pr is a probability function and $Pr(E) \neq 0$. The quantity on the left side of Theorem 1 is the *posterior probability* of H, given E, in contrast with $Pr(H)$, the *prior probability*. Note that the posterior probability is a conditional probability, and for personalist probabilities, Bayes's theorem relates this quantity (an agent's fair betting rate for a wager on H, with the bet to proceed only on condition that E is true) to an agent's probability for H independently of E, her probability for E independently of H, and her probability for E conditional on H.

To address the question about how beliefs should change, we need to add a rule that relates this conditional probability $Pr(H|E)$ to the unconditional probability the agent should attribute to H once she *has* learned that E is true. For the Bayesian, change in belief is a matter of replacing one probability function with another. Rules for doing so are called *updating* rules, and Bayesians rely on *updating by conditionalization.*

Principle 2 (updating by conditionalization): *At time $t = t_0$, let S's state of belief be represented by a probability function Pr. If, at $t = t_1 > t_0$, S acquires the information that E is true (and no other new information is acquired), then at t_1, S should adopt the probability function Pr′, such that for every proposition ϕ that is a subject of belief for S, $Pr'(\phi) = Pr(\phi|E)$.*

Of course, $Pr(\phi|E)$ is precisely the posterior probability of ϕ conditional on E that we can calculate using Bayes's theorem. Although updating by conditionalization is central to the Bayesian approach, it cannot be derived from the probability axioms via any 'Dutch Book' argument (though see Teller, 1973). Howson and Urbach conclude that the rule is valid only for an agent who leaves her conditional probabilities unchanged when she acquires additional information E. In other words, $Pr'(\phi) = Pr(\phi|E)$ provided $Pr'(\phi|E) = Pr(\phi|E)$. But the probability axioms do not require an agent to keep the same conditional probabilities over time.

Nonetheless, assuming that we may attribute degrees of belief to agents engaged in inquiry under conditions of uncertainty, and assuming that those agents maintain stable conditional probabilities, the Bayesian now has a theory of how such agents must revise their beliefs if they are to remain coherent.

8.5.1 Example: Fossil Origins

To see how this might illuminate scientific reasoning, consider an artificial – though historically inspired – example.[6] For a long time *fossils* posed a

[6] The example that follows draws heavily on Martin Rudwick's authoritative study (1985). Kyle Stanford (2010) offers a fascinating (non-Bayesian) discussion of the changing nature of the evidence for the organic origins of fossils from the period discussed here to contemporary times.

great mystery. They *looked* somewhat like organisms,[7] but they were not made of organic material; they were found inside of rocks, in places far from where one would expect to find organisms of the type they resembled (such as marine-looking forms found in the Alps); many of them corresponded in form to no living organisms. From the Renaissance – when serious study of the issue seems to have begun – until well into the eighteenth century, the idea that fossils resulted from the death and fossilization of organisms – so familiar to us today – was at best a problematic proposal and, for most of that time, was justly regarded as less convincing than competing explanations.

To simplify greatly, we will pretend that there were just two such competitors to the hypothesis of organic origins: Renaissance Neoplatonists believed that all of nature was related by a network of affinities and correspondences, rooted in an "ontological analogy between Man [microcosm] and his external world [macrocosm]," "which might be made manifest by resemblances not only between microcosm and macrocosm, but also between the heavens and the Earth, between animals and plants, and between living and non-living entities" (Rudwick, 1985, 18–19). Neoplatonists considered the creation of organic-looking forms in inorganic material the work of "the very same 'moulding force' or *vis plastica*" that directed "the growth and development of living organisms themselves" (Stanford, 2010, 222). Aristotelians developed an alternative approach. Since they held that spontaneous generation could give rise to living organisms when a living form acted on nonliving matter, it was natural to suppose that similar structures could be generated in rocks "by the form present in their characteristic seeds acting on the materials of the Earth, carried there by such natural processes as the percolation of groundwater" (Stanford, 2010, 223; see Rudwick, 1985, 34–35).

These two views we shall attempt to codify with the hypotheses – *NP*: 'Fossils are formed through the action of the *vis plastica* on inorganic materials to manifest their affinity with organisms;' and *A*: 'Fossils are formed by seeds of living things imparting their forms to inorganic materials.' Finally,

[7] Some of them did, at any rate. The term 'fossil' was for centuries used "to describe *any* distinctive objects or materials dug up from the earth or found lying on the surface" (Rudwick, 1985, emphasis in original).

we have the organic origins hypothesis – O: 'Fossils are formed by processes that transform the remains of organisms into mineralized forms.'

Such views were well known to renowned anatomist Niels Stensen (1638–1686) in 1665, by which time he had arrived in Florence from his native Copenhagen, Denmark, via Leiden and Paris. (The anglicized version of his name 'Steno' appears in most scholarly literature.) While in Florence, Steno undertook to compare *glossopetrae* ('tongue-stones') with the teeth of living (more precisely until-quite-recently-living) sharks, which tongue-stones had already been noted to resemble. In this comparison Steno found support for the hypothesis of organic origins. Steno's observations identify two salient facts beyond the mere resemblance of fossils to sharks' teeth with regard to their shape (Neoplatonists and Aristotelians alike claimed the ability to account for this): (1) tongue-stones show signs of *decay* and (2) tongue-stones show no signs of distortion of the sort seen in tree roots when they grow in rock crevices, but are "the same shape, whether they come to light from softer ground, or are chipped out of rocks" (Steno, [1667] 1958, 15). The first point is significant, Steno argues, because it implies "that they were not being formed at the present time but were relics of an earlier period." The second fact indicates that "the surrounding 'earth' must have been soft when it first enclosed them" (Rudwick, 1985, 50).

Steno's argument seems to be this: According to both Neoplatonist and Aristotelian theories, the processes that produce fossils act on a relatively short time scale (the time scale of the growth of individual living organisms) and are not restricted to any particular period of history. We should not expect to find indications of decay in all such tongue-stones, then, but should find some to be intact, just as in the case of recently formed teeth in living sharks. Moreover, both Neoplatonist and Aristotelian theories suppose that fossils grow inside rocks. But when we see other examples of things growing in rocks (as in the case of tree roots), we find that the difficulty of displacing the surrounding rock causes the form of the growth to be contorted. Neither of these conclusions is compatible with Steno's observations, so he concludes that these facts all point 'unanimously' to the conclusion that tongue-stones formed where sharks had died, followed by a process of sedimentation during which the surrounding rocks formed.

Bayesians have a fairly straightforward way to understand these arguments (with a little bit of reconstruction): Steno is arguing that these observed characteristics of tongue-stones are of a sort that hypothesis O

renders probable. We should expect to see just such traits if *O* is true. The hypotheses *A* and *NP*, by contrast, make it improbable that we would observe such facts. Moreover, the plausibility of *O* prior to the consideration of these facts is not insignificant, while it is not highly probable that there is some other hypothesis that no one has thought of yet that would make sense of these observations.

The argument thus characterized in qualitative terms can be connected more strongly with the Bayesian apparatus if we agree to engage in a little historical fiction and attach numbers to these judgments. The numbers are not likely to represent any of Steno's actual beliefs. However, if we grant the Bayesian claim that beliefs formed under conditions of uncertainty can be represented as susceptible to degrees, then, at least for the purposes of illustration, we might postulate a fictional, Steno-like character – let's call him 'Schmeno' – whom we assume to be coherent in the Bayesian sense and whose degrees of belief the probabilities given below might approximate. We will use *S* to refer to Schmeno, *E* to represent the proposition 'A collection of tongue-stones was examined and found to show signs of decay and not to show signs of distortion,' and *CA* to represent the catch-all hypothesis 'Fossils form as the result of some process other than those referred to in *NP*, *A*, and *O*.'

We will start with the prior probabilities:

$$Pr_S(NP) = 0.25, \quad Pr_S(A) = 0.25,$$
$$Pr_S(O) = 0.25, \quad Pr_S(CA) = 0.25.$$

In effect, we are supposing that Schmeno begins in a state of indifference with regard to the hypotheses of Neoplatonism, Aristotelianism, or organic origins as regards the explanation of fossils. He is also indifferent – relative to each of those hypotheses – with regard to the possibility that there is some other explanation of which he remains unaware.

Now for the probability of the evidence under these various hypotheses:

$$Pr_S(E|NP) = 0.1, \quad Pr_S(E|A) = 0.1,$$
$$Pr_S(E|O) = 0.8, \quad Pr_S(E|CA) = 0.5.$$

Here we are supposing that Schmeno regards the observations he made as rather improbable if either *NP* or *A* is the correct explanation for fossils, since, under both of those hypotheses, he believes he probably would not find signs of decay and the absence of contortions in all his samples. But the organic origin hypothesis makes such observations probable, in the sense that one should expect to find them. (Note that the purely fictional value $Pr_S(E|O) = 0.8$ simply expresses the vague assumption that Schmeno considers his observations to be pretty much expected but not guaranteed by the organic origins hypothesis. In those cases where a prediction can be *deduced* from a hypothesis, then the probability of the prediction, conditional on that hypothesis, is equal to 1. In many other cases, the probability of a particular observation can be calculated from a quantitatively expressed hypothesis, perhaps with the help of auxiliary assumptions.) Meanwhile, if some other explanation of these observations of fossils is correct (i.e., the catch-all hypothesis is true), then it is unclear whether one should expect these observations or not. Schmeno considers E in that case to be an even–odds proposition.

Updating by Bayesian conditionalization will lead Schmeno to a set of posterior probabilities. But it will help here if we use a slightly different form of Bayes's theorem. In the following we take advantage of the fact that, if $H_1, H_2, \ldots, H_n$ constitute a set of mutually exclusive propositions such that for some probability function Pr, $Pr(H_1 \vee H_2 \vee \ldots \vee H_n) = 1$ (i.e., with probability 1, at least one of the propositions $H_1, H_2, \ldots, H_n$ is true), then for any proposition Q,

Theorem 2:

$$Pr(Q) = Pr(Q|H_1)Pr(H_1) + Pr(Q|H_2)Pr(H_2) + \cdots + Pr(Q|H_n)Pr(H_n).$$

We apply Theorem 2 (sometimes called the *total probability theorem*) to the denominator in the right-hand side of Bayes's theorem (Theorem 1) to get the following:

Theorem 3 (Bayes's theorem expanded):

$$Pr(H|E) = \frac{Pr(E|H)Pr(H)}{Pr(E|H_1)Pr(H_1) + Pr(E|H_2)Pr(H_2) + \cdots + Pr(E|H_n)Pr(H_n)},$$

where $H_1, H_2, \ldots, H_n$ constitute a set of mutually exclusive propositions such that $Pr(H_1 \vee H_2 \vee \cdots \vee H_n) = 1$ (i.e., the H's exhaust the possibilities) and $Pr(E|H_i)Pr(H_i) > 0$ for some $1 \leq i \leq n$ (that is, the denominator is not equal to 0).

This form of Bayes's theorem provides some guidance for estimating the expectedness of E, which otherwise seems inscrutable. The prior probabilities of the hypotheses and the probabilities of E under those hypotheses provide us the means to determine the expectedness of E.

Applying Theorem 3 to Schmeno's probabilities, then, we arrive at the following posterior probabilities:

$Pr_S(NP|E) = 0.07$, $Pr_S(A|E) = 0.07$,
$Pr_S(O|E) = 0.53$, $Pr_S(CA|E) = 0.33$.

We can see here how the probabilities have been redistributed in light of E. The probability of O has increased significantly, to a point where Schmeno regards it as more probable than not. Meanwhile, NP and A have significantly decreased in probability. The probability of CA has also increased slightly.

Of course, were Schmeno to begin with different prior probabilities, then his posterior probabilities would also differ, and nothing in the Bayesian account constrains prior probabilities beyond the requirement that they be coherent.

Although this may appear to be a problem, Bayesians consider it a strength of their account. Recall how Kuhn emphasized that an account of the development of scientific thought should allow scientists to rationally disagree. Allowing coherent Bayesian agents to have different prior probabilities accomplishes that.

For example, the naturalist Martin Lister (1638–1712) disagreed with Steno's conclusions. When confronted with Steno's arguments, Lister found them unpersuasive not because he disagreed with the observations Steno had made, or their relevance for the hypothesis in question, but because he found the hypothesis of organic origins for fossils highly implausible on the basis of other considerations drawn from his own study of shell-like fossils: (1) the substance of these shell-like fossils did not include any actual shell; (2) fossils were found only in certain kinds of rock formation, suggesting that they preferentially grow in those kinds of formations in the same way that plants grow preferentially in certain kinds of soils; (3) although such fossils resemble shells, they do not resemble the shells of any existing species. (The idea of widespread extinctions of species would have been entirely foreign to Lister.) So Lister rejected the general claim that fossils could be understood as the result of some process whereby organic remains retained their shape but became "petrified" (Rudwick, 1985, 62–63).

We can avail ourselves again of the Bayesian framework to see how Lister, while not rejecting Steno's observations, might reject his conclusions. Again, in the absence of sufficient historical accuracy with regard to Lister's actual beliefs, we will have to invent a Lister-ish character to whom we may attribute a set of Bayesian probabilities. Let's call him 'Listerio.' We will add L to our existing notation to refer to him.

We can begin by stipulating that Listerio and Schmeno are in complete agreement over the probabilities imparted to Schmeno's observations by the hypotheses under consideration:

$$Pr_L(E|NP) = 0.1, \quad Pr_L(E|A) = 0.1,$$
$$Pr_L(E|O) = 0.8, \quad Pr_L(E|CA) = 0.5.$$

But to reflect Listerio's skepticism about the hypothesis O, we should attribute a different set of prior probabilities. Whatever Lister may have made of the Neoplatonist and Aristotelian viewpoints, we can pretend that Listerio found them neither more nor less plausible than did Schmeno. If he is to begin with a lower prior probability for O than Schmeno, he will need to attribute a greater probability to the catch-all CA. So we will stipulate the following priors:

$$Pr_L(NP) = 0.25, \quad Pr_L(A) = 0.25,$$
$$Pr_L(O) = 0.05, \quad Pr_L(CA) = 0.45.$$

This leads to the following posterior probabilities:

$$Pr_L(NP|E) = 0.08, \quad Pr_L(A|E) = 0.08,$$
$$Pr_L(O|E) = 0.13, \quad Pr_L(CA|E) = 0.71.$$

Note here that Listerio, like Schmeno, judges Schmeno's observations to raise the probability of O and to reduce the probabilities of NP and A. Those relative changes are forced by the probabilities conferred on those observations by the hypotheses, on which Schmeno and Listerio agree. But for Listerio, the posterior probability of O remains rather low, while the probability of the catch-all has increased significantly.

There are two important points to emphasize in our fake historical example. The first is that, by Bayesian standards, Schmeno and Listerio are both (according to our stipulations) being perfectly rational in spite of their disagreement. The second point is that Lister based his rejection of Steno's argument on *his observations* of shell-like fossils found in England. What we have depicted as a low prior probability for O could also be regarded as a low *posterior probability* relative to Listerio's observations. As deployed by personalist Bayesians, the notions of prior and posterior probability are always relative to some particular set of data. (The flip side of this observation is that Bayesian updating by conditionalization does not apply to the problem of how to redistribute probabilities in the face of a newly invented hypothesis.)

8.6 Problems for Bayesianism

8.6.1 Subjectivity and the Catch-All Hypothesis

I have been rather glib about the phoniness of this reconstruction. Certainly one could more conscientiously take historical figures' actually expressed beliefs into account when attempting a Bayesian reconstruction of some historical episode.[8] The point of the illustration is not the actual numbers we arrive at but the positive or negative *effect* that evidence has on the probability of a given hypothesis – and how that effect compares to what happens with regard to competing hypotheses – for agents starting from different prior probabilities.

Nonetheless, a little reflection on this reconstruction highlights a difficulty for Bayesianism quite distinct from the problem of coming up with realistic probability assignments to reflect scientists' beliefs: the dependence of the Bayesian calculation on the catch-all hypothesis.

Critics of Bayesianism often point to the fact that it does not dictate prior probabilities, leaving them as a 'subjective' matter. Subjectivity, such critics argue, should be avoided in a theory that purports to tell us how we should draw conclusions from scientific data. John Norton sees the problem as inherent in the personalist Bayesian shift that merges the notions of 'degree of support' and 'degree of belief' into a single representation in terms of probabilities. The result of this merger is that "[t]he evidential relations that interest us are obscured by a fog of personal opinion" (Norton, 2010, 507).[9]

[8] See Dorling (1979) for an example of this that seeks to use Bayesianism to address the underdetermination problem, and see Mayo (1997a) for a critique.

[9] Norton argues that Bayesianism cannot distinguish cases where the evidence *fails to support* some hypothesis (such as when evidence is completely neutral with regard to a class of competing hypotheses) from cases where the evidence *supports the denial* of a hypothesis. This conflation results, he argues, in 'spurious' Bayesian inferences.

We have already noted the standard Bayesian replies to criticisms of the subjectivity of prior probabilities: First, most prior probabilities are posterior to some previously considered evidence. Second, it is only sensible that plausibility judgments should make a difference to how scientists evaluate the evidence supporting theories; this just clarifies in what sense scientists can rationally disagree about the status of a theory. The catch-all points to a subjectivity problem that goes deeper than these responses acknowledge.

Catch-all hypotheses introduce subjectivity into Bayesian inference in two ways. Recall that in order to calculate the denominator when applying Theorem 3 to our example, we needed *both* $Pr_S(CA)$ and $Pr_S(E|CA)$. The former, of course, is a prior probability and hence subjective to the extent that any prior probability is. Estimating the latter quantity – the probability of the evidence under the hypothesis *CA* – seems an elusive task. In general, probabilities of particular data or observations conditional on hypotheses under consideration are treated by Bayesians as unproblematic and agreed upon by scientists, because they can often be calculated from the hypotheses in question (perhaps with the help of auxiliary hypotheses). But asking for the probability of the data under the catch-all seems to ask us to do the impossible: to determine the probability of some specific data or observation, conditional on a completely unspecified potentiality: '*Some other hypothesis is correct.*'

In our example, we specified $Pr_S(E|CA) = 0.5$, but why? That assignment seems to reflect a judgment that if none of the hypotheses considered is correct, then *E* is just as likely to be true as not. But that judgment has no justification. Indeed, it is hard to imagine what *would* justify anyone in assuming this or any other value for the quantity at issue. If Schmeno had a belief such that $Pr_S(E|CA) = 0.9$, there would be no Bayesian grounds for criticizing him, nor if he held beliefs such that $Pr_S(E|CA) = 0.1$. If the former were true his posterior probability for *O* would be only 0.42, which would be less than the posterior for *CA*, and if the latter were the case, then the posterior for *O* would be 0.73. The probability of *E* under the catch-all can, thus, have a significant influence on the posterior probability.[10]

[10] One way to avoid this problem while using Bayesian probabilities is to decline to calculate a posterior probability and instead compare hypotheses H_1 and H_2 using the 'Bayes factor,' evaluated by taking the ratio $\Pr(E|H_1)/\Pr(E|H_2)$. This approach does not involve determining the probability of *E* under the catch-all hypothesis, but it also does not allow one to determine a posterior probability for the purposes of updating by conditionalization.

8.6.2 Relevance and Normativity

Even if we grant, however, that the Bayesian account allows for the articulation of a probabilistic theory of rationality, it remains somewhat unclear how the notion of rationality that emerges relates to the aims of scientific inquiry. For example, we might think that the point of scientific inquiry is to discover theories that are true, or approximately true. Or, we might consider the point to be the discovery of theories that are generally reliable. According to the pragmatic approach to the Dutch Book argument, the person who holds incoherent degrees of belief is irrational in the sense of being open to a combination of bets that guarantee a loss. How is this relevant to the scientist who seeks to assess the merits of a scientific theory in the face of a body of evidence? The scientist cannot be regarded as even implicitly making a wager on the truth, approximate truth, or general reliability of a general theory, but at most on that theory's reliability in some *particular* series of applications. This is because any agreement to a betting contract (even a hypothetical one) must include agreement to a criterion by which to settle the bet in some finite time. But the truth, approximate truth, and general reliability of theories are all open-ended in the sense that any judgment regarding them is provisional. If we regard any of these as the aim of scientific inquiry, then scientists' preferences regarding wagers are irrelevant to the aims of science, which concern matters not subject to wagering.

A further problem for Bayesians concerns the nature of their project. One aspect of the logical empiricist project that both Bayesian and non-Bayesian methodologists have inherited is an interest in developing accounts of scientific reasoning that not only describe how scientists reason but that are *normative* in the sense that they can provide guidance to investigators seeking to reason *well*. This normative aspect of the Bayesian project is evident from their focus on rationality. Yet the Bayesian notion of rationality is arguably inappropriate for the purposes of a normative philosophy of science, being either so weak as to be trivially satisfied or so demanding as to prevent its being applicable to the situations encountered in practice by scientists.

Consider the situation of a scientist S who, acquainted with certain phenomena E in a given domain, entertains some number of competing hypotheses $H_1, \ldots, H_n$ as potential explanations of the phenomena in that domain. S wishes to know what inferences regarding those hypotheses might be warranted. The first weakness of Bayesianism arises at just this point: S might think that the situation calls for the collection of further data, perhaps disclosing new phenomena, to test these hypotheses. The Bayesian approach does not obviously provide guidance as to what kinds of procedures *S* ought to employ to produce data that will help discriminate among those hypotheses. It only tells *S* that the data will not make a difference to their prior rankings of these hypotheses if they are equally probable under each of them. That is not much help.

The Bayesian might counter that giving more concrete methodological guidance is not the task of a theory of rationality. The theory need only allow for the discrimination between beliefs that are jointly held reasonably and those that are not thus held. So let us consider the question of whether *S* is rational to hold a certain set of degrees of belief regarding $H_1, \ldots, H_n$ in light of *E*. According to the Bayesian approach, we should think of these degrees of belief as the result of an updating, in which *S*, having gained knowledge of *E*, replaced one belief function with another. Then this belief function would be rational in the Bayesian sense if that updating were to be carried out by applying the rule of conditionalization, that is, the new degrees of belief are equal to posterior probabilities calculated by applying Bayes's theorem to a prior belief function, where that prior belief function was coherent (i.e., satisfied the probability axioms).

In one sense, this requirement is very *easy* to satisfy. Whatever one's current degrees of belief, so long as they are presently coherent, there will be *some* prior probability function that, by Bayesian updating, would yield precisely that posterior probability function. Thus, it seems that if we are committed to viewing *S*'s updated beliefs as rational, then (assuming *S* is presently coherent) we can vindicate that judgment by attributing to *S* the requisite prior beliefs, which, after all, are not directly observable. The apparent triviality of its requirements renders doubtful the value of Bayesianism even for the limited task of rationally reconstructing historical episodes (Mayo, 1997a).

Note, however, that this argument did require that *S*'s present beliefs must be synchronically rational, meaning that they must be coherent in the Bayesian sense. As long as we confine our attention to a small number of propositions, it is easy enough to come up with a plausible probability assignment that meets these requirements. But Bayesian rationality requires that *all* of *S*'s degrees of belief must jointly satisfy the probability axioms. Because of the large number of beliefs one must hold at any given time and the subtle ways in which different beliefs are related to one another, giving a consistent explicit assignment of subjective probabilities to some small number of hypotheses under consideration might, nonetheless, yield incoherence in one's overall set of beliefs (Senn, 2001, 2011). Thus, even when a reconstruction appears to show the rationality of a set of beliefs in the face of some body of evidence, a real person holding such beliefs might be incoherent because of other, off-stage beliefs.

8.7 Bayesian Philosophy and Bayesian Statistics

Our discussion thus far has considered Bayesianism as a theory of rational belief under conditions of uncertainty, to be used to examine real or hypothetical episodes of scientific reasoning and – through a process of reconstructing the inference in Bayesian terms – display its rationality or lack thereof. Can Bayesianism be used by scientists to help them answer questions?

For many years, philosophical regard for Bayesian theory far exceeded the respect afforded it by statisticians, who found it computationally difficult to apply to their data. Applied statistics was dominated by methods rooted in the frequentist approach to probability that Chapter 9 discusses, while Bayesianism has dominated the philosophy of science (and percolated into other areas of philosophy such as epistemology[11] and the philosophy of religion).

While much statistical analysis of data across the sciences remains frequentist in orientation, the field of *Bayesian statistics* is growing. Although there is no rigorous definition of the term, statisticians tend to regard data analysis as Bayesian just in case it makes use of a prior probability distribution over the range of hypotheses of interest. Bayesian statistical analyses need not be rooted very deeply in the personalist Bayesian philosophical theory, in that the prior probabilities they use may not be intended to reflect some person's predata state of opinion regarding the hypotheses at stake. Bayesian statisticians often choose instead to use a *noninformative* prior distribution that distributes probability evenly among all competing

[11] Of particular significance is the centrality of Bayesian concepts and techniques in formal epistemology, which seeks to employ formal methods to solve a range of problems philosophers have traditionally approached by theorizing about knowledge in terms of concepts such as evidence and the justification or warrant of beliefs (Bovens & Hartmann, 2003; Lin, 2022; Weisberg, 2021).

hypotheses (Jaynes, 2003). Another approach advocates the use of *reference priors*. These are prior probability distributions chosen so that the distance (according to a particular measure) between the posterior distribution and prior distribution will be maximized (Berger & Bernardo, 1992; Bernardo, 1979). Many Bayesian statistical analyses concern problems in which there is a prior assumption that a certain parameter has a numerical value, possibly within some determinate range, and the hypotheses under consideration concern the value of that parameter. Provided that the prior assumption is correct (often not a trivial issue), the bite of the worry previously raised about the catch-all hypothesis is considerably mitigated, since all of the hypotheses consistent with the assumption can be stated explicitly and often yield determinate probabilities for the data under consideration.

The idea that the usefulness of Bayesian statistical methods is entirely independent of the status of Bayesian philosophy has a pragmatic appeal. Many scientists who have used them value the flexibility of Bayesian methods. But philosophical issues, particularly concerning *interpretation*, lurk wherever we find statistics used. Regardless of how the prior distribution is chosen and justified, the point of a Bayesian analysis is to generate a *posterior* distribution, which serves as the output of the analysis.[12] For a personalist Bayesian, the meaning of this posterior distribution is clear: It represents the state of belief a rational agent should have with regard to the relevant hypotheses, in light of the data, if that agent starts out with a state of belief represented by the prior distribution. If the prior distribution does not represent a state of belief, but is chosen because it is noninformative, or affords ease of computation, or for some other practical reasons, then the posterior distribution no longer has this same meaning. But then, what does it mean? David Cox and Deborah Mayo have stressed the need for objective Bayesian statistics to clarify how its posterior probabilities fulfill the goal of measuring the strength of the evidence supporting a given inference (Cox & Mayo, 2010).[13]

[12] Some statistical techniques that employ Bayesian ideas and techniques do not generate a posterior distribution, such as the use of Bayes's factors. Setting aside the terminological dispute over what counts as a legitimate Bayesian statistical analysis, any statistical method has to find answers to the questions of whether it is appropriate to the aim for which it is adopted and how to its product should be interpreted.

[13] In an interesting effort to address such foundational issues, Jon Williamson has pursued an interesting variant of Carnap's objective version of Bayesian inference (Williamson, 2010).

8.8 Conclusion

Does Bayesianism, in the personalist form that has been the primary focus here, provide the kind of neutral framework for the evaluation of evidence that Carnap and other logical empiricists sought? In one sense it does not. Carnap sought a measure of the degree to which any given data confirm any given hypothesis that was *objective*, in the sense that any two inquirers who agreed on the language to be used to describe the hypothesis and the data would also agree on the degree of confirmation. But personalist Bayesian inquirers can disagree on this if they start out with different prior probabilities, or if they disagree on how probable the data are given the competing hypotheses.

In another sense, however, the Bayesian framework does exhibit some features of the logical empiricists' vision. Although disagreements over prior probabilities can yield discrepant posterior distributions, Bayesian investigators will all share the same procedure for updating from the former to the latter. In other words, they will agree on how the data are relevant to the evaluation of the hypotheses at issue. Moreover, as Bayesians stress, if agents with differing prior probabilities continue to update by conditionalizing on the same accumulating body of data (and they agree on the probability of the data, including the probability conditional on the catch-all hypothesis, and they agree about which hypotheses have probabilities of 0 or 1), their repeated updating will yield a *convergence* of their probability functions (see Doob, 1971; Gaifman & Snir, 1982; and Earman, 1992, 144–149 includes a philosophical assessment). Thus, although Bayesians attribute subjectivity to scientific reasoning, they provide in principle a framework for reducing that subjectivity.

Noting these points, Wesley Salmon proposes that Bayes's theorem provides a means for building a bridge between the logical empiricist project and Kuhn's postpositivist insights (Salmon, 1990). Salmon argues that we can interpret the kinds of paradigm-dependent judgments that lead scientists to evaluate theories differently in Kuhn's account as yielding different prior probabilities representing plausibility judgments before some body of data had been taken into account, while Bayes's theorem provides a paradigm-independent device for bringing those data to bear on the theories in question. Nonetheless, Salmon worries that Bayesianism in its usual form allows too much subjectivity into scientific reasoning, because of the difficulty of determining the probability of the evidence under the catch-all hypothesis. Salmon proposes a solution – using a Bayes's factor approach – that does not determine a posterior probability, but allows one to compare hypotheses and decide, for any given pair, which of them ought to be preferred, given the data in hand. His approach eliminates the need to refer to the catch-all hypothesis, but at the cost of greatly limiting the strength of one's conclusions: Salmon's approach leads to conclusions only about which of the theories under consideration we ought to prefer to the others under consideration, but not to conclusions about the truth, probability, or truthlikeness of those hypotheses.

Debates between defenders and critics of Bayesianism are among the most hotly contested in the philosophy of science as well as in statistics, where the arguments at one time became so pugilistic that they became known as the 'statistics wars.' Here we have barely scratched the surface of the issues. A historically significant collection from the statistics wars is Savage (1962). Howson and Urbach (2005), on which I have drawn for the exposition here, give a good overview of many of the criticisms of Bayesianism in the context of a vigorous defense of the view. Deborah Mayo, whose frequentist-based approach to inference we discuss in Chapter 9, has voiced some of the most significant criticisms of Bayesianism (Mayo, 1996, 2018). Earman (1992) gives a mixed assessment of Bayesianism, surveying a number of debates that continue to develop.

Appendix: Solution to the Chevalier's Problem

Turning first to Game 2: That the die lands 6 up at least once in four tosses is equivalent to the denial of the claim that it does not land 6 up in any of four tosses. The latter claim is equivalent to the statement that the die lands with something other than 6 up on each of four tosses. To calculate the probability of Game 2 we can therefore calculate the probability of getting not-6 four times in a row, and then subtract that number from 1. (Recall that the probability of a logical truth such as 'p or not-p' is (by Axiom 1) equal to 1, so for any proposition p, $Pr(p) = 1 - Pr(not - p)$.) The probability of getting something other than 6 on a single toss is 5/6, and that probability for any given toss is independent of what happens in any other toss. So (applying the definition of conditional probability) the probability of not-6 on four consecutive tosses is $5/6 \times 5/6 \times 5/6 \times 5/6 \approx 0.482$. The probability of getting at least one 6 is then approximately $1 - 0.482 = 0.518$.

Calculating the probability of winning Game 3 uses a similar technique. There are thirty-six possible, equally probable outcomes for a toss of two dice, only one of which consists of two 6s. The probability of *not* getting two 6s on a single toss is 35/36, and on twenty-four tosses is $(35/36)^{24} \approx 0.509$. The probability of getting a pair of dice to land 6 up at least once in twenty-four tosses is, therefore, approximately $1 - 0.509 = 0.491$.

9 Reasoning with Probability

Frequentism

9.1 Introduction

In Chapter 8, we explored that aspect of probability ideas concerned with "the degree of belief warranted by evidence." Now we turn to the other aspect: "the tendency, displayed by some chance devices, to produce stable relative frequencies" (Hacking, 1975, 1). Our use of examples involving games of chance to introduce the mathematics of probability at the beginning of Chapter 8 relied implicitly on this idea. But are dice games relevant to science? As we will see, thinking of *scientific experiments* or *measuring procedures* as chance mechanisms with tendencies 'to produce stable relative frequencies' holds the key to the potential value of relative frequency ideas for the philosophy of science.

Frequentism understands probability statements as statements about the relative frequency with which a certain outcome would occur under a repeated execution of some process. Frequentist ideas have been central to the development of theoretical and applied statistics and, thus, have had a significant influence on how scientists analyze data and report their results. Some of these statistical practices have been subjected to considerable criticism, however, and philosophers of science have tended to regard Bayesianism as based on a more coherent set of principles. Recently, the *error-statistical philosophy* has attempted to respond to philosophical criticisms of frequentist statistics while remaining founded on the idea of experiments as chance mechanisms.

Since, however, error statistics arises from an attempt to reform and rehabilitate the frequentist tradition in statistics, we will start with a brief introduction to that tradition. Statisticians have long emphasized the importance of experience in the use of statistical methods and analysis of real data to a good understanding of statistics, and with good reason. What follows is too brief and superficial to underwrite a useable understanding of statistics. Nonetheless, we may hope that some grasp of the conceptual underpinnings of frequentist statistics will convey a sense of the philosophical orientation and import of these ideas.

9.2 Relative Frequency 'In the Long Run'

Suppose that, following the suggestion floated early in Chapter 8, we interpret the statement 'the probability of winning Game 1 is 1/6' to mean 'if you play Game 1 over and over again, then, on average, you are going to win one out of six games.' Although this interpretation has an intuitive appeal, we have taken an apparently definite mathematical statement and given it a vague and potentially misleading interpretation. The statement 'over and over again' is vague. How many times is that? If we think playing over and over again is something that can be achieved by playing some finite number of times, then the statement 'if you play Game 1 over and over again, then, on average, you are going to win one out of six games' might not be *true*, even if 'the probability of winning Game 1 is 1/6' is true. As we will see, you might win either more or less than that. (The name we have for that phenomenon is 'luck.') So, the interpretation might also be misleading. Let's see if we can do better.

As Hans Reichenbach notes, the frequency interpretation of probability has two distinct functions in probability theory. One is concerned with providing an account of the *meaning* of probability statements, while the other is concerned with accounting for the *substantiation*, or support, of probability statements (Reichenbach, 1938, 339). We will begin with the first function.

Suppose that you do undertake to play Game 1 repeatedly, recording after each roll of the die the number of times you have played N and the number of times you have won N_w. I did this.[1] Table 9.1 gives my recorded values for the fraction N_w/N – the *relative frequency* of winning – after twelve rolls.

[1] No, I didn't. Being too lazy to sit and roll a die over and over again and lacking the nerve to ask for funding to pay a student to roll a die over and over again, I used a random number generator that produced outputs of whole numbers from 1 to 6 to simulate my data. No conclusion about any actual die can be drawn from my fake data.

Table 9.1 *Data from the rolling of a die*

Roll	1	2	3	4	5	6	7	8	9	10	11	12
Result	5	6	1	3	2	5	6	6	6	5	3	6
N_w/N	0	0	0	0	0	0	0	0	0	0	0	0

Such bad luck! I continued rolling the die (let's pretend), and after thirty-six rolls, the relative frequency of wins was still only $N_w/N = 1/36$. At $N = 216$, I had $N_w/N = 23/216 \approx 0.11$ (note that $1/6 \approx 0.17$). At $N = 3{,}600$ (surely enough to make the point, so that I could take a break from the feigned tedium of repeatedly rolling that die!), the relative frequency of wins was looking much more like the 'right' number: $N_w/N = 613/3{,}600 \approx 0.17$.

In our dice game example, the relative frequency approach to probability interprets $Pr_M(X_1 = 4) = 1/6$ as a statement not about what happens in any finite sequence of rolls of the die but about the value of N_w/N as N becomes infinitely large, or *in the limit as* $N \to \infty$. More generally, according to the frequentist view:

Definition 3 (frequentist probability): *If one is able to repeat a trial under identical conditions and note whether some event E occurs as an outcome of the trial, and if* N_E/N *is the relative frequency with which E occurs, then*

$$Pr(E) \equiv lim_{N\to\infty}\frac{N_E}{N}. \tag{9.1}$$

Of course, this simple definition only applies if the probability that a certain kind of trial has a certain kind of outcome is the same from one trial to the next and is independent of how many trials have been carried out. Even repeatedly throwing a die would seem not to *exactly* satisfy this requirement: As we extend the number of throws to infinity, we will at some point undoubtedly notice the die to start showing signs of wear, however long that might take, and the relative frequency of 4s might not remain the same as the slow erosion of the die continues. To suppose that it will remain the same constitutes an *idealization* in our probability model of Game 1. To be an appropriate idealization, it needs to be the case that, although it may not be strictly true that $lim_{N\to\infty}\frac{N_W}{N} = 1/6$, we will not be misled in any of our real-world reasoning if we pretend that it is.

The die-rolling data exhibits a pattern that, intuitively, seems typical: Although small amounts of data show a behavior that we might not have been able to predict, the more data we collect, the more the data align with what we think *ought* to happen (based on our theory about the die). The mathematical elaboration of frequentist probabilities allows us to make precise our intuitive sense of what is *typical.* Probability distributions are crucial to this clarification. They convey mathematically both what outcome is most to be *expected* and the degree to which the outcomes of a given process are *variable.*

To see how, think of the 3,600 throws of the die as a series of 100 *experiments,* in each of which the die has been thrown thirty-six times. In any given experiment, there will be some number of wins, so let's define a random variable Y_i that – for a given trial i consisting of thirty-six throws – will take the value $Y_i = y_i$, where y_i is the number of wins in the ith trial.

Our theoretical assumption that the probability of a win on any individual throw of the die is 1/6, and that this value is the same for each throw of the die and independent of the outcome of other throws (in statistical parlance, the outcomes of individual throws are 'Independent and Identically Distributed,' or IID) has implications about both facets of our experiment via the *binomial distribution,* which is a mathematical function of the number of opportunities (here $N = 36$) and the probability of winning in a single opportunity (here $p = 1/6$). We will forego the mathematical details here (this is covered in any elementary text on statistics or probability) and simply display the results for the possible numbers of wins of Game 1 in a trial consisting of thirty-six throws of the die. The distribution, given in Figure 9.1, reveals that the mean value or *expectation value* for Y_i, $\langle Y_i \rangle_T = 6.0$ for each i (the subscript T indicates that this is the theoretical value). In other words, for each trial, the expected number of wins is six. The distribution also yields a useful measure of variability, or the degree to which the results are 'spread out' around the expectation value. This is given by the *standard deviation* σ, which for the binomial distribution is given by the expression $\sigma = \surd\{Np(1-p)\}$ (≈ 2.24 for Y_i). For comparison, Figure 9.2 shows the data from the 100 experiments by tabulating how many of them yielded particular values for Y. The observed mean of these experiments is $\langle Y \rangle_o = 6.13$.

Our example of the dice game exhibits simply and obviously a feature that is found throughout the sciences in more complex and subtle ways:

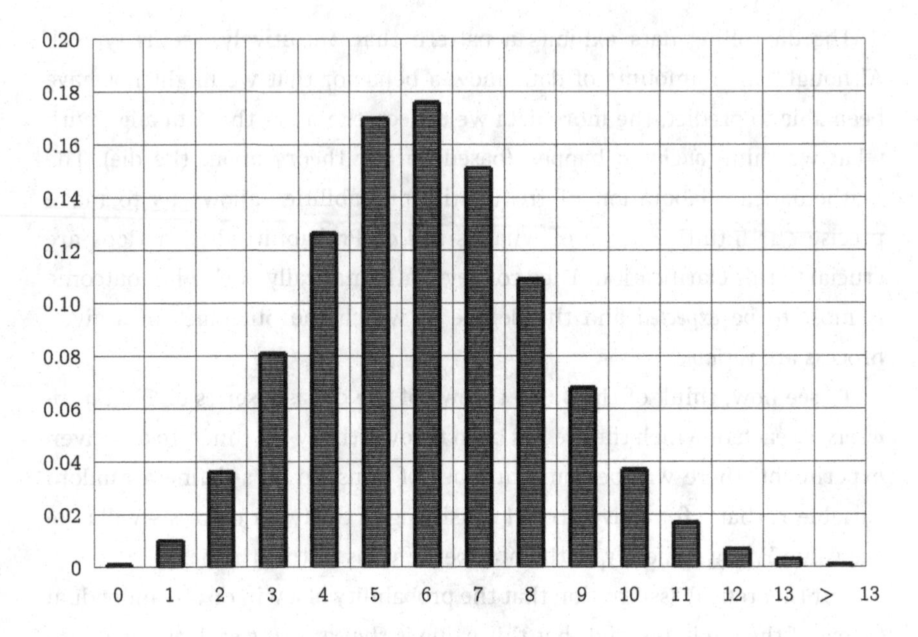

Figure 9.1 The binomial distribution for $Yi(p = 1/6, N = 36)$.

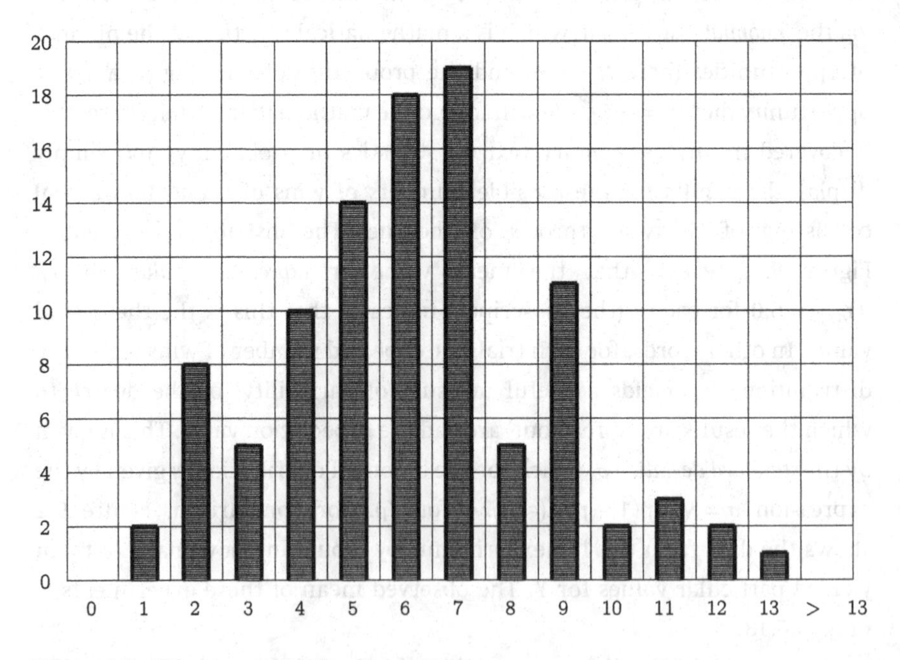

Figure 9.2 Dice game results, $N = 100$.

Although what happens in any *individual* case is unpredictable, the data *in aggregate* exhibit an understandable pattern that can serve as a basis for forming expectations about the future. Many advances in statistics have come through the efforts of mathematicians and natural philosophers to duplicate the successes enjoyed when dealing with games of chance in the analysis of astronomical data, in the prediction of the behavior of markets and other economic phenomena, and in many other domains in which patterns emerge in the aggregate from noisy behavior at the level of individual observations. Returning to Reichenbach's two functions of frequentism, probability ideas and the statistical science based upon them are so useful because they help both to *represent* such patterns and to facilitate their *discovery*. Thus far, the emphasis was on the representation function; next we will consider how the concepts of frequentist statistics facilitate the discovery of such patterns.

9.3 Two Traditions in Frequentist Statistics

In our die-throwing example, we have assumed that we know from the outset that the probability of winning Game 1 on a single throw of the die is 1/6, that is, that the die is fair. If, however, we did not know that the die is fair, we would face quite a different situation. If we *knew* that the die is fair, then it is reasonable to respond to my failure to win any games in the first twelve throws – and winning only one game in the first thirty-six – by saying 'what bad luck!' If, however, we did not already know that the die really is fair, then we might be inclined to respond by thinking that the data indicate not that my luck is bad but that the die is loaded, that is, not fair. But would such data constitute good evidence that the die is loaded?

In this section, we will consider two important schools of thought within the frequentist tradition in statistics to see how they might address such questions. One tradition emphasizes the search for *statistically significant* discrepancies from some target hypothesis. The second tradition advocates the use of statistics to test among a class of hypotheses all belonging to some more general model. We will note the genuine differences between these approaches, but also consider a philosophical approach to scientific reasoning that draws on insights borrowed from both camps.

9.3.1 Fisherian Significance Testing

The kind of problem that interests us confronts not only those who like to gamble with dice but also their kindred spirits – research scientists. Let's consider a recent example.

Example: The Higgs Boson?

In July 2012, scientists at the Large Hadron Collider (LHC), a particle accelerator at the CERN (Conseil européen pour la recherche nucléaire) laboratory near Geneva, Switzerland, announced they had discovered a new particle very similar – and perhaps identical – to the previously hypothetical Higgs boson (additional data allowed them later to conclude unambiguously that it was the Higgs). A discussion of what exactly the Higgs boson is, and why it is such an exciting discovery, would take us far afield of our immediate concerns (Weinberg, 2012). (Suffice it to say that in the prevailing Standard Model of the elementary constituents of the physical world, the Higgs boson is implicated by an ingenious hypothesis that seeks to explain why particles like electrons and quarks are not massless.) If the Higgs boson exists, then it will be produced out of the energy released in collisions between super-high-velocity protons and then decay into other particles according to the characteristic patterns dictated by well-established principles of the Standard Model. Two groups of physicists at the LHC use massively complex particle detectors to measure the properties of the particles produced in proton–proton collisions in order to see how well they match the 'signature' pattern expected from Higgs boson decays.

Two features of this scientific endeavor make statistics relevant. First, the theoretical description of the production of Higgs bosons rests in Quantum Field Theory, an inherently statistical theory yielding predictions in terms of probability distributions rather than certain outcomes. Second, *background* processes that will occur even if the Higgs boson does not exist can mimic the signature of Higgs decays, and these processes are also best described by probability distributions.

The two teams of LHC physicists, in announcing their discoveries, both described it as a '5-sigma (σ)' effect (Aad et al., 2012; Chatrchyan et al., 2012).[2] Here σ refers to our just-introduced statistical concept – the *standard deviation.*

[2] You can find philosophical reflections on the reasoning employed in this episode – including the statistical features discussed here – in a special issue of *Synthese* on "Evidence for the

Here is how that is relevant: Even in the absence of the Higgs boson, they knew they would find some number of Higgs-like 'signatures' in their data, a number they could only predict with probability. Part of their problem (a very difficult part) was to understand this *background distribution* so that they could say – for any given number of Higgs-like signatures found in their data – what the probability is that they would get as many or more of such signature decays if only background processes were present. Their background distribution, just like the binomial distribution we used to characterize the behavior of a fair die, had an expectation value – the average number of signature events that would be expected from background alone – and a standard deviation – a measure of the variability of the background in terms of the *width* of the distribution, describing how much the number of signature events could be expected to vary from one dataset to another. By describing their result as a '5σ' effect, the LHC physicists were indicating how far out in the background distribution their result is: It is five standard deviations greater than the expectation value assuming that only background processes are present.

In this way, they followed the methodology advocated by Ronald Fisher (1890–1962), who made significant contributions to evolutionary biology and genetics as well as statistics. The essence of Fisher's methodology can be described in fairly simple terms (see Cox & Hinkley, 1974, ch. 3 for a careful introduction), but it requires certain things. Table 9.2 gives a shopping list for a Fisherian test. You begin with a *question*. To learn something that will help answer that question, you need a substantive hypothesis that is a possible answer to that question, which you can test. This is the *null hypothesis*. The hard part comes next: figuring out how to generate *data* for such a test. You will need to be able to define some quantity that is a function of the data, called a *test statistic*, which will have a known probability distribution if that hypothesis is true, such that larger values of the test statistic should indicate stronger evidence of departure from what is expected if the null hypothesis is true. The probability distribution of the test statistic under the null hypothesis is the *null distribution*. The null distribution serves as a mathematical *model* of the null hypothesis, and what we test directly is the

Higgs Particle." See especially Beauchemin (2017), Cousins (2017), Dawid (2017), Franklin (2017), Staley (2017), and Wüthrich (2017). Mayo (2018, 202–217) discusses this episode from the severe testing standpoint discussed later in this chapter.

Table 9.2 *Shopping list for a Fisherian test*

A question	Based on what you want to learn about
A null hypothesis	A possible and testable answer to that question
Data	From observations or experiment
A test statistic	Some function of the data
	Known probability distribution if the null hypothesis is true
	Larger values indicate stronger evidence against null
The null distribution	The distribution of the test statistic assuming the null is true

statistical hypothesis that the data are generated by a process characterized by the null distribution. We test that claim by looking at the value of the test statistic: How probable is it that one would get a value equal to or greater than the observed value, assuming the null hypothesis is true?

For example, if we wish to test whether a die is fair, then our statistical hypothesis will state that the data are drawn from a binomial distribution with $p = 1/6$ and $N = 36$. We could use the average number of 'wins' in our previously described 100 experiments by defining our test statistic to be $T \equiv |\langle Y_i \rangle - \langle Y \rangle_T|$, the observed value of which is denoted by $t \equiv |\langle y_i \rangle - \langle Y \rangle_T|$. The search for the Higgs boson required much more complicated data to test the hypothesis that those data are drawn from the probability distribution that characterizes 'background physics.' The relevant random variable in that case relates to the difference between the number of signature events observed and the expectation value of the background distribution (the test statistic used has a slightly more complicated definition). In reporting the results of the Higgs search, the shorthand '5σ' stands for a certain probability statement: Were the null hypothesis to be true, the probability of getting as large an excess of signature events as they did (or more) is about 3×10^{-7}. This kind of probability statement, based on the observed value of the test statistic, is called a *p-value*, or sometimes the *observed statistical significance* of the result.[3]

[3] In the die-tossing experiment, we test for divergence from the null hypothesis due to differences in either direction; a significant excess, *or* deficiency, relative to the expectation value might lead us to reject the null hypothesis, and this is reflected in our calculation of the *p*-value. This is called a *two-sided* test. In the Higgs boson case, only excesses are considered relevant to the calculation of *p*-values, because the test is designed

To appreciate the role p-values play in reporting an experimental result, compare the values cited in the Higgs announcements to what we should report were we to use the data from die-throwing, reported above, to test whether our die is fair. Recall that the observed mean $\langle y \rangle_o = 6.13$, whereas theoretically we should expect a fair die to yield an average of six wins out of every thirty-six throws. Does the difference of $t = 0.13$ mean anything? The p-value of a difference of 0.13 is approximately 0.61.[4] That means that if you do an experiment of 100 trials with thirty-six tosses each, and you *are* using a fair die, you will get values for the random variable T as large as 0.13 or more *quite often* – on average, about 61 out of 100 times. The LHC physicists cite their very small p-values to communicate that their data strongly indicate the negative claim that the null hypothesis of background-only processes does *not* yield a distribution that does a good job of describing their data. (That the data *do* indicate the existence of a Higgs boson rests on further features of the data beyond what is captured by the p-value.) Our rather *large* p-value for the die-throwing experiment, by contrast, clearly does not indicate that the null distribution is incompatible with the data. The result we obtained seems to be pretty typical of a fair die.

Does that mean that these data are good evidence that the null hypothesis is true? Fisher declared that "the null hypothesis is never proved or established, but is possibly disproved, in the course of experimentation. Every experiment may be said to exist only in order to give the facts a chance of disproving the null hypothesis" (Fisher, 1949, 16). Note that the null hypothesis makes a very exact statement: The probability of winning Game 1 in a single toss of the die is *exactly* 1/6. Hypotheses that attribute exact values to quantities are often called *point hypotheses*. The slight asymmetry among the six faces of the die (each does have a different number of dots) provides some reason to suspect that, construed as an exact statement, the null hypothesis is false, but whatever discrepancy from the exact truth of the null hypothesis might obtain is likely to be too slight to be of any practical interest to us and is, in any case, not of a sort that can be revealed by the experiments we performed.

for departures from the null with respect to the appearance of nonbackground processes. This is a *one-sided* test.

[4] This is based on what statisticians call a two-sided t-test.

Despite their widespread use among scientists, *p*-values are subject to various misunderstandings and misuses.[5] This has even given rise to proposals that journals should adopt policies of not publishing papers that rely on or report *p*-values (i.e., "don't say 'statistically significant'") (Wasserstein, Schirm, & Lazar, 2019; see Mayo, 2022 for a critical response). A common misunderstanding is to think of the *p*-value as the probability that the null hypothesis is false. Frequentists deny that probabilities can be applied to hypotheses (unless the hypothesis can appropriately be regarded as the output of some chance mechanism). Hypotheses such as 'This die is fair' or 'Humans first used pottery for cooking approximately 20,000 years ago' are either true or false. No sense can be made of the relative frequency with which they are true. (Bayesians, of course, are perfectly happy to assign probabilities to hypotheses such as these, but would not endorse the *p*-values delivered by Fisherian significance tests as the correct means to calculate them.) Rather than the probability that the null hypothesis is false, a *p*-value reports the probability that one would observe a value for the random variable being considered that fits the null hypothesis as poorly or worse than the value actually observed, *assuming the null is true*.

It is also sometimes alleged that in significance testing the null hypothesis is 'presumed true' or 'given the benefit of the doubt.'[6] But being chosen as a target for a significance test does not in itself indicate any judgment regarding the truth of a hypothesis. (We already have noted that a null hypothesis might be useful to test in spite of our having a good reason to suspect that it is false, at least strictly speaking.) Point hypotheses yield well-defined probability distributions (*sampling distributions*) for the random variable of interest, and thus the means for calculating *p*-values. Typically the aims of the test – that is, what the investigator seeks to find out from the data – dictate the choice of null hypothesis. If you are interested in learning whether the die is loaded, it is more useful to test the hypothesis that it is not than to test one that says that it is loaded to some specific degree. If you are interested in finding out whether the Higgs boson exists, it is more useful to test the hypothesis that it does not than the one that attributes particular

[5] See Morrison and Henkel (1970) for a historically significant collection of perspectives and Ziliak and McCloskey (2008) for a more recent jeremiad.

[6] Even well-known and mathematically sophisticated social scientists can fall into this error (Silver, 2012).

characteristics to it (though testing the latter kind of hypothesis might also prove important to understanding the import of your data).

Finally, one should bear in mind that a result that is *statistically* significant may be *scientifically* or *practically* insignificant. One way in which this may happen comes from the fact that as the sample size increases, the sampling distribution tends to become more sharply peaked around the expectation value, so that smaller discrepancies become more improbable assuming the null hypothesis is true. If we were, for example, to obtain $t = 0.13$ from our die-rolling data on the basis of 10,000 experiments rather than 100, we would find that our result has a statistical significance of about 3×10^{-7} (just like the data from the search for the Higgs boson!). At that point we might justifiably claim to have evidence that the die is not *exactly* fair, but we might not care about such a small departure from fair odds. Indeed, with enough data, any slight discrepancy from the null expectation can become very statistically significant. With $N = 1{,}685{,}000$, a $p = 3 \times 10^{-7}$ result would be obtained with $t = 0.01$! (Partly for this reason, it is often recommended that *p*-value statements be accompanied by estimates of 'effect size,' a quantity that can be variously calculated, but is intended to convey the magnitude of the difference on which the *p*-value is based.)

None of this means that *p*-values are valueless. The statistician David Cox summarizes the point nicely:

> The essential point is that significance tests in the first place address the question of whether the data are reasonably consistent with a null hypothesis in the respect tested. This is in many contexts an interesting but limited question. (Cox, 2006, 42)

Cox goes on to note that a more developed methodology will allow us to formulate "more informative summaries of what the data plus model assumptions imply" (Cox, 2006, 42). What might such a more developed methodology look like?

9.3.2 Neyman–Pearson Testing

Fisherian hypothesis testing aims to discriminate between the null hypothesis and *every other* possible probability distribution. If a Fisherian test provides evidence against the null, this entails only that it provides a reason to believe the data to be incompatible (statistically) with the null but compatible with *some other* hypothesis, without the latter being further specified.

Model Assumptions

Consider again our Fisherian test of the hypothesis that the die is fair. The null hypothesis is well defined, but if we found the data statistically incompatible with the null hypothesis, we might want to know what we can reliably conclude from the discrepancy we do find. For example, intuitively you might think that if our 100 experiments of thirty-six trials each yielded a value of $\langle Y \rangle_0 = 18.0$, then we could infer that $p \approx 1/2$ and more generally that, for whatever value we obtain for $\langle Y \rangle_0$, $p \approx \frac{\langle Y \rangle_0}{36}$. This intuitive judgment is based, however, on an easily overlooked assumption that needs to be made explicit.

The quantity p is a parameter in the binomial distribution that we used (by letting $p = 1/6$) as a mathematical model of the hypothesis 'the die is fair.' That model, along with the information about our testing procedure, was used to derive a sampling distribution for the random variable Y, *under the assumptions of the null hypothesis.* But if the data indicate that we have a reason to *reject* the null hypothesis, what happens to those assumptions? In particular, may we assume that the behavior of the die is best described by the binomial distribution? If not, we are not warranted to draw conclusions about the value of p, which has meaning only as a parameter in that distribution.

As already noted, the binomial distribution assumes that the outcomes of individual throws are IID. The hypothesis that the die is fair not only involves attributing a probability to winning the game on a single throw of the die but also asserts that this probability of winning, for a given throw, is independent of the outcome of previous throws and remains the same for each throw in the sequence (the possible outcomes of each throw are described by the same – identical – probability distribution). So, the null hypothesis that picks out the value $p = 1/6$ belongs to a family of hypotheses. We can use the notation $Y \sim B(n, p)$ to mean 'random variable Y is distributed according to the Binomial distribution with parameters n and p.' Then we can express the statistical null hypothesis as $H_0 : Y \sim B(36, 1/6)$.

The intuition that we can infer, at least approximately, the value of p from the observed value $\langle Y \rangle_0$ really rests on the intuition that most dice satisfy the IID assumptions, even if we cannot always trust them to be fair. (Whether this intuition is warranted is, of course, another matter.) If we assume that there is some value of p for which $Y \sim B(36, p)$ is true (we control the value of n, so we just *make* $n = 36$ true when we perform the experiment), then we can use the results of this test to draw conclusions about the range of values for p that are *supported* by those data. In other words, we assume the model $Y \sim B(36, p)$ and use that model to guide our inferences about the value of p. But in doing so, all of our inferences are hostage to the adequacy of the assumed model.

The question of *adequacy* of the assumed model in model-based inferences raises important issues about how such models are specified and tested (Mayo & Spanos, 2004; Spanos, 1999; Staley, 2012; Tukey, 1960). Here we will focus on the question of how to use that model so as to ensure that we are drawing reliable inferences from the data in an optimal manner.

9.3.3 Error Probabilities

In the late 1920s and early 1930s, Jerzy Neyman (1894–1981) and Egon S. Pearson (1895–1980) laid out a mathematical framework for choosing an *optimal* approach to testing a given hypothesis (Neyman & Pearson, 1928, 1933). By invoking the kind of model assumptions just discussed, their framework allows for more definite statements regarding the comparison between the null hypothesis and its alternatives than a Fisherian test would support.

Again, we will need to forego a full mathematical discussion of Neyman–Pearson (NP) theory and settle for a characterization that will allow us to grasp the basic stance this approach takes toward the problem of statistical inference (Cox & Hinkley, 1974, ch. 4 again provides a helpful systematic treatment).

Note that model assumptions such as IID delimit a class of possibilities. If our test procedure assumes at the outset that the die-tossing experiment outcomes are IID, then we will not be able to use that procedure to make inferences about hypotheses that are incompatible with those assumptions, such as a hypothesis according to which winning the game on one toss of the die makes it 50 percent less probable that you will win on the next toss. In the NP approach, model assumptions determine the class of hypotheses among which the test can discriminate. If the statistical null hypothesis is $H_0 : Y \sim B(36, 1/6)$, then the *alternative* against which we are testing it is $H_1 : Y \sim B(36, p)$; $p \neq 1/6$. In other words, all hypotheses under consideration in this test assume that Y is binomially distributed with $n = 36$, but the null specifies one value for p, and the alternative says that p has some other value. As is often the case in an NP test, the null hypothesis here is a point hypothesis ascribing a precise value to a parameter in the statistical distribution, whereas the alternative *compound hypothesis* simply says that the value of that parameter lies *somewhere* within a range of numbers – in this case the interval [0, 1], exclusive of the point 1/6.

Size, or Significance Level

Traditionally NP tests are described as having two possible outcomes: accept the null H_0 or reject it. As Neyman and Pearson noted, having two possible outcomes means the test is susceptible to two distinct kinds of error:

> If we reject H_0, we may reject it when it is true; if we accept H_0, we may be accepting it when it is false, that is to say when really some alternative $[H_1]$ is true. These two sources of error can rarely be eliminated completely; in some cases it will be more important to avoid the first, in others the second. (Neyman & Pearson, 1933, 296)

To specify an NP test for discriminating between H_0 and H_1, one begins by choosing the level at which one would like to limit the probability of the first kind of error, called a *Type I* error. For example, we may decide to design our test so that it will reject H_0 when it is true not more than once in twenty times. Then we may determine the *critical value* t_c for the test statistic T such that the probability of getting a value for T that is greater than or equal to t_c, assuming the null hypothesis is true, is .05,[7] that is,

$$Pr(T \geq t_c; H_0) = .05. \tag{9.2}$$

[7] The appearance of a semicolon in Equation (9.2) introduces some new notation. Expressions of the form $Pr(A; B)$ should be read as 'the probability of A assuming B is true.' This expression might not seem substantively different from the conditional probability statement 'the probability of A given (or conditional on) B,' which we notate as $Pr(A|B)$. However, mathematically speaking, the difference is significant. Recall that conditional probabilities are defined such that $Pr(A|B) = Pr(A \wedge B)/Pr(B)$. (This is the version of Definition 2 (conditional probability), introduced in Chapter 8, for probabilities applied to sentences. The symbol $\wedge$ represents logical *conjunction*, which connects two sentences into one that is true if and only if both of the conjoined sentences are true.) For frequentists, however, the conditional probability expression $Pr(T \geq t_c|H_0) = Pr(T \geq t_c \wedge H_0)/Pr(H_0)$ would be meaningless. There is no relative frequency with which H_0 is true, making both the numerator $Pr(T \geq t_c \wedge H_0)$ and the denominator $Pr(H_0)$ undefined.

More generally, one specifies an NP test to have a type I error probability limited to a certain value α (called the *size* or *significance level* of the test) by choosing a *critical value* t_c for the test statistic so that the condition

$$Pr(T \geq t_c; H_0) = \alpha \qquad (9.3)$$

is met.

Power

The choice of a critical value t_c, then, ensures a certain kind of reliability for the test. By choosing a large enough value to make α – the Type I error probability of the test – very small, the investigator ensures that her test will only very rarely reject the null when it is true. One could, of course, ensure the same with a Fisherian test. The LHC physicists did just this in searching for the Higgs boson. They set a standard such that they only announce a discovery of a new phenomenon when the probability of such a large divergence from the null is ~ 3×10^{-7} or less, assuming the null hypothesis is true. But what about the other kind of error, in which one accepts the null hypothesis (or fails to reject it[8]) although the alternative is true? This is called *Type II* error. One of Neyman and Pearson's innovations was to introduce a quantity called the *power* of a test, which is 1 minus the probability β of a Type II error. This enabled them to address what they saw as a shortcoming in Fisher's approach – namely, that Fisher did not provide a systematic rationale for the choice of test statistic.

Neyman and Pearson tackled, in a statistical framework, a general problem of inquiry. On the one hand, we do not want to acquire a large stock of false beliefs. We would like to reason *reliably*. On the other hand, we also do not want to fail to learn things that are true. One could always avoid the acquisition of false beliefs by simply not forming beliefs. But having *true* beliefs is valuable to us for many reasons.

[8] Conceived as a dichotomous test, NP tests pose two exclusive and exhaustive options: (1) accept H_0 and reject H_1, or (2) reject H_0 and accept H_1. On this view, for a scientist to decline both options and suspend judgment is to abandon the test entirely. Another option would be to reformulate the test to explicitly allow 'suspension of judgment' as a possible test outcome. For an important discussion of this option, see Levi (1967) .

So, a test should not only have a small size or significance level; it should also have a high power, which is to say, a high (enough) probability of detecting an effect of interest, if it is present.

Assuming, at the outset of our test, that we are discriminating between the null hypothesis H_0 and the alternative H_1, you might think that the power of a test can be defined as $1 - \beta$, that is, $1 - Pr(T < t_c; H_1)$. This would indeed be the case were one to test a null point hypothesis against a *point* alternative hypothesis, because then the probability distribution needed for determining the value of $Pr(T < t_c; H_1)$ would be specified. Rarely, however, do real scientific problems lend themselves to such a treatment. Much more typical are cases where the target null hypothesis is contrasted with a continuum of alternative possibilities, just as in our die-rolling example. In such cases, the power of the test varies according to the possible alternatives in H_1.

What Neyman and Pearson's work allows investigators to do, then, is to define a test – both in the choice of the critical value of the test statistic and in the choice of the test statistic itself – that is optimal in a certain sense: For a given size (Type I error probability), it is in many cases possible to choose a test that also maximizes the power (minimizes the Type II error probability) for *all* possible alternatives in H_1 (often called a *uniformly most powerful test*). Just how one goes about determining the optimal test is an important problem for the specification of NP tests, but it would take us too far afield of our present concerns. We now turn instead to the questions of how to understand the aims of NP testing and interpret the outcomes of NP tests. We will consider some traditional interpretations and their shortcomings before turning to Deborah Mayo's error-statistical philosophy that seeks to overcome those limitations through a reconsideration of both the aims and methods of frequentist statistical appraisal.

9.3.4 Statistical Tests: What Are They Good For?

A Behavioristic Interpretation

In their 1933 paper, Neyman and Pearson articulate a rationale for NP testing that emphasizes the usefulness of such tests in limiting the rate of erroneous decisions in the long run. Having asserted that "no test based upon the theory of probability can by itself provide any valuable evidence of the truth or falsehood" of a particular hypothesis, they go on to note:

> But we may look at the purpose of tests from another view-point. Without hoping to know whether each separate hypothesis is true or false, we may search for rules to govern our behaviour with regard to them, in following which we insure that, in the long run of experience, we shall not be too often wrong ... Such a rule tells us nothing as to whether in a particular case *H* is true when [the rule says to accept it] or false when [the rule says to reject it]. But it may often be proved that if we behave according to such a rule, then in the long run we shall reject *H* when it is true not more, say, than once in a hundred times, and in addition we may have evidence that we shall reject *H* sufficiently often when it is false. (Neyman & Pearson, 1933, 290–291)[9]

Here Neyman and Pearson express a *behavioristic* understanding of the rationale for NP testing, according to which the results of an NP test do not indicate whether the data provide good support or evidence for or against a given hypothesis. Rather, on this interpretation, the NP tester begins with a choice between two courses of action that provide a practical signification of the outcomes *accept H* and *reject H*, as well as some idea of the costs of making a choice between those actions erroneously. Following the decision rule specified as part of the NP test – thus allowing for the limitation of error rates – allows the NP tester to contain costs. Indeed, by facilitating the design of *optimal* tests, Neyman and Pearson allow one to maximize a certain kind of *efficiency*.

Critics such as R. A. Fisher have noted that, whatever usefulness NP tests thus understood might have for, say, the pursuit of economic gains in manufacturing (a typical example would be the sampling from batches of manufactured goods to see whether they meet quality assurance standards before shipping them to buyers), the relevance of such considerations to science is doubtful. Scientists seek answers to exactly the kinds of questions Neyman and Pearson say they cannot answer. They would like to have 'valuable evidence of the truth or falsehood' of the hypotheses they investigate!

[9] In later works Neyman continued to advocate an interpretation of NP tests in terms of "inductive behavior" (Neyman, 1950, 1–2), but Pearson, argues Mayo, became a "heretic" to the NP philosophy and advocated a view closer to Mayo's own (Mayo, 1996, ch. 11).

A Measure-of-Evidence Interpretation

An alternative to such a behavioristic interpretation would be to interpret the error probabilities of a test as indicators of the strength of evidence provided by the test with regard to the hypotheses tested. An approach of this kind, which we might call a *measure-of-evidence* interpretation, was attempted by Allan Birnbaum (Birnbaum, 1977). In Birnbaum's approach, a rejection of the null hypothesis in favor of the alternative (or vice versa) would be interpreted in terms of the strength of evidence it provided ('strong,' 'very strong,' 'conclusive,' 'weak,' or 'worthless') in favor of the alternative as against the null (or vice versa). This interpretation would be based simply on the error probabilities of the test that led to that rejection. For example, if a test yielded a rejection of H_0 in favor of H_1 with error probabilities $\alpha = 0.06$ and $\beta = 0.08$ (recall that H_1 must be a point hypothesis in order to yield a single value for β), then Birnbaum would interpret that as "strong statistical evidence" for H_1 as against H_0 (Birnbaum, 1977, 24).

As Birnbaum acknowledges, statistical evidence in the form of outcomes of statistical tests alone do not suffice to form reasoned judgments regarding substantive scientific questions. Even as an account of statistical evidence strictly, however, Birnbaum's approach is inadequate. Two different tests with the same accept/reject outcome and the same error probabilities need not yield statistical evidence of the same import. For example, in an NP test, one designates a single cutoff value t_c, and the test rejects the null hypothesis if the data yield any value for T that is greater than t_c. Neither the outcome nor the error probabilities a and β are sensitive to the magnitude by which the observed value of T exceeds the cutoff value. For example, setting a cutoff value $t_c = 0.5$ for testing a fair die hypothesis on the basis of the 100 trials of thirty-six tosses yields a size of $\alpha \approx .05$. The null hypothesis H_0 would then be rejected by data yielding observed values of $t = 0.6$ as well as by data yielding $t = 3.6$, but the latter result gives much stronger evidence against the null than the former. Also, as stated previously, as the sample size grows, the sampling distribution tends to become more sharply peaked around the expectation value. As a consequence, the null will be more easily rejected at a given size by a smaller effect if the sample size is large than if it is small. This is another way in which factors beyond the accept/reject outcome of a test and its error probabilities have a bearing on the interpretation of the statistical evidence (see Cousins, 2017; Mayo, 1985, 1996, 2018; Pratt, 1977).

9.4 Error-Statistical Philosophy of Science and Severe Testing

Deborah Mayo has developed an alternative approach to the interpretation of frequentist statistical inference (Mayo, 1996).[10] But the idea at the heart of Mayo's approach is one that can be stated without invoking probability at all. Let's start with this core idea and then see how and why probabilities enter the discussion.

Mayo takes the following 'minimal scientific principle for evidence' to be uncontroversial:

Principle 3 (minimal principle for evidence): *Data* x_0 *provide poor evidence for H if they result from a method or procedure that has little or no ability of finding flaws in H, even if H is false.* (Mayo & Spanos, 2009, 3)

This minimal principle is weak but not vacuous. It states neither a criterion the satisfaction of which suffices for the data to constitute *good* evidence, nor a criterion the satisfaction of which is necessary for the data to constitute *poor* evidence. However, it does commit one to regarding the method by which data are generated, and not the data alone, as *relevant* to the data's evidential import. Moreover, it indicates that the relevant aspect of that method or procedure concerns the 'ability' it has to reveal a flaw in *H*, if there is such a flaw (such as being false). Mayo thus considers even this weak principle philosophically and methodologically significant.

Philosophical accounts of scientific reasoning have, however, generally failed to satisfy this principle.[11] Philosophers of science have constructed theories of evidence that divorce the evidential import of data from a

[10] Mayo's 1996 book *Error and the Growth of Experimental Knowledge* served to bring the main ideas of Mayo's philosophical approach to attention of many philosophers of science. Her 2018 *Statistical Inference as Severe Testing* delves more deeply into statistical science and its philosophical foundations while also introducing further developments of her view and applications to current debates. The introduction to a collection edited by Mayo and her collaborator Aris Spanos includes an incisive summary of the error-statistical philosophy (Mayo & Spanos, 2009).

[11] An exception would be the account of Charles S. Peirce (1839–1914), who is typically associated (along with William James and John Dewey) with the origins of philosophical pragmatism. Peirce insisted that in *probable reasoning* one must take account of "the manner in which the premisses have been obtained." Peirce's theory of probable reasoning anticipates some features of the NP approach and Mayo's error-statistical elaboration of it (Peirce, 1883, 128; see also Kyburg, 1993; Mayo, 1993).

consideration of the methods used to generate those data. Mayo argues that we should *reject* the logical positivists' project of finding a way to allow one, for any data or observations *E*, to calculate the degree of support or confirmation afforded to any hypothesis *H*. We should not, however, abandon the ideals of *neutrality* and *objectivity* themselves. An account of scientific reasoning that respects Principle 3 will better promote these ideals by emphasizing that reliable inferences from data require a consideration of the properties of the method that produced such data.

Principle 3 tells us that the properties of methods that matter to the evidential appraisal of data are those relevant for a method's 'ability of finding flaws' in the hypothesis under consideration. Error probabilities enter into Mayo's account because she maintains that (1) probability distributions provide useful models of the operating characteristics of the methods of hypothesis appraisal, and (2) the operating characteristics that best reflect a method's 'ability of finding flaws' in relevant hypotheses are its error probabilities calculated from such a probability distribution (assuming that distribution is, in fact, an adequate model of the method).

9.4.1 Severity and Evidence

The use of probabilities as models for the flaw-finding capacities of tests leads to a refinement of Principle 3 that addresses the question: 'When do data x_0 provide good evidence for or a good test of hypothesis *H*?'

Principle 4 (severity principle): *Data* x_0 *provide a good indication of or evidence for hypothesis H (just) to the extent that test T has severely passed H with* x_0. (Mayo & Spanos, 2009, 21)

We need to understand what the expression 'severely passed' means, but first note that this expression relates *three items:* (1) the hypothesis under test, (2) the data with which the hypothesis passes, and (3) the test itself. Non–error-statistical accounts of evidence or confirmation typically only concern themselves with a relation between the data and the hypothesis.

Mayo gives the following account of what it means for a hypothesis, given certain data, to pass a severe test or to pass a test with severity:

Definition 4 (severity): *A hypothesis H passes a severe test T with data* x_0 *if*

(S-1) x_0 *agrees with H (for a suitable notion of 'agreement') and*

(S-2) with very high probability, test T would have produced a result that accords less well with H than does x_0*, if H were false or incorrect.* (Mayo & Spanos, 2009, 22)

Evaluating the 'agreement' in condition S-1 will often use probabilities: The data fit or agree with the hypothesis H well if H confers a probability on the observed value of the test statistic (or whatever data-dependent quantity serves as a measure of distance from what is expected if H is true) that is relatively high compared to the probabilities conferred by competing hypotheses.

As a *first pass* at appreciating the relevance of condition S-2, consider the context where the hypothesis that is considered to pass the test with data x_0 is the alternative H_1 in an NP test of the fair die hypothesis. If (1) our test has a cutoff fixed at t_c that results in significance α (= .05, let's say), (2) the data yield a value of $t = t_c$, and (3) we take this to be a passing result for the alternative hypothesis H_1, then perhaps we can use the significance level of the test to evaluate how well the condition S-2 has been met. Specifically we know that $Pr(T \geq t_c; H_0) = \alpha = .05$. In other words, assuming the null hypothesis is true (and hence the alternative false), rarely – once in twenty times on average – would one get a result that agrees so poorly (or even worse) with H_0. It follows then that $Pr(T < t_c; H_0) = 1 - \alpha = .95$. In other words, assuming the null hypothesis to be true and the alternative false, one would almost always obtain a value for T that agrees better with the denial of H_1 (i.e., with H_0) than does the result in hand. So can we say that conditions S-1 and S-2 have been met by H_1 and it has passed a severe test with such a result?

Answering requires us to go beyond the standard NP testing approach invoked so far in this example. Recall that H_1 is usually (as in the die example) a *compound* hypothesis. In our example, H_1: $Y \sim B(36, p); p \neq 1/6$. Although this hypothesis says something definite about the kind of distribution that describes Y, all that it says about the value of parameter p is $p \neq 1/6$, which is not saying much at all! This makes the idea of 'agreement' with H_1 so poorly defined as to preclude the application of severity. If we want to know what hypotheses might have passed with severity in our example, we will have to scrutinize the various possibilities buried within the compound hypothesis H_1.

Recall that the original rationale for NP tests appeared to be a behavioristic interest in limiting the rate of erroneous decisions, whereas we now seek an account of what we can learn about hypotheses from experimental data. Mayo insists that the NP emphasis on the size and power of a test remains relevant, particularly for *pre-data planning*. Such considerations help investigators devise most efficient ways to gather data to facilitate learning about the hypotheses under test. But given that size and power do not translate directly into a measure of the strength of evidence for or against a hypothesis, one must use a broader consideration of error probabilities to discriminate among the possible inferences one might draw with regard to the entire family of hypotheses under consideration *after the data have been gathered.*

Two points deserve emphasis here. *First:* probabilities entered into the account as a means for representing the error-detecting capacities of test procedures. Very strong arguments based on severity considerations can often be given without appealing to quantitative probability measures at all. *Second:* Mayo proposes that we may regard learning from experimental data as a matter of learning to construct *arguments from error*, the structure of which "is guided by the following thesis":

Principle 5 (arguing from error): *It is learned that an error is absent when (and only to the extent that) a procedure of inquiry (which may include several tests) having a high probability of detecting the error if (and only if) it exists nevertheless fails to do so, but instead produces results that accord well with the absence of the error.* (Mayo, 1996, 64)

We can thus apply error-statistical reasoning without explicit appeals to error probabilities. Recall the example from Chapter 1, in which Newton investigated the differential refrangibility of sunlight. An argument from error would require us to ask, first, what are the ways in which it might be erroneous to conclude – as Newton does – that light from the Sun consists of rays that are differentially refrangible? We would then want to ask whether Newton's method of inquiry had a high probability of (or strong ability for) revealing such errors, supposing them to be present. In doing so, we can determine which conclusions are warranted by having been well probed for possible errors (e.g., refraction with a prism reveals sunlight to be analyzable into light of of different colors, with each color refracted at a different angle; the refracted light is not further analyzable by the same means into further components), as well as which conclusions are not warranted because the method is incapable of revealing them as erroneous even if they are. (Refraction does not warrant any conclusions about whether light is composed of particles or waves, for example. If something is wrong with the idea that light is composed of waves (or particles), Newton's experiment did not have the capacity to reveal that.)

Before turning to another example, it will be helpful to emphasize a distinctive feature of Mayo's philosophical approach and its relationship to frequentism. In practice, frequentist statistical tests – especially those for statistical significance – often are applied in a somewhat mechanical way. Worse, they can be manipulated in a variety of ways to create the appearance of evidence by crafting a selection of data to produce a calculated significance at a desired level, such as $p < 0.01$. (This is sometimes called 'p-hacking.') This motivated some to call for abandoning significance testing, as previously mentioned.

Mayo's severe testing approach emphasizes a principle of frequentist statistical reasoning that has not always been honored in practice: The error probabilities of a test depend on the procedure actually used in arriving at a test outcome, which may not be well captured by a brute calculation if an investigator has manipulated or just been careless in the selection of data, the choice of test statistic, or the choice of test parameters (like the point at which data collection stops). The 'nominal' error probabilities (those calculated based on standard mathematical formulas in statistics textbooks or software packages) may not be the same as the actual error probabilities, that is, the rate at which errors would result in a long run of test repetitions, including all factors relevant to the outcome.

In her recent book, Mayo has emphasized this point in advocating for the importance of *auditing* as an element in scientific inquiry (Mayo, 2018). The idea behind auditing is that conducting a scientific inquiry (including its statistical aspects) requires that the investigator maintain a critical scrutiny of every aspect of their procedure. This includes testing the statistical model relied upon. More generally, auditing is important for securing severity in a scientific inference. Mayo writes, "Because they alter the severity," certain considerations "must be taken account of in auditing a result, which includes checking for (i) selection effects, (ii) violations of model assumptions, and (iii) obstacles to any move from statistical to substantive causal or other theoretical claims" (Mayo, 2018, 269).

9.4.2 Example: Evidence That Formaldehyde Is Carcinogenic

Let us now develop an example in more detail.

Formaldehyde is a colorless flammable gas present in a wide range of products, including cosmetics, cleaning products, and carpets, as well as in cigarette smoke and industrial exhaust. If you have ever dissected a frog or other specimen preserved in formalin (a solution of formaldehyde gas in water), then you know its distinctive odor. The US Environmental Protection Agency (EPA) in its Integrated Risk Information System presently lists formaldehyde as a "probable human carcinogen, based on limited evidence in humans, and sufficient evidence in animals" (US Environmental Protection Agency, 1990). The National Toxicology Program of the US Department of Health and Human Services (DHHS), in the twelfth edition (2011) of its *Report on Carcinogens*, revised the status of formaldehyde from "reasonably anticipated to be a human carcinogen" (as it had been listed since 1981) to "known to be a human carcinogen," citing "sufficient evidence of carcinogenicity from studies in humans and supporting data on mechanisms of carcinogenesis" (US Department of Health and Human Services, 2011, 195).[12]

Much of the evidence that formaldehyde can cause cancer in humans comes from epidemiological studies, that is, studies that compare rates of various types of cancer among humans known to have significant exposure to formaldehyde (pathologists and embalmers, for example) with rates among humans without such exposure. Such studies require great care to ensure that the selection of subjects and interpretation of data is unbiased and should take into account possible confounding causes (other causes of cancer, exposure to which may differ among the two groups studied). Moreover, it is difficult to measure actual rates of exposure to formaldehyde, even among those whose jobs involve constant exposure to formaldehyde-containing substances. Recent determinations that formaldehyde is indeed a human carcinogen, as reflected in the DHHS *Report on Carcinogens*, arise from significant advances in the ability of epidemiologists to tackle these problems, as well as from the accumulation of data over many years, allowing the long-term effects of formaldehyde exposure finally to emerge.

In the early 1980s, the situation was rather different. The few existing epidemiological studies were generally considered inconclusive. Experiments on rats and mice, however, yielded more definite results. Our discussion of these experiments will focus on how scientists used frequentist statistics in the evaluation of their evidence, and how an error-statistical approach would interpret and possibly expand upon those assessments. (We return to this

[12] The fifteenth edition of the report, released in 2021, retains the language quoted here (US Department of Health and Human Services, 2021).

research in Chapter 12 as an example of the complicated relationship between scientific research and public policymaking.)

These experiments and the epidemiological studies that had been carried out at that time were central to a controversy that arose over the safety of formaldehyde in the early 1980s. Although several scientific panels and the EPA's own scientists had recommended the substance be classified as a carcinogen, EPA administrators then newly appointed by the Reagan administration declined to do so. Critics charged the EPA administrators with having neglected scientific evidence to advance a political agenda of industrial deregulation (see the discussions in Ashford, Ryan, & Caldart, 1983; Jasanoff, 1987; and Mayo, 1991).

In experiments conducted by researchers at the Chemical Industry Institute of Toxicology, male and female rats and mice were randomly assigned to four groups subjected to different levels of exposure to formaldehyde gas: 14.3, 5.6, 2.0, and 0 parts per million (ppm). Each exposure group had about 120 members of each of the two species and both sexes. Those exposed to 0 ppm were the *control group*. Exposures were scheduled to last for twenty-four months, followed by six months of nonexposure. At intervals of several months, researchers would 'sacrifice' preestablished numbers of animals; they would then conduct pathological examinations on these and other deceased study animals.

We will not concern ourselves with the subtleties of the statistical analysis of data from these studies, but will content ourselves with a rough characterization of the hypotheses under investigation and the nature of support they established for the inferences they drew. Although the researchers report a variety of pathological observations, we will focus on results that became central to debates over the safety of formaldehyde: the incidence of squamous cell carcinomas (SCCs) in the various groups.

The researchers tested hypotheses concerning the *difference* that exposure to formaldehyde makes to the risk of developing SCCs in rats and mice. A null hypothesis regarding this difference would state that formaldehyde makes no difference to the risk of developing SCC. Let us define a parameter RI_{RE} (or RI_{ME}) to represent the difference between the risk of SCC among rats (or mice) exposed to formaldehyde at a level of exposure E and the risk among those not exposed to formaldehyde. Then the null hypothesis would for rats be $RI_{RE} = 0$.

Among the rats in the 14.3 ppm exposure group, 103 were found to have SCCs, as compared to 2 in the 5.6 ppm group, and none in either the 2.0 ppm or 0 ppm group. Among the mice, investigators found two SCCs in the 14.3 ppm exposure group, and none in other groups. How should we interpret these results?

Following a widespread convention, the researchers did not report the actual *p*-values for these, but instead reported their results as either surpassing or failing to surpass certain thresholds of statistical significance. They noted that the 103 SCCs in the 14.3 ppm exposure rats are significant at a level of $p < 0.001$. As is common practice in biomedical sciences, the researchers reported *p*-values >0.05 as 'not significant.' The two SCCs in the 5.6 ppm exposure rats and the two in the 14.3 ppm exposure mice fall into this category.

Most researchers use 'significant' as a technical term, but it has been easily misunderstood. It does not mean the same thing as 'important,' either when affirmed or when denied. To say that 103 SCCs among the 14.3 ppm exposure group is *significant at a level of* $p < 0.001$ is to say that, assuming $RI_{R14.3ppm} = 0$, one would expect to discover 103 or more SCCs in a similar sample of rats in less than once in 1,000 trials. (This does *not* mean that there is a less than 1 in 1,000 chance that these SCCs were due to chance or that there is a less than 1 in 1,000 chance that formaldehyde does not cause SCCs in rats exposed at a level of 14.3 ppm. Such statements would concern the probability of a particular inference from given data, which is not licensed in a frequentist framework.) To say that the two SCCs in the 5.6 ppm exposure rats or the two SCCs in the 14.3 ppm mice are 'not significant' simply means that, assuming $RI_{R5.6ppm} = 0$, finding two or more SCCs would happen more often than once in twenty times, on average. These latter two results may not be significant in a technical sense, but they are *relevant*, as we will see.

Regarding the much greater incidence of SCCs among rats compared to mice exposed to 14.3 ppm, the researchers noted, "It is of interest to note that the incidence of squamous cell carcinoma was similar in mice exposed to 14.3 ppm and rats exposed to 5.6 ppm of formaldehyde" (Kerns et al., 1983, 4388). Although neither number is statistically significant, the similarity between the two rates *suggests* a hypothesis that would, if correct, explain the discrepancy in the effects of formaldehyde between the two species – namely, that the "differences in response ... may be related to differences in their physiological responses to formaldehyde inhalation" that would result in rats receiving, at a given level of exposure, a much greater "dose" of formaldehyde than mice exposed to the same level (Kerns et al.,

1983, 4388). The researchers cited independent support for such a hypothesis on the grounds of physiological differences in the nasal cavities between the two species, as well as from data on "cell turnover in the nasal cavities of rats and mice" (Kerns et al., 1983, 4388).

From the standpoint of the Severity Principle, we might at this point wish to say that the data provide good evidence for the hypothesis that $RI_{R14.3ppm} > 0$, since the difference in rates of SCCs among the control group and the 14.3 ppm exposure group fits well with the hypothesis that formaldehyde exposure at that level increases the risk of SCCs, and it is highly improbable that one would observe as great a difference (or greater) were it the case that formaldehyde exposure did not have such an effect.

Metastatistical Severity Analysis of the Fates of Rats and Mice

Mayo and Spanos (2006) have emphasized that an assessment of the severity with which a hypothesis has passed a given test cannot be read off directly from *p*-values, or the outcome of an NP test, but requires a *metastatistical severity analysis* of the results of such tests. Under the behavioristic rationale discussed above, Fisherian or NP tests have their value in the ability the investigator has to limit her rate of erroneous inferences in repeated applications of the tests. But the rationale for a severity analysis is to determine what warranted inferences the investigator may draw from a *particular* set of data.

Mayo and Spanos specify three requirements for success at that task: (1) The analysis must concern inferences about the presence or absence of errors, rather than (merely) decisions about a course of action to pursue. (2) One must use a test statistic (such as the rate of SSCs) to serve as an 'appropriate measure of accordance or distance' with respect to the hypotheses under consideration. Roughly this is simply the idea that larger values of a quantity used to determine whether the test rejects or accepts a hypothesis H must indicate larger discrepancies from (or worse fit with) what one would expect were H to be true. (3) The assessment must be "sensitive to the *particular outcome* x_0; it must be a *post-data assessment*" (Mayo & Spanos, 2006, 329–330, original emphases).

In our example, a severity analysis would not, therefore, rest with simply reporting that finding 103 SCCs among the rats exposed at 14.3 ppm was significant with $p < 0.001$. One would next proceed to ask about the hypotheses asserting specific sizes of increased risk, relying on the very same data. For example, do such data support the inference that $RI_{R14.3ppm} > 0.10$ (i.e., exposure at 14.3 ppm adds *more than 10 percent* to the probability a rat will get an SCC)? The answer to this depends on how probable it is (how often it would happen) that one would obtain *no more than* 103 SCCs in their sample were it the case that exposure added *just* 10 percent to that risk. What about a 20 percent increase? 30 percent? and so on ... In this way, one could calculate the severity – for a given possible inference about $RI_{14.3ppm}$ – with which that hypothesis passes given the data in hand. Although Mayo and Spanos do not propose any particular fixed severity threshold above which one should take the data to provide good support for a hypothesis, and below which one should not, the idea is that, assuming the conditions for a severity analysis have been met, one can distinguish well-warranted from poorly warranted inferences based on whether the inferred hypotheses passed with high or low severity, respectively (Mayo & Spanos, 2006, 341–344).

Such an analysis can be especially useful, according to advocates, in case a test fails to reject the null hypothesis, or if a result is found not to be statistically significant. What, for example, follows from the fact that among the mice exposed to 14.3 ppm, only two SCCs were found, which was judged to be not statistically significant? If we consider this to be a passing result for the hypothesis $RI_{M14.3ppm} = 0$ (in effect, 'formaldehyde is safe for mice'), does this constitute good evidence for that hypothesis? A severity analysis would direct our attention beyond simply the absence of a statistically significant departure from what we would expect were this null hypothesis to be true. We would first ask how probable it is that the difference between the expected value under the null and the observed value of the test statistic would have been *larger* than it is, if $RI_{M14.3ppm} > 0$. If that probability is *very high*, then the null hypothesis passes with high severity. But here again, the calculation of this probability will depend on *which* value of $RI_{M14.3ppm}$ we choose that is greater than 0. Presumably, if $RI_{M14.3ppm}$ exceeds 0 by only a very small amount, such as 0.001, then it is *not* the case that it is highly probable that more than two SCCs would have been observed (Mayo & Spanos, 2006, 337–339).

In case a test fails to reject a null of the form $\mu \leq \mu_0$ (or $\mu = \mu_0$), for some numerical value of μ_0, a severity analysis will focus on the severity of inferences of the form $\mu \leq \mu'$ for various values of μ' greater than μ_0. In our example, the question to ask of the mice results is what inferences they warrant of the form 'the increased risk in mice of SCCs from exposure to formaldehyde at 14.3 ppm is *not greater than r*'? Error statisticians thus acknowledge Fisher's cautions against accepting a null hypothesis based on a failure to reject it, but they go beyond Fisher insofar as they allow for a postdata analysis about the extent to which the data support inferences that the null is not wrong by more than a given amount. (This goes beyond orthodox NP testing as well, though Mayo and Spanos have argued that aspects of this approach can be found in the writings of Neyman himself (Mayo & Spanos, 2006, 334–336).)

A severity analysis requires one to distinguish between errors that the results of a given test do and do not rule out (in the sense that, were those errors to be actual, the test would have the capacity to tell us so). Such a perspective would be potentially relevant to the broader significances of these results. What do they suggest about the health effects of formaldehyde on human beings?

One of the EPA administrators involved in the early 1980s decision against regulating formaldehyde as a carcinogen noted that, although "formaldehyde is a carcinogen in the rat by the inhalation route . . . its carcinogenic potential appears to vary significantly with species and route." He furthermore stated that epidemiological data supported "the notion that any human problems with formaldehyde carcinogenicity may be of low incidence or undetectable" (Ashford et al., 1983, 327). As we already noted, the investigators who performed the experiments on rats and mice suspected that mice might suffer lower rates of SCCs because of a difference in the dose of formaldehyde they actually received, rather than (or perhaps in addition to) a difference in the carcinogenic potential of formaldehyde between mice and rats. In other words, the differences in the results between the two species could *not* be reliably interpreted as supporting a substantive claim about the difference in carcinogenicity between the two species, a claim not well tested by this study. Moreover, from the standpoint of a severity analysis, the absence of statistically significant results from epidemiological studies would only warrant the claim that the carcinogenicity of formaldehyde in humans is 'of low incidence or undetectable' were it to be the case that those studies had a very high probability of finding an effect if the carcinogenicity of formaldehyde in humans exceeded the threshold of 'low incidence.' There was at that time good reason to think that this criterion was not met by these studies (Mayo, 1991, 271–273). In other words, even were it to be the case that formaldehyde posed a significant carcinogenic potential to humans, it was not improbable that these studies would have failed to find a significant effect.

9.5 Conclusion

In this chapter, we have considered the idea of relative frequency as a way of understanding probability statements, as a foundation for some important approaches to statistical data analyses, and as a jumping-off point for an approach to the philosophy of science that regards the error-probing power of tests as crucial to the determination of what it is that we learn from scientific inquiry.

The error-statistical approach we explored has been subjected to considerable criticism from other philosophers of science (Achinstein, 2010; Howson, 1997; Musgrave, 2010), and Mayo has responded to such criticisms in turn (Mayo, 1997b, 2018; Mayo & Spanos, 2009). Here I discuss two critical issues.

Some critics argue that the error-statistical approach is committed to regarding some hypotheses as passing severe tests, which nonetheless have low posterior probabilities (Achinstein, 2010). Suppose that we are interested in the hypothesis H_T: Tatyana is well qualified to participate in eating contests. The contrary hypothesis H'_T states that Tatyana is not thus qualified. To test H_T against H'_T, we ask Tatyana to eat ten saltine crackers in 60 seconds without drinking any fluids. If she can do so, she will pass the test. Only one out of twenty people who are not well qualified for competitive eating can pass this test, whereas nearly everyone who is qualified would pass it. However, in the city where Tatyana lives, only one person in a thousand is well qualified for competitive eating. From the test we will obtain the result E: Tatyana passes the test. We have the following probabilities:

$$Pr(H_T) = 0.001, \qquad Pr(H'_T) = 0.999,$$

$$Pr(E|H_T) \approx 1, \qquad Pr(E|H'_T) = 0.05. \tag{9.4}$$

Suppose that we regard H_T as having passed a (somewhat) severe test. The severity principle would then have us regard this as (somewhat) good evidence for H_T. But a Bayesian calculation reveals that $Pr(H_T|E) \approx 0.02$, which is still quite low, while $Pr(H'_T|E) \approx 0.98$.

Mayo's response first points out that this criticism begs the question by invoking a criterion of appraisal rejected by error-statistics. That approach does not appraise a scientific claim by considering its posterior probability, but asks how well it has survived scrutiny by methods with a good capacity for revealing errors, should they be present. Indeed, a test based on the criterion of posterior probability would use the results of Tatyana's performance as an indication that she is not well qualified for competitive eating regardless of what she did, simply on the basis of the city where she lives, and would thus be committed to "innocence by association" (Mayo & Spanos, 2009, 192–201). (Spanos, 2010 offers another criticism of such arguments.)

The error-statistical approach does not merely offer a different answer to a traditional question in the philosophy of science: 'How do we evaluate the degree to which a given body of observations confirm (make probable) or disconfirm (make improbable) a theory?' Rather, the error-statistician seeks to *replace* this question with another: 'How do we learn about scientific claims by subjecting them to tests designed to reveal such errors in those claims as we might be interested to discover?'

A second challenge for error-statistics has been to show how it can shed light on scientific claims that appear to be less directly connected with particular bodies of experimental data, what philosophers sometimes call 'high-level theories.' Granting that a severity analysis provides a helpful perspective on the use of data from rats to evaluate the carcinogenicity of formaldehyde, what can it tell us about the status of such broad theoretical frameworks as Neo-Darwinian Evolutionary Theory, Keynesian Economic Theory, or the General Theory of Relativity (GTR)? These theories do not lend themselves to severe testing directly.

Although much work on this challenge remains to be done, the basic approach that defenders of error-statistics have taken is to argue (analogously to their response to the first challenge) that an understanding of how severe tests are relevant for high-level theories requires one to first *give up* the assumption that the nature of the problem concerns *how probable* such theories are or the assignment of a *degree of confirmation*. Instead, we should ask of our theories *the extent to which we have effectively probed them for errors and how well they have survived such probing.*

Mayo's writings on this topic have focused on the experimental testing of theories of gravity (Mayo, 2002; Mayo & Spanos, 2009; see Staley, 2008, 2013 for a discussion and extension; see Harper, 2011 for an extended philosophical and historical discussion that relates the testing of modern gravity theories to Isaac Newton's methodology). She has focused on the use of a framework in which Einstein's GTR can be put into a common mathematical representation with other theories of gravity and tested – in a piecemeal way – for how well they describe a broad range of phenomena in the domain where relative velocities are not large and gravitational fields are not extraordinarily strong. In this way, data from many different observations and experiments can be used to discriminate the various ways in which GTR *might* be in error (and other theories more correct), so that, although GTR might not be subjected in its entirety to a single severe test, we can severely test claims that it makes about, for example, how much a given quantity of mass curves the space around it – one of the features that distinguishes GTR from some of its alternatives (Will, 1993a, 1993b).

A significant feature of this case is that the theories to which this framework compares GTR are neither those that are logically possible (which would include many that are silly or incomprehensible), nor only those we are clever enough to have thought of (which might not include the correct theory). The framework allows one to test among an entire class of theories, all of which satisfy certain weak assumptions.[13] This includes GTR, all of its rivals that physicists have regarded as serious, and (infinitely) many other rivals that no one has explicitly entertained but which are mathematically represented within the framework. Whether a similar error-statistical approach can (or should) be applied to theories regarding aspects of nature other than gravitational physics (particularly in domains such as biological or social sciences) remains an open question.

However defenders of an error-statistical approach in the philosophy of science grapple with these issues, we can see that the frequentist tradition in statistics has given rise to an interesting alternative to the Bayesian approach that challenges long-standing conceptions of the task facing philosophers of science.

[13] Weak but not trivial. The framework, called the Parametrized Post-Newtonian (PPN) framework, includes all theories that satisfy the Einstein Equivalence Principle; equivalently they are all *metric* theories of gravity. See Will (1993a) for details, and Staley (2008, 2014) for a discussion of the empirical grounds for the assumption of Einstein Equivalence in this context.

10 Realism and Antirealism

10.1 Introduction: Success, Change, and Empiricism

Scientists pose questions about the world and, in some sense, succeed in arriving, however tentatively, at answers. In this chapter we ask how we should understand that success. In particular, should we think of scientific inquiry as a project aimed at revealing a *true* description of the world, even when that description involves things we cannot possibly see, hear, touch, taste, or smell? Or should we consider scientists to be engaged in a search for theories that are simply in some way *useful*?

On the one hand, the ways in which scientists are sometimes successful would be, some philosophers argue, inexplicable or miraculous unless it were the case that they have accepted some theories that are (at least approximately) true. The approximate truth of some of their theories is the *best explanation* of certain successes scientists have enjoyed. *Scientific realism* maintains that we should regard empirically successful theories in the *mature* sciences (like physics, chemistry, and biology) as at least approximately true *because* this is the best explanation of the success of those theories.

On the other hand, we have seen how theories can change in dramatic and unexpected ways. A theory might even seem to enjoy tremendous success, only to be replaced by a theory with a radically different conception of the basic entities responsible for the phenomena in its domain. A historically informed caution then seems appropriate with regard to successful theories, calling into doubt the close connection between success and truth that scientific realism postulates. *Theory change* poses a *historical* challenge to scientific realism.

Another criticism of scientific realism alleges it to be inconsistent with a thorough-going scientific attitude. Such an attitude, according to this challenge, should consider the ultimate arbiter of questions of truth to be nothing other than the observable data. But, the argument goes, theories that involve terms referring to unobservable entities, such as electrons, necessarily involve statements that go beyond what can be tested with observable data. So, we should understand any commitment scientists make to such theories not in terms of belief in the truth of those theories but as a pragmatic decision. This is the *philosophical* challenge of *empiricism*.

10.1.1 Example: The Hole Theory and Antimatter

In 1928, the physicist Paul Dirac published a theory that for the first time described the quantum mechanics of electrons in a way that was consistent with Albert Einstein's Special Relativity (Dirac, 1928). Dirac's theory, however, included something peculiar: electrons with *negative* energy. Such things had never been observed in any experiment.[1]

Dirac valued the mathematical elegance of a theory very highly, and his own theory exemplified this attribute. So, he sought to convert the oddity of negative energy states into a virtue by using them to explain already known particles. This interpretation became known as the 'hole theory.' According to hole theory, there really is an ever-present 'sea' of negative-energy electrons. Because its density, although infinite, is uniform, the net electromagnetic field from it will be zero everywhere except where there are local divergences from this uniform density. The existence of this sea then helps to ensure that positive-energy electrons will stay positive. According to a principle of quantum physics called the exclusion principle, no two electrons can occupy exactly the same state. Since all negative-energy states are already occupied, electrons in positive-energy states cannot jump into them.

In an effort to unpack all the consequences of his theory and solve problems threatening it, Dirac then asks what it would be like if there were a *hole* in this sea of negative-energy electrons, a location at which no negative-energy electron were to be found. His surprising answer is that such a hole

[1] Theoretically, negative-energy electron states were not entirely new. The classical theory of the electron also included negative-energy states, but physicists ignored them as 'unphysical.' This was not possible in a *quantum* theory, because for any electron in a positive-energy state, there would be a nonzero probability of that electron undergoing a transition to a negative-energy state.

would behave just like a particle, but one with an electric charge opposite that of electrons. At first he concluded that his theory had just predicted protons, but Hermann Weyl showed that, if that were the case, protons would have to have the same mass as electrons, which they do not. Dirac thus took the radical step of proposing an entirely new kind of particle. He made the prediction with great diffidence:

> A hole, if there were one, would be a new kind of particle, unknown to experimental physics, having the same mass and opposite charge to an electron. We may call such a particle an anti-electron. We should not expect to find any of them in nature, on account of their rapid rate of recombination with electrons, but if they could be produced experimentally in high vacuum they would be quite stable and amenable to observation. (Dirac, 1931, 61)

Meanwhile, a young physicist at Cal Tech named Carl Anderson was beginning a new series of experiments on cosmic rays in a strong magnetic field using an improved version of the *cloud chamber*, a device that allows physicists to take pictures of the tracks charged particles leave in a vapor. Anderson, who had no knowledge of Dirac's prediction, noticed in late 1931 that some of his photographs seemed to depict positively charged particles with masses considerably less than that of a proton. They had to be charged because their paths curved (as particles carrying a charge should do when passing through a magnetic field), and the charge had to be positive. (Anderson had inserted a lead plate in his cloud chamber, allowing him to determine whether the particles entered the chamber from below or above. He could then infer the sign of the charge by noting whether the track curved to the left or the right.) Moreover, the particles could not be protons, but had to have a mass similar to that of an electron because of the very thin tracks they left. A proton, with mass about 1,800 times that of an electron, would leave a much thicker track. Based on a number of photographs such as the one shown in Figure 10.1 (Anderson, 1933, 492), Anderson proclaimed the discovery of a "positive electron," which he dubbed the "positron" (Anderson, 1933).[2]

[2] Anderson did not set out to find new particles but to investigate the origins of cosmic rays, a question of great interest to the head of Cal Tech, Robert A. Millikan (Anderson, 1999). C. T. R. Wilson had originally developed the cloud chamber in order to investigate the formation of clouds, not to study cosmic rays, or to discover new particles (Chaloner, 1997; Galison & Assmus, 1989). The unfolding of scientific discovery cares little for the expectations of those who pursue it!

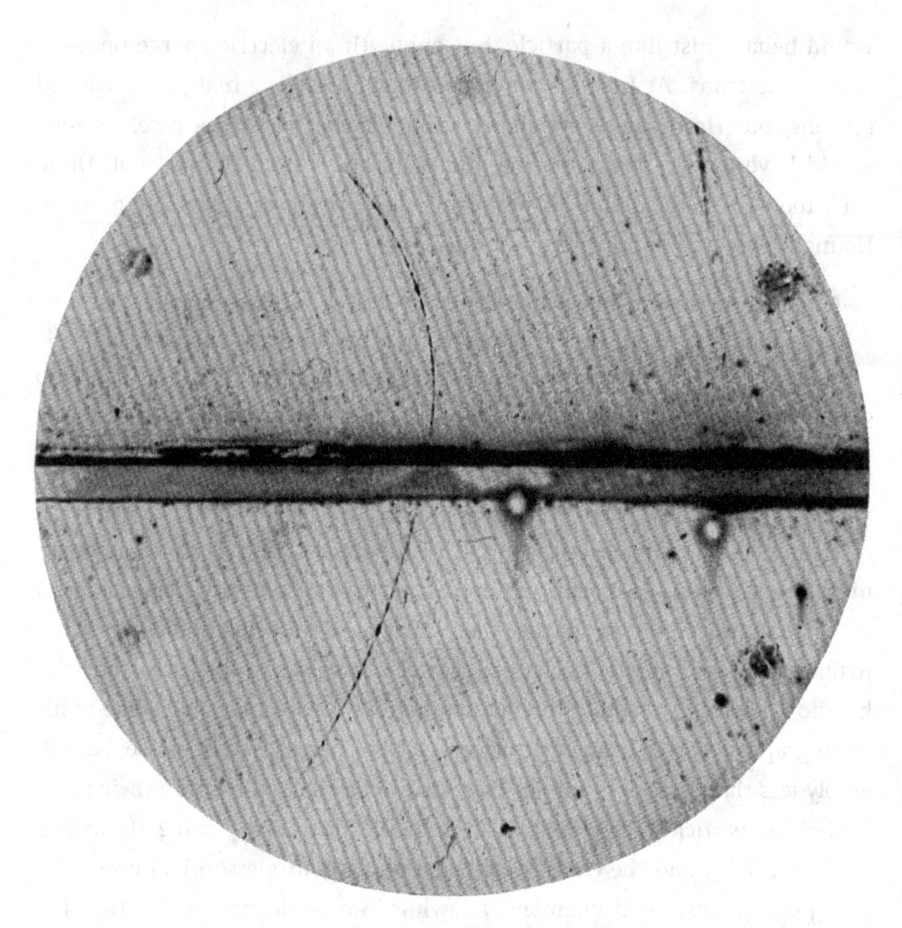

Figure 10.1 A cloud chamber photograph of a track left by a positron. The particle enters from below and passes through a 6-mm lead plate across the center. The subsequent loss of momentum results in greater curvature of the track in the upper region. Reprinted figure with permission from Anderson (1933). Copyright 1933 by the American Physical Society.

Physicists expressed skepticism at first, but experiments by P. M. S. Blackett and G. P. S. Occhialini provided more detailed measurements. Unlike Anderson, Blackett and Occhialini did discuss their findings in light of Dirac's theory, concluding that although their data were not sufficient to test it *directly*, Dirac's theory "predicts a time of life for the positive electron that is long enough for it to be observed in the cloud chamber but short enough to explain why it had not been discovered by other methods" (Blackett & Occhialini, 1933, 716).

Further experimentation strengthened the positron's status as an elementary particle and as the first of a whole class of antiparticles. Dirac had effectively predicted this, too. In the same paper he argued that the proton – the only other particle known at the time – would also have a sea of negative-energy states – holes that would constitute antiprotons. In spite of its seemingly odd way of viewing the world, Dirac's theory was strikingly successful.

10.2 'No Miracles': An Argument for Realism

When a theory is able to get things right in this way, it seems intuitively as though the scientists who developed it must be onto something. That intuition lies at the heart of one argument for *scientific realism.* This argument (sometimes called the 'no miracles argument') asks *what is the best explanation* of the fact that certain scientific theories provide a basis for scientists to predict successfully the outcomes of experiments never before performed and for engineers to extend our ability to intervene in and control natural processes? Either Dirac got tremendously – even miraculously – *lucky*, or else he had come up with a theory that was, if not *exactly* true, at least close to the truth. As stated in an often-quoted phrase from Hilary Putnam, "realism is the only philosophy that does not make the success of science a *miracle*" (Putnam, 1979, 73, emphasis in original).

Although statements of the realist thesis about scientific theories vary somewhat, the basic idea is this:

Thesis 2 (realism): *Scientific theories that achieve a certain level of success in prediction and experimental testing are (probably) approximately true.*

What is the alternative?[3] Antirealists regard theories, even successful ones, as somehow able to perform their function without being true, or at least without being literally true. One version of this idea is *instrumentalism*, which holds that theories should be regarded as tools (or instruments).

[3] In Chapter 6, we encountered one alternative in constructionist views that treats scientific knowledge strictly as a matter of consensus or some other social state of affairs. In the present chapter the focus will be on philosophical arguments for and against scientific realism. These will occasionally intersect with considerations taken on by social constructionists.

Different versions of instrumentalism emphasize different uses for these instruments: making predictions, exerting control over natural processes, representing phenomena (Stein, 1989), or facilitating our understanding of observable phenomena (Rowbottom, 2019). What is crucial to the instrumentalist point of view is that theories can serve their function without being even approximately true. Thus, from the fact that a theory serves its function successfully, we should not infer that it is even approximately true. Advocates of scientific realism contend that the argument they offer is of the same sort that scientists use to defend their theories: They consider scientific realism to be superior to alternative ways of thinking about theories such as instrumentalism because it provides the best explanation of the success of certain scientific theories. Because, they argue, scientists themselves accept, in any given domain, the theory that offers the best explanation of the phenomena in that domain, scientific realism earns the appellation 'scientific' not simply because it is a thesis about science but because it uses a scientific argument in defense of that thesis.

Scientific realists thus adopt a model of the reasoning by which scientists defend scientific theories called *Inference to the Best Explanation* (IBE). Another term sometimes used for this kind of inference is *abduction*.[4] To infer the best explanation is to infer "from the premise that a given hypothesis would provide a 'better' explanation for the evidence than would any other hypothesis, to the conclusion that the given hypothesis is true" (Harman, 1965, 89). Although philosophers disagree over which features make an explanation the *best*, some prominent considerations have included plausibility, consilience, simplicity, and the extent to which a proposed explanation is analogous to explanations already known to be true (Lipton, 2004; Thagard, 1978).

More specifically, the argument for scientific realism focuses on the ways in which scientists rely on methods that depend on theoretical assumptions. Such assumptions, they argue, enter into the design of experiments, the analysis of data, the use of instruments, and the evaluation of evidence. "All aspects of scientific methodology are deeply theory-informed and theory-laden" (Psillos, 1999, 78). Moreover, scientists succeed in conducting

[4] Charles S. Peirce introduced the term 'abduction,' but the kind of inference he intended for it to characterize is not quite that which contemporary philosophers typically have in mind when they use the term (Peirce, 1883).

experiments and making accurate predictions while relying on these theory-laden methods. The best explanation for such success is that "the theoretical statements which assert the specific causal connections or mechanisms by virtue of which scientific methods yield successful predictions are approximately true" (Psillos, 1999, 78; see also Boyd, 1989).

10.2.1 Circularity?

Because scientific realism relies on IBE as a mode of argument, critics have alleged that the argument for scientific realism is viciously circular. They point out that instrumentalists *reject* IBE as a means of justifying the belief that a theory is true. Were they to accept the IBE principle that the best explanation of the phenomena in a given domain should be regarded as true, they would already be realists about scientific theories.

Since skepticism about IBE is a chief motivation for instrumentalism, the most an instrumentalist would conclude from the fact that scientific realism is the best explanation of empirical success in the sciences is that scientific realism is instrumentally useful (Fine, 1986; van Fraassen, 1980). (They may not concede even that much. Unlike the scientific theories alluded to in the scientific realist argument, scientific realism itself has not provided us with any successful predictions; nor has it enabled anyone to do a better job of analyzing data, devising instruments, or performing experiments (Laudan, 1981, 46).) Since the argument for scientific realism assumes the reliability of a mode of reasoning that is rejected by those who reject the conclusion of that argument, the defense of scientific realism assumes that which is at issue and hence begs the question.

Defenders of scientific realism such as Stathis Psillos respond to this charge by drawing an important distinction between two different ways in which an argument can be circular. An argument is *viciously* circular, or *premise-circular* (Braithwaite, 1953, 274–285), if it includes among its premises a claim that is identical or equivalent to the conclusion of that argument (and would not be valid without that premise). It is unreasonable to take such an argument as a reason to adopt a belief in its conclusion. If the proposition expressed by the conclusion is not one that you accept, then the argument depends crucially on a premise you do not accept. If it is already among your beliefs, then accepting the conclusion of the argument does not involve adopting any new belief. Note that what is at issue is not the validity of the argument but whether the argument provides a rational basis for accepting its conclusion.

Defenders of scientific realism then point out that the defense of scientific realism by appeal to IBE is not circular in this way, because it does not include *as a premise* the claim that IBE is reliable.

That distinction matters because, although it is true that a person must accept the premises of an argument for that argument to be rationally persuasive, it is not (at least not obviously) true that for an argument to be rationally persuasive, a person must believe that the kind of argument used is reliable. It is only necessary that the form of argument used *is* reliable. So scientific realists can argue cogently for their position without first convincing those who are skeptical of their position to accept that IBE is reliable, provided that IBE is *in fact* reliable.

You may be thinking "that seems like a bit of a dodge," but recall from Chapter 1 how defenders of induction run into a similar difficulty. Justifying the use of induction in an inference requires another inductive argument.[5] Even deductive inferences are subject to this difficulty. Could you imagine defending the reliability of deductive reasoning without using deductive reasoning? So long as deductive reasoning is *in fact* reliable (in the sense of being truth-preserving), this need not lead to a debilitating skepticism. So, in this respect, abductive reasoning that scientific realists employ is no worse off than inductive or deductive inference (Psillos, 1999, 89).

We should approach this 'no worse off' reply carefully, however. It shows that the form in which this circularity arises should not *by itself* cause us to reject the kind of reasoning being used, as that would lead to the self-defeating conclusion that we should also reject deductive and inductive inferences. That might defeat any attempt to reject IBE as unreliable were such skepticism to be based *only* on the charge of circularity against attempts to justify IBE. Those skeptical of IBE, to be taken seriously, need also to provide some kind of evidence that IBE is *not actually* reliable.

Can critics of scientific realism produce such evidence? Let's consider the historical record.

[5] Gilbert Harman argues that all inductive inferences really are inferences to the best explanation (Harman, 1965). For the purposes of the present argument, I will assume induction and IBE are distinct.

10.3 The Problem of Theory Change

Scientific theories have come and gone over the years. Newton's emission theory of light enjoyed quite a few years of success before nineteenth-century physicists abandoned it in favor of the theory that light consists of waves in an aether filling all of space. The idea of the aether as a mechanical medium for light vibrations proved fruitful for a time, but was found ultimately unworkable, while the electromagnetic theory of James Clerk Maxwell provided an alternative conception that identified the waves with fluctuations in an electromagnetic field. In light of twentieth-century developments in quantum theory, physicists now decline to regard light as *fundamentally* either a wave or a particle.

In an influential 1981 essay, Larry Laudan argues that successful theories that were later rejected in favor of conceptually quite distinct successors pose a significant challenge to scientific realism by calling into question the connection between the truth or even approximate truth of a theory and its empirical success.

According to Laudan, the scientific realist is committed to the claims that (1) "[s]cientific theories (at least in the 'mature' sciences) are typically approximately true and more recent theories are closer to the truth than older theories in the same domain" and that (2) "[t]he observational and theoretical terms within the theories of a mature science genuinely refer." Laudan aims his arguments at a "convergent" version of scientific realism committed to further claims that (3) "[s]uccessive theories in any mature science will be such that they 'preserve' the theoretical relations and the apparent referents of earlier theories" and (4) "[a]cceptable new theories do and should explain why their predecessors were successful insofar as they were successful" (Laudan, 1981, 20–21).

The scientific realist argues that we should accept these claims because they explain the empirical success of the scientific theories in question. For such an explanation to succeed, Laudan argues, the following two principles must be true:

Principle 6 (reference connection): *If the central terms in scientific theories genuinely refer, those theories will generally be empirically successful.*

Principle 7 (truth connection): *If scientific theories are approximately true, they will typically be empirically successful.*

Laudan offers counterarguments to these principles on several fronts, drawing on the history of science. For example, he argues against Principle 6 by first distinguishing two claims that might provide support for it:

Principle 8 (reference to success): *A theory whose central terms genuinely refer will be a successful theory.*

Principle 9 (success to reference): *If a theory is successful, we can reasonably infer that its central terms genuinely refer.*

Laudan's attack on Principle 8 points to theories with (what we currently regard as) genuinely referring terms that were *not* empirically successful (such as the original version of tectonic plate theory, as articulated by Alfred Wegener). Principle 9 must contend with historical examples of theories whose central terms we currently regard as *nonreferring* that nonetheless were empirically successful (such as the phlogiston theory of combustion, or the theory that light is a vibration in the 'luminiferous aether'). Moreover, Laudan points out, Principle 9 is not strong enough by itself to support the scientific realist's contention that scientific realism *explains* the success of science. Principle 9 tells us not that we can expect genuinely referential theories to be successful but only that we can expect successful theories to be genuinely referential. The scientific realist might be committed to the latter claim, but the *argument for* scientific realism depends on the former claim.

The argument Laudan offers against Principle 7 builds on these points. Parallel to the previous distinction with regard to reference, we can point to two commitments of the scientific realist with regard to truth:

Principle 10 (approximate truth to success): *If a theory is approximately true, then it will be empirically successful.*

Principle 11 (success to approximate truth): *If a theory is empirically successful, then it is probably approximately true.*

Laudan acknowledges that Principle 10 enjoys a certain air of plausibility, but that, he claims, is due to its resemblance to the quite distinct and 'self-evident.'

Principle 12 (truth to success): *If a theory is true, then it will be successful.*

Of course if a theory is true, we should expect it to succeed, but scientific realists cannot take much comfort from this, because their thesis does not assert only that being *true* explains the success of theories but that being *approximately true* does so. Absent a theory of approximate truth that would underwrite Principle 10, the realist can issue only a promissory note here.

Moreover, the history of science provides ample reasons to reject Principle 11 if we accept that theories postulating nonexistent entities are not approximately true. Laudan expands on his earlier discussion of theories with nonreferential terms that were empirically successful. His long list of examples includes the "crystalline spheres of ancient and medieval astronomy," "the effluvial theory of static electricity," "the caloric theory of heat," "the vital force theories of physiology," and the electromagnetic and optical aethers.

Indeed, as Thomas Pashby has argued (Pashby, 2012), Laudan might have added Dirac's 'sea' of negative-energy electrons to this list. Although Dirac's theory enjoyed genuine empirical success, including its remarkable prediction of positrons, his proposal for understanding the theory did not survive. Through the middle decades of the twentieth century, physicists such as Sin-Itiro Tomonaga, Julian Schwinger, Freeman Dyson, and Richard Feynman developed mathematical techniques to handle the problem of infinities in Dirac's theory (Schweber, 1994). The result was an approach to quantum electrodynamics that describes particles and antiparticles in terms of *fields* rather than seas of negative-energy particles.

10.3.1 The Pessimistic Meta-induction

The list of successful, not-even-approximately-true theories can, Laudan asserts, "be extended ad nauseam" (Laudan, 1981, 33) and is so extensive that scholars have taken it as the basis for the *pessimistic meta-induction*. Authors have used this term to refer to two distinct arguments. The first argues (in a *reductio ad absurdum*) that if we assume our current theories are true, then we must regard past theories as false, which provides us with an inductive reason to expect our current theories to turn out also to be false (Putnam, 1978). We will not concern ourselves further with this version of the pessimistic meta-induction, which relies on a weak inductive argument (Lewis, 2001).

Laudan's version of the pessimistic meta-induction begins by assuming, as the realist believes, that most currently successful theories in mature sciences are approximately true. But from this, his argument seeks to show not (directly) that we should believe our current theories will turn out to be false but only that the empirical success of a theory is not a reliable indicator of its truth or even its approximate truth. Look at how many empirically successful theories have not even been right about what kinds of things exist!

Even accepting Laudan's historical claims about the prevalence of empirically successful theories that are not even approximately true by the light of currently accepted science, it remains unclear what follows from this regarding the relation between success and truth. For example, suppose we think of empirical success as a kind of diagnostic test for truth. The fact that it has often given a positive result for false theories does not by itself prove that it is unreliable. Suppose it is reliable in the sense that it rarely yields a negative result when applied to a true theory (the *false-negative rate* is low) and rarely yields a positive result when applied to a false theory (the *false-positive rate* is low). We might still see numerous examples of positive results from this test as applied to false theories if most of the theories to which it has been applied are false. As Peter Lewis has pointed out, this leaves a way for the realist to explain Laudan's historical evidence: The population of *past* theories to which the realist's test for truth has been applied consists of mostly false theories, thus yielding a significant number of false positives. But the kind of realist Laudan is criticizing believes scientific progress entails that we now have many more true theories than we have had in the past (Lewis, 2001; see also Magnus & Callender, 2004).

Applying such statistical ideas to the history of science is problematic on a number of levels. Do theories form a *population*, from which scientists have sampled? Does the realist employ success as a diagnostic test to be applied to such a population? One sure conclusion is that historical change complicates any attempt to reason inductively from the historical record.

Nonetheless, Laudan's argument poses at least this challenge to the realist: What supports the claim that IBE is a reliable form of argument, given that it has so often misled us in the past? The rule-circularity of the IBE argument for scientific realism may not by itself lead us to doubt its cogency, but the historical record of apparently strong but erroneous IBEs suggests that we are owed an argument in support of the reliability of IBE.[6]

[6] A different kind of historical argument has been proposed by Kyle Stanford. In his argument against scientific realism, the historical record shows us the occurrence not merely of empirically successful theories that are not approximately true but of the repeated failure of scientists to conceive of relevant alternatives to their favored explanatory theories. In this way Stanford's argument makes use of history to support a claim about the cognitive limitations of scientists, and it is the latter that provides us with a reason to reject realism (Stanford, 2006).

10.4 The Empiricist Challenge

The other significant challenge scientific realism has confronted has its source not in the history of science but in a philosophical orientation historically associated with a scientific outlook: *empiricism*. Empiricist philosophers share a commitment to accounting for human knowledge in terms of experience gained through the use of our senses.[7] Empiricist critics of scientific realism argue that believing in the truth of a theory that refers to unobservable things goes beyond what experience can warrant.

Logical empiricists provide one example (see Chapter 4). In their attempt to distinguish between cognitively meaningful discourse and meaningless metaphysics, logical empiricists sought to characterize the responsible use of theoretical language in terms of the use of *correspondence rules* that would link theoretical terms (like 'electron') to terms describing observable things (like tracks in a cloud chamber photograph). The correspondence rules were to provide the *meaning* of theoretical terms. So, a physicist who avows belief in a theory about electrons would not need to believe in the existence of subatomic particles zooming around with their miniscule mass and electric charge; she would commit herself instead to the existence of a theory in which expressions using such terms are deductively linked, via correspondence rules, to descriptions of observations and experimental outcomes.

10.4.1 Constructive Empiricism

In his 1980 book *The Scientific Image*, Bas van Fraassen set forth such a bracing challenge to realists that it established much of the framework in which subsequent discussion of the issue has taken place. Van Fraassen advocates a position called *constructive empiricism*:

Thesis 3 (constructive empiricism): *Science aims to give us theories that are empirically adequate; and acceptance of a theory involves as belief only that it is empirically adequate.* (van Fraassen, 1980, 12)

[7] This statement suffers from an unavoidable vagueness. Empiricists must regard the nature of experience and its role in knowledge production as empirical questions that cannot be settled at the outset by philosophical fiat. Bas van Fraassen concludes that empiricism cannot be characterized by a belief in any particular thesis, but must be understood as a *stance* or orientation in one's theorizing (van Fraassen, 2002).

Van Fraassen's rough account of empirical adequacy states that a theory is empirically adequate "exactly if what it says about the observable things and events in this world, is true" (van Fraassen, 1980, 12). For greater precision, recall the contrast in Chapter 7 between syntactic and semantic views of scientific theories. Van Fraassen endorses a semantic view. For him, to give a scientific theory is to specify a family of structures: the models of the theory. Such a specification will include the *empirical substructures* of those models, which potentially represent observable things and events. An empirically adequate theory, then, is one for which all of the data ("the structures which can be described in experimental and measurement reports," or what van Fraassen calls *appearances*) correspond to elements of the empirical substructure (van Fraassen, 1980, 64). By "all the data" van Fraassen means *all* the data: not just all observations recorded so far but all those that will be in the future.[8] To state it in more intuitive but less precise terms, an empirically adequate theory is a collection of models into which all the appearances fit.

Unlike the logical empiricist, the constructive empiricist does not seek to interpret away a theory's apparent reference to unobservable things. If a theory uses language that refers to electrons and is true, then it entails the existence of electrons. However, the constructive empiricist regards such a theory as contributing to science, not by providing a true description of an unobservable microworld but rather by describing a set of models into which all the data might possibly fit. Those models might also refer to things and events that do not correspond to any possible data, but the acceptance of a theory does not involve belief in *those* things, only in the claim that the theory is empirically adequate, that is, the data will fit into its empirical substructure.

Thesis 3 makes no descriptive claim about the beliefs of scientists (perhaps most of them are realists) but a *prescriptive* claim about the aims of scientific inquiry. It tells us what *should* count as a successful contribution to science. Note also that to say that a theory is empirically adequate requires an inductive leap beyond what the current data state, for it entails that the theory will be consistent with all *future* data.

[8] Of course, the data might have noise in them. Moreover, some data may be erroneous due to instrumental flaws or errors in experimental design or execution. Even the distinction between data and models of data is not so clear and, as some have argued, may be a matter of context (Leonelli, 2019). A more careful statement of the definition of empirical adequacy would need to take account of these realities of data collection.

10.4.2 What Is Observable?

Because van Fraassen defines empirical adequacy in terms of what is observable, the exact meaning of the constructive empiricist thesis depends on what it means to be observable. Two questions confront the constructive empiricist: (1) Can one state a consistent and defensible criterion of observability? and (2) Does observability *matter* in a way that would justify giving it such a central role in our understanding of the aims of science?

Beginning with the first question, consider an argument due to Grover Maxwell against the possibility of a principled distinction between what we can and cannot observe. Acts of observation range from a direct and 'unaided' observation of an object, through such simple aids as a magnifying glass or a low-power optical telescope or microscope, to cases involving more technological mediations, such as the use of radio telescopes, or electron microscopes, or bubble chambers. This range constitutes a continuum that leaves us, according to Maxwell, "without criteria which would enable us to draw a non-arbitrary line between 'observation' and 'theory'" (Maxwell, 1962, 7).

In response, van Fraassen proposes the following criterion of observability, commenting, "This is not meant as a definition, but only as a rough guide to the avoidance of fallacies" (van Fraassen, 1980, 16):

Principle 13: *X is observable if there are circumstances which are such that, if X is present to us under those circumstances, then we observe it.*

Maxwell's argument indicates that we can draw no sharp line along the continuum separating the certainly and unequivocally observable from the certainly and unequivocally unobservable. But this only indicates that 'observable' – like many other perfectly meaningful and useful words – is a *vague predicate*. By pointing to clear cases and counter-cases, we can provide evidence for the meaningfulness of such predicates – like 'bald,' 'blue,' or 'bilingual' – even if in other cases we are not sure whether they apply or not. The moons of Jupiter are observable because we not only can observe them through the telescope but "astronauts will no doubt be able to see them as well from close up" (van Fraassen, 1980, 16). A subatomic particle passing through a bubble chamber is clearly not observable. You may see the track of bubbles that it leaves, but under no condition can you observe the particle itself.

Such a response suffices to establish that the distinction between what is and is not observable is meaningful, but leaves our second question, which concerns "what ontological ice" such a distinction cuts (Maxwell, 1962, 8). One aspect of this problem concerns the relevant range of possible circumstances in which we might observe a thing. For example, a mutation that enabled a person to form visual images using radiation of wavelengths different from those that human vision normally utilizes would allow one to observe many things we cannot currently observe (Maxwell, 1962, 11). This suggests that the delineation of the observable from the unobservable is *indeterminate*, since we cannot predict just what possible modes of sensory experience we might develop in the future.

Van Fraassen's reply is to insist that we should understand the possibilities of observation in terms of actual, present human capabilities: "The human organism is, from the point of view of physics, a certain kind of measuring apparatus. As such it has certain inherent limitations ... It is these limitations to which the 'able' in 'observable' refers – our limitations, *qua* human beings" (van Fraassen, 1980, 17). Constructive empiricism thus pins the aims of science to something of a moving target. To the extent our cognitive capacities change, so will the meaning of empirical adequacy. Van Fraassen, however, accepts this as an entirely reasonable commitment, for the alternative is to require that "our epistemic policies should give the same results independent of our beliefs about the range of evidence accessible to us" (van Fraassen, 1980, 18).

Does that answer our second question, though? We want to know why observability should make a difference to what we should believe regarding an empirically successful theory.

10.4.3 Underdetermination

Constructive empiricism is grounded in a certain view about what empiricism requires: Only the appearances should matter to the epistemic commitments of science. Since theories with different nonempirical substructures can share the same empirical substructures, a decision to believe one such theory rather than another would have to attribute an epistemic import to such *nonempirical* virtues as simplicity (in one or more of its guises), mathematical elegance, fruitfulness, and so on. Van Fraassen does not deny that scientists value such features in a theory, but insists that an empiricist must value them only as pragmatically *useful*, not as signs of truth. Dirac may have considered his theory of the electron to be mathematically beautiful, but to the extent that he took that aesthetic value as a reason to believe his theory to be true (as opposed to making a pragmatic commitment to develop and defend it), he would be adopting an epistemic attitude based on something other than appearances. A consistent empiricist would not believe in Dirac's 'sea' of negative-energy electrons, or in holes in that sea, which after all are absent from later formulations of quantum theories of electrons. Nor should one believe in electrons and positrons themselves, which some future theory making all the same experimental and observational predictions might eliminate.

Constructive empiricism, therefore, relies strongly on an appeal to a kind of *underdetermination*. That a currently embraced theory that refers to unobservables gives the best explanation of the phenomena – even if it does so with respect to all *future* phenomena – is no warrant for the truth of that theory because of other empirically equivalent theories that differ only with respect to theoretical virtues that an empiricist regards as no reason to believe those theories that possess them. One might, however, resist this argument if one could make the case that such theoretical virtues are in fact reliable indicators of the truth of a theory (McAllister, 1999; Musgrave, 1982).

Van Fraassen employs a different underdetermination argument to defeat one strategy for defending scientific realism. Scientific realists sometimes argue that scientists should use IBE as a means of inferring the *truth* of scientific hypotheses because we do this in ordinary circumstances and we should consistently apply the same kinds of reasoning in both ordinary circumstances and in science. To adapt van Fraassen's own example: Zig has left an (obviously) unwanted horseradish doughnut on the kitchen counter overnight. When he wakes up, he finds no doughnut, but observes in its place a few crumbs and some small black pellets. Zig infers H_M: A mouse has taken up residence (and has a higher opinion of horseradish doughnuts than does Zig). Of course, he has not seen the mouse, but the best explanation of the evidence he has seen is that H_M is true. It would be silly to insist that Zig infer only that things are *as if* there were a mouse, but to remain agnostic about the existence of an actual mouse. Likewise, it would be silly to ask the scientist to remain agnostic about positrons in the face of the evidence compiled by Anderson and Blackett and Occhialini.

This argument interprets Zig's reasoning as an instance of IBE in which Zig infers the truth of the hypothesis. But van Fraassen points out that one could equally well describe Zig's inference regarding the mouse as an inference to the empirical adequacy of H_M. Mice being observable, H_M's truth and its empirical adequacy amount to the same thing. All of its consequences are observable, so it is true exactly if it is empirically adequate, and vice versa. Van Fraassen thus offers a rival hypothesis to that of a scientific realist regarding how we reason: "We are always willing to believe that the theory which best explains the evidence is empirically adequate." This hypothesis, he claims, "can certainly account for the many instances in which a scientist appears to argue for the acceptance of a theory or hypothesis, on the basis of its explanatory success" (van Fraassen, 1980, 20). The empirical evidence of how people do reason underdetermines the choice between realist and antirealist philosophies of science.

We now turn to some of the ways realists and antirealists have elaborated or modified their positions and arguments over years of debate. Various forms of realism and arguments for it have emerged from these efforts.

10.5 Strengthening Scientific Realism

In the face of Laudan's historical challenge, scientific realists have tightened their argument by both refining their standards for what counts as success and by distinguishing those aspects of theory that must be regarded as approximately true to explain such success. In this way, scientific realists have refined their IBE argument by modifying their explanation. They have modified both the *explanans* (the propositions doing the explaining) and the *explanandum* (the proposition describing what is being explained).

10.5.1 Maturity and Success

We can start with their modification of the idea of success. Laudan's argument appeals to a long list of theories that enjoyed periods of empirical success, although we now regard them as false and their central terms as nonreferential. In defense, scientific realists have sought to strike items from this list by invoking a more stringent standard of empirical success than Laudan's.

One such standard requires theories to yield successful *novel predictions* to qualify for a realist attitude (Leplin, 1997; Psillos, 1999). In Chapter 5 we discussed Lakatos's use of a novel prediction criterion to distinguish progressive from degenerating research programs. Philosophers have variously interpreted this standard as requiring a theory to predict some new result that: (1) was not known before, (2) is not predicted by any of the theory's rivals, (3) would be very surprising or unexpected independently of that theory, or (4) was not used in the construction of that theory. In his defense of scientific realism, Stathis Psillos employs the concept (4) of *use novelty*, as advocated by John Worrall (Psillos, 1999; Worrall, 1985, 1989a). Jarrett Leplin defends realism with an analysis of novelty that combines and refines elements of both Worrall's and Lakatos's conceptions (Leplin, 1997).

However conceived, the novel prediction standard significantly reduces the number of episodes in the history of science in which a successful theory turned out to be incompatible with our current best theories. Psillos argues in addition that many of Laudan's successful-but-false theories emerged at times when their respective disciplines were not yet *mature*. They had not yet produced "a body of well-entrenched beliefs about the domain of inquiry which, in effect, delineate the boundaries of that domain, inform theoretical research and constrain the proposal of theories and hypotheses" (Psillos, 1999, 107). By thus invoking higher standards, scientific realists can thus escape any threat from such putative successes as the humoral theory of medicine, the crystalline spheres theory of medieval astronomy, or the effluvial theory of static electricity.

10.5.2 Divide and Conquer

Raising the standards for maturity and success does not, however, suffice to address the entirety of Laudan's challenge, for some significant episodes remain in which a genuinely mature science seems to have succeeded in producing novel successes from a theory with nonreferring terms. Both realists and antirealists repeatedly turn their attention to the case of the nineteenth-century wave theory of light and the luminiferous or optical aether, which physicists used to derive some strikingly successful novel predictions, although the aether plays no role in our current understanding of optical phenomena.

Psillos summarizes the strategy for answering this challenge: "[I]t is enough [to defeat Laudan's argument] to show that the success of past theories did not depend on what we now believe to be fundamentally flawed theoretical claims" (Psillos, 1999, 108). He calls this strategy *divide et impera* (divide and conquer). The strategy has important consequences for the *methods* used by the defenders of scientific realism and for the *content* of the scientific realist position.

The divide-and-conquer strategy requires the defenders of scientific realism to conduct historical investigations to determine just what theoretical claims *were* essential to a given successful prediction and how scientists evaluated the contributions of various elements of the theory to such a prediction. Based on his reading of the contributions of various nineteenth-century physicists, Psillos argues at length that we should distinguish between their mathematical description of the dynamics of light propagation – the essentials of which carried over from wave optics into the later reconceptualization of light in terms of the electromagnetic field – and their attempts to construct a mechanical model of an elastic aether, the vibrations of which could be understood as light waves. Whereas the former played an important *explanatory* role (and were preserved in later theories), the latter mechanical models were merely *illustrative*, and later theories of light excluded them precisely because physicists could not give them a description from which they could predict the behavior of light. He summarizes the situation thus: "The parts of 'luminiferous aether' theories which were taken by scientists to be well supported by the evidence and to contribute to well-founded explanations of the phenomena were retained in subsequent theories. What became paradigmatically abandoned was a series of models which were used as heuristic devices for the possible constitution of the carrier of light-waves" (Psillos, 1999, 140).

In this way, scientific realists seek to turn the problematic case of the luminiferous aether into a success story for a more nuanced realist position. The case of Dirac's sea of negative-energy electrons might pose a more difficult challenge. Thomas Pashby has argued that the early successes of Dirac's theory depended crucially on the sea and could not have been derived without it (Pashby, 2012).

The divide-and-conquer strategy requires the scientific realist to add nuance to their thesis, as expressed in Thesis 2. We should not regard a theory in its *entirety* as 'approximately true' simply because that theory successfully predicts novel phenomena; we should adopt a realist attitude only toward those "constituents which contribute to successes and which can, therefore, be used to account for these successes" (Psillos, 1999, 110).

10.6 Experimental Routes to Realism

Thus far, our discussion of realism and antirealism has focused on the interpretation of *theories* that provide explanations making a reference to unobservable entities. Debates over realism, thus construed, share a preoccupation with theories and their status typical of twentieth-century philosophy of science up until the early 1980s. Then, a few philosophers began to argue that such 'theory-dominated' philosophy of science resulted in a distorted understanding of scientific knowledge that privileged abstract logical or quasi-logical relations between theory and evidence over scientific practices involving material engagement and interaction with experimentally induced phenomena. These 'New Experimentalists' (Ackermann, 1989) argued that valuable philosophical lessons were to be learned by attending to the practices of experimentation. Scientists, they insisted, do not only record observations so they can formulate observation statements to put into some formal relationship with theoretical statements. They intervene in the world, tinkering with technological systems to *make* them convey information about the phenomena that interest them (Ackermann, 1985, 1989; Franklin, 1986; Galison, 1987; Gooding, 1990; Hacking, 1981, 1982, 1983; Hon, 1987, 1989; Mayo, 1996). In the words of Ian Hacking, "Experimentation has a life of its own" (Hacking, 1983, 150).[9]

This turn toward experimental practices in the philosophy of science has led to a focus on questions about the nature of data and the uses of data in science (Antoniou, 2021; Leonelli, 2016, 2019). James Bogen and James Woodward (1988) proposed a distinction between *data* and *phenomena* as part of a critical response to the idea that scientific theories should explain what we observe. Suppose we take data to refer to records of scientific observations. Theories, they argue, do not in general explain data, because the latter are too closely tied to the concrete details of their production. They are in a sense too idiosyncratic to be explained by theories. What theories instead explain are phenomena, such as the differential refraction of light according to its wavelength or the evolution of a new species from an ancestral population under conditions of geographic isolation. These "occur in a wide variety

[9] Here I have treated 'the New Experimentalism' simplistically. An informative discussion that casts doubt on the narrative alluded to in this paragraph can be found in Potters and Simons (2023).

of situations or contexts" (Bogen & Woodward, 1988, 317). Bogen and Woodward propose to be "ontologically noncommittal" regarding phenomena beyond affirming that they "think of particular phenomena as in the world, as belonging to the natural order itself and not just to the way we talk about or conceptualize that order" (Bogen & Woodward, 1988, 321). This expresses a kind of realist orientation. We shall see that the distinction they invoke can play an important role in articulating different options for arguments regarding realism.

10.6.1 Entity Realism

Ian Hacking made some of the earliest and most influential of contributions to the New Experimentalism. Hacking argued that the basis for a realist attitude about unobservable entities such as electrons lay not in the fact that our explanatory theories make apparent reference to them but in the fact that we can *use* them to make things happen – especially in the course of investigating things other than electrons. Hacking described an experiment that used an 'electron gun' to investigate the phenomenon of parity violation in what physicists call 'weak neutral currents.' We will pass over an explanation of the physics of weak neutral currents here – Hacking himself gives only the briefest discussion – because the crucial aspect of the experiment for Hacking's point is not the physics being investigated but the *engineering* enterprise of building an electron gun that makes an experimental investigation of that physics possible. He summarizes his point thus: "We are completely convinced of the reality of electrons when we regularly set out to build – and often enough succeed in building – new kinds of devices that use various well understood causal properties of electrons to interfere in other more hypothetical parts of nature" (Hacking, 1983, 77). Hacking discusses an experiment in which physicists use positrons – the particle predicted by Dirac and discovered experimentally by Anderson. In the experiment, physicists 'spray' positrons on a niobium sphere to change its charge. What impresses Hacking is neither Dirac's theoretical prediction nor Anderson's experimental discovery. It's the much later *spraying* that indicates we should take positrons to be a part of the world. Hacking quips: "So far as I'm concerned, if you can spray them, then they are real" (Hacking, 1983, 23).

Hacking does not simply observe that experimenters, as a matter of fact, tend to believe in the existence of entities they manipulate to investigate phenomena. Instead he claims that the "enterprise" of using entities to experimentally investigate the world "would be incoherent without" a realist attitude toward the entities thus manipulated. Given that the experimental enterprise "persistently creates new phenomena that become regular technology," this enterprise is evidently not incoherent (Hacking, 1982, 73). Thus, "[E]ngineering, not theorizing, is the proof of scientific realism about entities" (Hacking, 1982, 86).

The realism Hacking endorses *only* concerns entities that experimentalists use to learn about other phenomena. He does not endorse realism about theories as such and denies that his argument relies on the kind of inference to the best-explanation argument scientific realists typically employ. (At around the same time, Nancy Cartwright arrived at a similar view, following a different path. In *How the Laws of Physics Lie*, she argues that the laws of physics have great explanatory power precisely because they do *not* describe actually occurring states of affairs ('the facts'). But, she insists, the laws of physics do truthfully describe *something*: They "describe the causal powers that bodies have" (Cartwright, 1983, 61).)

Hacking's writings on entity realism seem to point to two distinct arguments with different conclusions, but Hacking so intermingles them that he may not see them as distinct. One argument seeks to establish good evidence for the claim that not only do unobservable entities exist, but we make regular and reliable use of them in our experimental undertakings: "The 'direct' proof of electrons and the like is our ability to manipulate them using well understood low-level causal properties" (Hacking, 1982, 86). Other passages, however, point to a different argument. When Hacking declares that the fact that "experimenters are realists about the entities that they use in order to investigate other hypotheses or hypothetical entities" is "not a sociological fact" but a requirement of the evident *coherence* of the enterprise of scientific experimentation (Hacking, 1982, 73), his argument is not a 'direct' proof of the existence of electrons or anything else. It does not establish the truth of any scientific realist claim. It argues that experimental scientists should adopt realist *beliefs* about certain kinds of entities because doing so contributes to the coherence of their enterprise. A belief can contribute to the coherence of something without being true. The child who believes that the crumbs left on the plate by the Christmas tree are the consequence of Santa Claus's midnight snack contributes to the coherence of the story her parents tell her about the presents that have appeared overnight.

Independence from Theory?

The more philosophically substantial of Hacking's arguments, then, is the first one. David Resnik reconstructs that argument as follows:

> (1) We are entitled to believe that a theoretical entity is real if and only if we can use that entity to do things to the world. (2) We can use some theoretical entities, e.g. electrons, to do things to the world, e.g. change the charges of niobium balls. (3) Hence, we are entitled to believe that some theoretical entities, e.g. electrons, are real. (Resnik, 1994, 401)

Under this reconstruction, critics can either raise objections against premise 1 (e.g., Shapere, 1993) or premise 2. Resnik targets premise 2, not to dispute the soundness of Hacking's argument but to dispute Hacking's claim that the entity realist need not rely on inference to the best explanation, and thus can maintain an antirealist attitude to *theories*. Resnik's critique attempts to show that justifying premise 2 requires the entity realist to argue that the best explanation of the reliability of instruments experimenters use is given by "low-level generalizations (theories?) and other assumptions" that describe those instruments as taking advantage of theoretical entities (Resnik, 1994, 404). In other words, the realist about positrons cannot simply claim without justification that they are spraying positrons, they must warrant that claim by invoking an explanation of the functioning of their instruments that appeals to at least some theoretical claims.

Michela Massimi also disputes Hacking's claim to divorce entity realism from realism about theories. Her version of the argument does not saddle the theory realist with an appeal to inference to the best explanation, however. Massimi relies on an underdetermination argument to raise a worry for the entity realist that parallels the underdetermination problem for realism about theories: For any given claim about using a particular entity to produce certain effects in an experimental setting, distinct but *empirically equivalent entities* might produce the same effects. Massimi does not raise this possibility to underpin a skeptical argument based on underdetermination. Rather she invokes experiments in particle physics testing the *quark* theory of subatomic particles against the rival *parton* theory to argue that the very experimental evidence that suffices to resolve underdetermination regarding which entity the experimentalist is using also provides the grounds on which to choose one theoretical model over another. This makes theory-free entity realism an unstable position: One either foregoes appeals to theory, leaving one vulnerable to objections based on empirically equivalent entities, or one embraces theory, committing to realism about at least some theoretical claims in order to vindicate realism about entities (Massimi, 2004).

10.6.2 A Direct Experimental Argument for Realism?

Peter Achinstein has advanced a different argument for realism based on experiment (Achinstein, 2002, 2020). Drawing upon a suggestion by Wesley Salmon (Salmon, 1984, 213–227), Achinstein defends the claim that Jean Perrin's experimental investigations of Brownian motion, carried out in 1908, demonstrate scientific realism by giving an experimental demonstration of the existence of molecules.

Brownian motion, which derives its name from the English botanist Robert Brown who observed it in 1827, is an erratic and ongoing movement of microscopic particles suspended in liquids. Perrin carried out a variety of experiments in which he observed through a microscope the motion of particles of gamboge (a resin derived from certain trees) in an emulsion. Perrin derived an equation relating observable quantities by assuming, among other things, "that the motions of the visible Brownian particles are caused by collisions with the molecules making up the dilute liquid in which those visible particles are suspended" (Achinstein, 2002, 472). With this equation he was able to calculate *Avogadro's number*, which he understood as the number of molecules in 1 gram-molecule of hydrogen (i.e., the quantity of hydrogen whose mass in grams is numerically equal to the mean molecular mass of hydrogen). Perrin used a number of different experimental arrangements to produce data yielding distinct but convergent calculations of Avogadro's number N (~ 6×10^{23}). Perrin concluded:

> Even if no other information were available as to the molecular magnitudes, *such constant results would justify the very suggestive hypotheses that have guided us* [including that molecules exist], and we should certainly accept as extremely probable the values obtained with such concordance for the masses of the molecules and atoms . . . The objective reality of the molecules therefore becomes hard to deny. (Perrin, 1916, 105, quoted in Achinstein, 2002, 473, emphasis in original)

Achinstein's essay reconstructs the reasoning whereby Perrin arrives at this conclusion. He argues that Perrin's conclusion is indeed a realist one, in that he not only asserts the existence of unobservable entities (molecules); he does so in opposition to empiricist scientists of his day, such as Pierre Duhem, Ernst Mach, Wilhelm Ostwald, and Henri Poincaré, who "raise general methodological objections to inferences from what is observed to what is unobservable" (Achinstein, 2002, 493). According to Achinstein, Perrin used a *causal eliminative* argument, in which one begins by listing all of the causes of some observed phenomenon *E* that are possible 'given what is known' (alternatively, this is a list that – it is highly probable – includes the cause of *E*). If one can conclude of all but one of these possible causes that they do not in fact cause *E*, then one can justifiably conclude that (probably) the remaining possible cause of *E* is its actual cause. Perrin, Achinstein argues, had good reason to believe that he could – through a combination of experimental and statistical arguments – rule out all other possible causes of Brownian motion sufficiently to claim that it was at least more probable than not that it was caused by the motion of molecules. His subsequent derivation of Avogadro's number then raised this probability even higher.[10]

Achinstein attempts to anticipate and respond to several antirealist objections to his argument. One empiricist objection deserves particular attention. According to this argument, empiricism requires us not to draw inferences beyond what would be susceptible to either support or correction by possible future evidence. Claims about things that are unobservable cannot be supported or corrected by data concerning observable things because we cannot know that the conclusions we draw concerning observable things will also hold for unobservable things. Suppose that we are interested in sampling from things of type *A* and observing whether they have some property *B*. Even if we find that all of the members of our large and varied sample of *A*'s have *B*, we cannot infer that *all* things of type *A*, including those that are unobservable, have *B*, since it may be that being observable is a *biasing condition* with respect to *B*. That is, perhaps *B* is a universal property among observable *A*'s but is absent or rare among unobservable *A*'s (assuming the latter even exist).

[10] Achinstein employs a species of Bayesian probability; a frequentist would reject such probability statements. His interpretation of probability departs from the usual Bayesian one, however. For him, the probability of a hypothesis is a measure not of the degree of belief the agent does or should hold in the hypothesis but of the degree to which it is reasonable to believe the hypothesis. See Achinstein (2001) for an explanation and defense of this view and his general theory of evidence.

An argument like this would warrant denying empirical credentials to claims about unobservables: Such claims require an extrapolation from data regarding observable things to conclusions about unobservables that is unwarranted in principle. But Achinstein notes that the principle required to make this argument is either self-defeating or arbitrary.

Suppose that the antirealist argument from biased sampling is based on this principle: If all members of a sample of *A*'s share some property *P* that is not shared by all *A*'s outside of the sample, then no conclusion regarding all *A*'s is warranted. That would suffice to underwrite the antirealist conclusion about observability (observability would be property *P*), but it is so strong that *no* inference from a sample to a population would be warranted, since one can always find *some* such property for any sample (such as 'observed in Dr. Carlsson's lab' or 'located in the Northern Hemisphere').

If this very strong but self-defeating principle is not behind the antirealist appeal to biased sampling, what principle is? While being observable may be a biasing condition for some inferences, the antirealist argument here considered requires us to believe that it always is. That being observable *might* be a biasing condition should not by itself lead us to reject inferences concerning unobservables, since many other properties also might. Indeed, as Achinstein notes, we can go further and note how experimentalists deploy strategies for addressing worries about the warrant for extrapolating into the realm of the unobservable: Although one cannot observe the unobservable, "[o]ne can vary conditions or properties in virtue of which something is observable (or unobservable)" such as size, proximity, duration, and interactions with other bodies (Achinstein, 2002, 483). Perrin seems to have employed just such a strategy in his own experimental arguments. Speaking of the particles of gamboge he used, he wrote: "We *shall thus be able to use the weight of this particle, which is measurable* [and which Perrin did vary by preparing several different emulsions of gamboge], *as an intermediary or connecting link between masses on our usual scale of magnitude and the masses of the molecules*" (Perrin, 1916, 93–94, emphasis in original).

10.6.3 Experiment and Realism about Positrons

Although Hacking's attempt to drive a wedge between realism about theoretical claims and realism about entities may not succeed, Massimi's and

Achinstein's arguments support his insight that attention to experimental practice would add to the resources for vindicating a realist attitude in science in at least some instances. Moreover, even if experimental arguments for the existence of unobservable entities are not entirely theory-free, they often allow one to draw conclusions about unobservable entities at a stage when the theoretical description of those entities remains unresolved by evidence (or perhaps no candidate for a theoretical description has even emerged).

Consider Carl Anderson's arguments for the existence of a positive electron. Anderson's findings could not serve as a conclusive support for Dirac's theory. Even when Blackett and Occhialini developed more detailed evidence, they denied they could use their data to test Dirac's theory directly. But Anderson did provide an argument for the existence of *something* that, though unobservable, possessed particular attributes: It had a mass similar to that of an electron, and a charge opposite that of an electron. Anderson's argument drew upon some theoretical claims: particularly laws describing the effects of a magnetic field on a charged particle. He employed reasoning based upon previous experimental determinations such as "well established and universally accepted" descriptions of the range of protons of a given energy in a cloud chamber (Anderson, 1933, 491).

Indeed, drawing upon Chapter 9, we might construe Anderson's argumentation along the lines of an argument from severity in Mayo's sense: He considered what he took to be all of the alternatives to the conclusion he drew that were possible given what he knew, and showed how, under any alternative possibility, it would be very unlikely that one would observe an image that fit so well with the hypothesis of a positively charged electron-like particle (Staley, 1999a). Following Anderson's work, subsequent experimentation would both strengthen the argument for this fairly weak conclusion and provide a basis for stronger claims about the properties of positrons (Roqué, 1997).

We can certainly grant that in 1933 Dirac's theory gave 'the best explanation' of Anderson's observations, but did that provide good grounds for believing that theory? In accordance with Hacking, we can regard Anderson as having given an argument for the existence of positrons that we need not regard as an IBE. But Anderson's argument was also not based on being able to *use* positrons for some experimental purpose; it is not an engineering argument. By Hacking's criterion, a justified realist attitude toward positrons would have to wait for the advent of experiments using beams of positrons as probes of a subatomic structure. We would need to be able to spray them.[11]

10.7 Structural Routes to Realism

We have seen how different forms of realism arise from adopting a realist attitude toward different aspects of a theory. One version of this idea differentiates between the entities postulated by a theory and the structural relations between them. Beginning in late 1980s, John Worrall began advocating *structural realism* (though he regards Henri Poincaré and Pierre Duhem as earlier advocates). Structural realists adopt a realist attitude toward the *form* or *structure* of certain theories, but not to the kinds of *things* the theory postulates. This theory began in its *epistemic* form, in which ontology is considered a matter about which one ought to remain agnostic. A more recent *ontic* version has recently emerged claiming that when it comes to fundamental ontology, *structure is all there is*.

10.7.1 "The Best of Both Worlds": Epistemic Structural Realism

Surveying the terrain of long-standing disputes between realists and antirealists, Worrall finds that both sides have both a compelling positive argument and a fatal flaw. Moreover, each position's strength reflects the other's weakness. On the realist side, Worrall acknowledges that the no-miracles argument exerts a strong intuitive pull. It *does* seem that some of scientists' empirical successes are inexplicable unless the theories responsible for those successes are getting something right. Prevailing antirealist philosophies have no adequate alternative account for this great success. On the antirealist side, it seems that even in fairly recent history the mature disciplines have undergone upheavals that have overthrown understandings of phenomena that had seemed settled, replacing them with completely transformed interpretations of the natural world. It would seem hubristic to claim that all of that is behind us and that we will not be subjected to such revolutions in the future; rather we should *expect* such transformations to continue. No realist

[11] In the early 1960s an Italian collaboration built the first particle accelerator that used colliding beams of electrons and positrons, the AdA (*Annello di Accumulazione*) (Bernardini, 2004).

philosophy, even if we allow the invocation of 'approximate truth,' seems to do justice to this historical pattern.

Worrall proposes that epistemic structural realism promises "the best of both worlds" by articulating a view that gives due consideration to the no-miracles argument, but can accommodate conceptual revolutions in science. Consider, for example, the nineteenth-century wave theory of light. This theory generated some remarkable successes. For example, Augustin Jean Fresnel gave it a mathematical formulation in an essay he submitted to the French Académie des Sciences for a competition in 1819. One member of the committee, Siméon Denis Poisson, used Fresnel's theory to calculate the surprising result that, given the right configuration, one would find a bright spot at the center of a shadow cast by an opaque disc. The chairman of the committee, François Arago, decided to test this unexpected consequence of Fresnel's theory and found the predicted bright spot. Fresnel won the prize. But Fresnel's theory represented the propagation of light as a movement of vibrations in an all-pervading mechanical aether, which current physics denies. For the scientific realist, the wave theory of light poses a problem as an empirically successful theory using theoretical terms referring to things that do not exist. (We have already noted the 'divide and conquer' response to this problem in Section 10.5.2.)

But the history of the theory has an important feature for which a purely antirealist attitude seems inadequate: When the wave theory based on a mechanical aether was supplanted by Maxwell's theory of electromagnetic fields, the mathematical equations that theorists had developed to describe the undulations of aether *remained*, but they now described the vibrations of electromagnetic fields. The progression of theories of light thus exhibits the *preservation of structure* alongside a significant discontinuity in its description of entities. Worrall quotes a telling passage from Poincaré discussing the equations of Fresnel's theory:

> [T]hese equations express relations, and if the equations remain true, it is because the relations preserve their reality. They teach us now, as they did then, that there is such and such a relation between this thing and that; only the something which we then called *motion*, we now call *electric current*. But these are merely names of the images we substituted for the real objects which Nature will hide for ever from our eyes. The true relations between these real objects are the only reality we can attain. (Poincaré, [1905] 1952, quoted in Worrall, 1989b, 118, original emphases)

In other cases, the mathematical structure of an earlier theory might be subsumed within a later theory as a *special case*. We find this pattern in the replacement of Newton's dynamics by the dynamics of Einstein's Special Relativity. Kuhn argues (as discussed in Chapter 5) that the change from Newton's physics to Einstein's physics is so great that even though the same terms appear in both theories, their meanings changed. So, Newton's physics as a whole (with its understanding of mass as a quantity independent of one's frame of reference, for example) is not really a special case of Einstein's physics. But even if Kuhn is right about this, the structural realist's more modest claim remains valid: The mathematical form of Newton's laws can be derived from Einstein's laws as a special case where relative velocities are very small compared to the speed of light. This shows that in Einstein's theory, the relations between quantities that Newton's equations describe remain (within a certain domain) unchanged. Einstein's theory revises what those quantities signify, but Newton's theory gave the correct description – within a restricted domain – of how they relate to one another.

Dirac's Hole Theory, One More Time

In one respect, the case of Dirac's theory and its successful prediction of the positron seems to lend support to structural realism. If one consults a current textbook on electrodynamics, one will find equations with apparently the same mathematical form given by Dirac, although he regarded his theory as a description of the relations between particles that fill space even in a vacuum, whereas the current theory describes relations between fields.

Once again, however, the success of Dirac's peculiar theory challenges our attempts to craft a systematic philosophical understanding. As Pashby notes, interpreting the transition from Dirac's hole theory to our current understanding of quantum electrodynamics as a story of structural preservation tells only a part of the story. Although we will not delve here into the mathematics required for a full account, the crucial point is that in spite of the formal similarity between Dirac's own version of his equation and the modern understanding of it, the *mathematics* of the two equations are in fact quite different. In Dirac's theory, his equation describes the evolution of a *wave function* that determines the probability of finding a particle in a given location at a given time. But the modern equation applies instead to fields that assign *operators* to particular locations in spacetime. These operators then determine the probability that a particle will be *created* or *annihilated* at a given location. The difference here cannot be easily interpreted as simply a change from one kind of thing to another while keeping the relations between the things the same. Pashby concludes: "[T]heory replacement involves structural discontinuity as well as structural preservation, and since these discontinuities may reflect the problematic ontological or metaphysical shifts that the structural realist had hoped to avoid, she had better account for them too" (Pashby, 2012, 470).

10.7.2 Ontic Structural Realism

The structural realist position we have been discussing has come to be called *epistemic* structural realism to distinguish it from the more recently emerged *ontic* structural realism. Epistemic structural realism argues that we can safely regard our best theories as mapping an *aspect* of the world: its structure. But "the *nature* of the basic furniture of the universe" will remain forever elusive to us (Worrall, 1989b, 122).

Ontic structural realists, however, argue that the structural realist understanding of our best current theories in physics provides us with just such an understanding of the "basic furniture of the universe" (or, perhaps better, how the furniture has been arranged) and that what it tells us is that it is precisely *structure* that is fundamental to the makeup of the universe. In one version of this idea, at the most basic level of the makeup of the universe, there are no *things* that exist independently of the relations that hold between them. Rather, it is the relations (i.e., the structures) that are fundamental (Ladyman et al., 2007; Frigg & Votsis, 2011 gives an insightful critical overview; see Ladyman, 2023 for an extensive survey of structural realist views, criticisms, and responses).

Epistemic structural realism attempts to carve out a realist understanding of the success of scientific theories that disentangles them from metaphysics. Ontic structural realism invokes that same understanding as a way of *doing* metaphysics.

10.8 Perspectival Realism

10.8.1 Giere's Scientific Perspectivism

In his 2006 *Scientific Perspectivism*, Ronald Giere argues for a view of scientific knowledge that makes the perspective of the knower crucial to the knowledge they achieve. Giere's argument is not directed at the explanationist scientific realist or the constructive empiricist but is instead meant to respond to the kind of constructivist views discussed in Chapter 6 as well as to declared enemies of such views, such as scientists[12] articulating a kind of 'objectivism' that maintains that science is converging toward a single, uniquely correct description of reality.

Giere presents a sequence of three main arguments that aim to show that scientific knowledge is perspectival, in the sense that what we can claim to *know* on the basis of scientific inquiry is always going to be that something is so *from some perspective*.

The first argument concerns the perception of colors. Giere observes, drawing upon the scientific study of color vision, that the human visual experience of color is a product of a complicated sequence of electromagnetic and neurophysiological processes involving an "*interaction* between aspects of the environment and the evolved human visual system" (Giere, 2006, 31–32, emphasis in original). Other species have different color experiences because of differences in their visual systems, and humans themselves exhibit significant variability in their visual systems with regard to their sensitivity to different wavelengths of light. One aspect of this variability is so-called colorblindness that constitutes a significant group of variants upon the more common *trichromatic* structure that equips humans with cone cells in the retina that are sensitive to short-, medium-, and long-wavelength light. Color perception is a function of intensity differences among signals from these receptors. A person lacking cones sensitive to medium- or long-wavelength light has a *dichromatic* visual system and "experiences only somewhat faded yellows and blues" (Giere, 2006, 28). Giere concludes that color perception is perspectival in the sense that one's perception of the color of an object – a rug, perhaps – is always a perception from one's perspective, meaning that it is the result of an interaction between one's visual system and the environment. For individuals with different visual systems, that interaction – and the resulting perception of color – may be different, but it would be a mistake to say that, for example, a trichromatic person's judgment of the color of rug must be more true than that of a dichromatic person. Each of them is sensitive to some kinds of electromagnetic radiation and not others.

[12] Giere draws upon statements from theoretical physicist Steven Weinberg to illustrate this position (Giere, 2006, 4).

As Giere puts it, perspectives are "always partial" in this way (Giere, 2006, 35).

The argument for perspectivism about color perception forms a template for the next stage of argument: Scientific observation is also perspectival. This extension of the argument rests on features scientific instruments share with the human visual system: Scientific instruments "respond to only a limited range of aspects of their environment" and their response within that range "is limited" (Giere, 2006, 41). Using examples from astronomy and neurophysiology, Giere shows how the scientific study of a given object (the Milky Way, the human brain) benefits from using different kinds of instruments that interact in different ways with that object, yielding different kinds of observations. The Hubble Space Telescope yields images of objects using instruments that process electromagnetic radiation from particular parts of the spectrum, including infrared radiation. Image production involves complex data processing procedures. Giere shows a Hubble image made with an instrument called the Advanced Camera for Surveys (ACA – added to the Hubble by astronauts in 2002) in which a cluster of galaxies called Abell 1689, 2.2 billion light-years away, is said to function as a gravitational lens (a concept from General Relativity) to show light emitted about 13 billion years ago. Giere says that it does not simply show "the universe as it was 13 billion years ago" but instead constitutes an image of "the early universe from the perspective of the Hubble ACS system using Abell 1689 as a gravitational lens. The picture is the product of an *interaction* between light from the early universe and the Hubble telescope system" (Giere, 2006, 45, original emphasis). An entirely different kind of telescope system, the Compton Gamma Ray Observatory (CGRO), which orbited earth from 1991 to 2000, was sensitive to a quite different range of energies of electromagnetic radiation and thus produced quite different kinds of images from its multiple instruments. No single telescope conveys a complete image. Images from the Hubble ACS and images from CGRO instruments differ not in how accurate or veridical they are but rather in the aspects and manner in which they represent. They constitute different partial perspectives on objects in the cosmos. Giere shows how similar points apply to the various imaging technologies that have been used to study the human brain, such as computer-assisted tomography (CAT), positron emission tomography (PET), magnetic resonance imaging (MRI), and functional magnetic resonance imaging (fMRI).

The final step in Giere's argument for his claim that scientific knowledge is perspectival extends the arguments from color perception and scientific observation to scientific theorizing: "The basic idea is that conception is a lot like perception, or, that theorizing is a lot like observing ... [I]n creating theories ... scientists create perspectives within which to conceive of aspects of the world" (Giere, 2006, 59). Giere treats scientific theorizing as a practice in which scientists use theories to represent aspects of the real world for certain purposes (Giere, 2006, 60), and he understands scientific theories in terms of models. As discussed in Chapter 7, one way to approach scientific theorizing is to think of scientists as relying on the kind of general principles we find in, say, Newton's laws of motion or the Darwinian principle of natural selection in order to guide them – with the help of specific facts about an aspect of the world to be represented – in the construction of models that serve to represent that aspect. Theorizing, on this view, is a matter of representing selected features of the world by means of models constructed on the basis of "appropriate general principles and specific conditions" (Giere, 2006, 63). Like Nancy Cartwright (and in opposition to scientific realism in its explanationist form), Giere denies that the general scientific principles represent the world (it is in this sense, according to Cartwright, that the laws of physics "lie"). Instead, they define "very abstract models" that scientists then specify in "artful" ways, incorporating information about specific conditions of the world to construct a model that represents the intended features of the world to achieve some aim of the investigators. A perspective is defined by the selection of (1) features of the model, (2) features of the world to be represented by that model, and (3) aims to be achieved by such representation. Theorizing is a practice of representing from such a perspective. The model succeeds in representing when it has features bearing similarity in the right kind of way to the intended features of the world. What counts as 'the right kind of way' depends on the context. Bohr's model of the hydrogen atom was intended to resemble actual hydrogen atoms with respect to their discrete energy levels, but not with respect to, for example, their prevalence in the interior of the Sun.

The position that Giere seeks to defend against constructionist and objectivist alternatives is "perspectival realism," according to which "the strongest claims a scientist can legitimately make are of a qualified, conditional form" (Giere, 2006, 5). Giere describes that form as "something like: Given the assumed observational and theoretical perspectives, [representational model] M exhibits a good fit to the subject matter of interest" (Giere, 2006, 92). Because representational models face a pervasive problem of a realm of alternate possibilities that is "too rich" to be "described analytically," no stronger realist claims are ultimately defensible. This problem for realists dissolves once one accepts "relativized, perspectival conclusions rather than absolutely unqualified conclusions" and realizes that "perspectivally realistic conclusions are not unacceptably relativistic, but, rather, the most reliable conclusions any human enterprise can produce" (Giere, 2006, 92).

10.8.2 Massimi's Perspectival Realism

Michela Massimi extends and develops perspectival realism, drawing upon Giere's work – as well as other varieties of perspectivism advocated by Paul Teller (2019) and Alexander Rueger (2005) – in an ambitious recent work. Her *Perspectival Realism* (Massimi, 2022) presents a pluralistic philosophy of science that aims to "celebrate scientific knowledge as a distinctive kind of *social and cooperative knowledge* – knowledge that pertains to *us wonderfully diverse human beings* occupying a kaleidoscope of historically and culturally situated perspectives" (Massimi, 2022, 19, emphases in original). To this end, she offers the following 'working definition' of a scientific perspective:

> A scientific perspective *sp* is the actual – historically and culturally situated – scientific practice of a real scientific community at a given historical time. Scientific practice should here be understood to include: (i) the body of *scientific knowledge claims* advanced; (ii) the experimental, theoretical, and technological resources available to *reliably* make those scientific knowledge claims; and (iii) second-order (methodological-epistemic) principles that can *justify* the *reliability* of the scientific knowledge claims so advanced. (Massimi, 2022, 6, emphasis in original)

Massimi's project is motivated in part by criticisms of perspectival realism as "redundant, inconsistent, and unstable" (Massimi, 2022, 63). In particular, she seeks to respond to the 'Problem of Inconsistent Models' (PIM). For example, the atomic nucleus is represented as a set of concentric orbital shells in some contexts, and as "a bunch of valence quarks exchanging gluons" in others (see Morrison, 2011, 2015).[13] The model that represents the atomic nucleus as a set of orbital shells, and one that represents it as valence quarks exchanging gluons, attribute to the nucleus features that are incompatible with one another. Critics of perspectival realism (e.g., Chakravartty, 2010; Morrison, 2015) use the PIM to argue that, since some perspectival representations of a target system (like the orbital shell and quark/gluon representations of atomic nuclei) are incompatible with other representations of the same system, perspectival realism must either give up its realism or accept that our attempts to represent the world cannot ultimately succeed in perspectival terms alone.

Massimi provides a nuanced, multifaceted response to the PIM; a rough sketch might go as follows:

1. First, Massimi draws a distinction between two notions of perspectival representation, inspired by a distinction in art between representation that is perspectival in the sense of being drawn *"from a specific vantage point"* (perspectival$_1$) as opposed to being a *"representation drawn towards one (or more than one) vanishing point(s)"* (perspectival$_2$) (Massimi, 2022, 32, emphasis in original). In the case of scientific representation, distinct perspectival$_1$ representations of atomic nucleus may be thought of as representing it as having different features "viewed *from the point of view* of isotopic phenomena" that are incompatible with features represented "from the point of view of quantum chromodynamics" (Massimi, 2022, 40, original emphases). Massimi encourages us to consider how we might think of scientific models as also (or instead) providing us with perspectival$_2$ representations. A perspectival$_2$ representation "has a directionality ... *towards* one or more vanishing points that create the effect of a 'window' on reality extending beyond the boundaries of the representation itself." The content of a perspectival$_2$ representation "is not affected by the vantage point *from* which the representation takes place: it is not an instance of *representing-as* understood as ascribing alternative and incompatible attributes or properties ... It instead allows for a plurality of lines of inquiries and inferences about the target system" (Massimi, 2022, 41).
2. Massimi next identifies an assumption of PIM that she calls the *representationalist assumption*: "[O]ne of the main tasks of any scientific

[13] "There are over 30 different nuclear models based on very different assumptions, each of which provides some 'insight' into nuclear structure and dynamics" (Morrison, 2015, 179).

model M is to represent (in part at least) relevant aspects of a given target system S, and . . . claims of knowledge based on the scientific model are true (or approximately true) when the model provides a partial yet accurate representation of the target system" (Massimi, 2022, 44). The representationalist assumption is "in tension with" the *perspectivalist assumption* that "[s]cientific models offer perspectival representations of relevant aspects of a given target system," in a sense "often tacitly" assumed to be a perspectival$_1$ representation (Massimi, 2022, 44).

3. She reconstructs the PIM as obtaining when when (i) two representational models M_1 and M_2 both give a "partial yet *de re* representation[14] of selected relevant properties" of a target system, (ii) the properties represented by M_1 and M_2 are "real-qua essential properties of the target system in that they capture *essential* features of it," and (iii) the properties represented by M_1 are not only different but also inconsistent with those represented by M_2 (Massimi, 2022, 56, emphasis in original). Whatever the phase state of nuclear matter, for example, it cannot essentially be simultaneously a liquid, a gas, a semisolid, *and* a fourth state defined only for quarks (Morrison, 2015, 180). Critics contend that instances of PIM reveal a serious problem for perspectivism. Morrison expresses the problem this way: "If we take perspectivism seriously, then we are forced to say that the nucleus has no nature in itself and we can only answer questions about it once a particular perspective is specified" (Morrison, 2015, 160). Chakravartty takes the PIM to reveal the perspectival realist's position as holding the dubious position that "empirical reality itself consists of a hodgepodge of contradictory states of affairs, created (in part) by the human act of theory use and model construction" (Charavartty, 2010, 411).
4. Massimi's initial response to this criticism targets two assumptions not made explicit in the *representationalist assumption* but necessary for identifying a case of PIM as specified in the three conditions just stated. The first (*representing-as-mapping*) holds that an accurate, partial, *de re* representation involves establishing a "one-to-one mapping between relevant (partial) features of the model and relevant (partial) . . . states of affairs about the target system" (Massimi, 2022, 67). Massimi opts for an alternate "inferentialist" account of model representation (Suarez, 2004,

[14] To say that the representation of properties is *de re* (literally 'about the thing') means that it is a representation of particular properties of the target system itself, and not just a representation of a description of the target system. The latter would be a *de dicto* ('about what is said') representation.

2015). The second assumption (*truth-by-truthmakers*) holds that the states of affairs represented by a "partial yet accurate" representation "ascribe essential properties to particulars, and, as such, they act as ontological grounds that make the knowledge claims afforded by the model (approximately) true." This is why instances of PIM are described as cases in which the partial yet accurate representations "capture *essential* features" of the target system. To express the metaphysical character of this commitment, Massimi quotes from the metaphysician L. A. Paul, who writes that essential properties "give sense to the idea that an object has a unique and distinctive character, and make it the case that an object has to be a certain way in order for it to *be* at all" (Paul, 2006, 333, original emphasis).

5. Having exposed these assumptions behind the objection from cases of PIM, the obvious move for a defense of persectivalism is to reject the assumptions.

But what positive view of model representation should a perspectival realist adopt in their stead? Massimi's positive view draws on the account given above of a *scientific perspective*, emphasizing the importance of *perspectival$_2$* representation. She exemplifies her account through three case studies (models of the atomic nucleus, climate modeling, models of language development in children). Here we briefly consider Massimi's examination of the already mentioned plurality of models of atomic nucleus.

Early twentieth-century models drew from multiple disciplines, including geochemistry, earth sciences, and cosmochemistry. The shell model and the liquid drop model of the atomic nucleus grew out of a plurality of scientific perspectives that supported reliable inferences about such phenomena as neutron absorption and the stability of certain species of nuclei with particular numbers of protons and neutrons. An alternative "odd particle model" served as a "bridge" that "showed the epistemic limit of the liquid drop model" and paved the way for the development, in 1949–1950, of a "'unified model' that combined features of the liquid drop model and key insights of the shell model" (Massimi, 2022, 105). The history is too complicated and the cast of characters too large to permit a quick summary here, but let's consider how a perspectival realist should think about this episode, according to Massimi: "To be a perspectival realist about the atomic nucleus is ... to engage with an open-ended series of modally robust phenomena (e.g., nuclear stability, neutron capture, nuclear rotational spectra) at the experimental level and the many exploratory models that over time have allowed physicists to gain knowledge about *what is possible* concerning each of these phenomena" (Massimi, 2022, 106, original emphasis).

Several points regarding perspectival modeling deserve emphasis: (1) scientific models serve an *exploratory* function by enabling epistemic communities to make "relevant and appropriate inferences" (they are *inferential blueprints*); (2) they do this by also serving a "*social and collaborative function*," enabling scientists "to work together over time, make changes to and tweak an original model"; (3) such models consequently evolve over time (they are *dynamic*), making them "often the battleground for scientific rivalries and questions about co-authorship"; (4) perspectival modeling is "an integral part of a scientific perspective in being embedded into historically and culturally situated scientific practices" (Massimi, 2022, 106–108).

Whereas perspectival$_1$ representations provide static, map-like representations of their targets, Massimi emphasizes the ways in which the models of atomic nuclei she discusses function as perspectival$_2$ representations: They serve as resources for enabling differently situated epistemic communities to draw relevant and appropriate inferences regarding phenomena, which she likens to 'opening a window' on those phenomena.

A realist who holds the *truth-by-truthmakers* assumption will articulate realism in terms of the way a scientific discourse asserts true statements about the essential properties of the objects of that science. For Massimi, successful perspectival representation enables investigators to draw "relevant and appropriate" inferences (on the basis of data) about phenomena, and specifically about *modally robust phenomena*. The reference to phenomena in contrast to data invokes the distinction between the two introduced by Bogen and Woodward (Bogen & Woodward, 1988), discussed in Section 10.6, but with some important distinctions. Massimi regards phenomena as indexed to particular domains of inquiry. For example, the phenomenon of "large-scale ocean heat uptake" occurs in the domain of "the hydrosphere" (Massimi, 2022, 208). By calling these phenomena "modally robust," Massimi is also emphasizing that these are "phenomena that do not just occur, but *could occur* under a range of different experimental, theoretical and modelling circumstances and across a variety of perspectival data-to-phenomena inferences." The term *modal* marks the philosophical distinction between what is the case or does happen from what might, must, could, or would happen. Describing phenomena as modally *robust* emphasizes that there are "many ways in which epistemic communities *infer* the relevant phenomenon by connecting often diverse datasets to the stable event" (Massimi, 2022, 211). Examples include "the bending of cathode rays, the decay of the Higgs boson, the electrolysis of water, germline APC [Adenomatous poplyposis coli] mutations, the pollination of melliferous flora, the growth of a mycelium" and many more (Massimi, 2022, 15).

Massimi's account of scientific perspectives treats them as "historically and culturally situated." Massimi uses perspectival realism to examine the historical and cultural dimensions of science and especially to inquire into the forms of epistemic injustice[15] that can arise in those dimensions. This inquiry distinguishes two ways in which scientific perspectives relate to one another: Two scientific perspectives *intersect* when enhancing the reliability of knowledge claims requires contributions from both. Two scientific perspectives *interlace* when encounters between the two lead to the exchange of tools, instruments, and techniques. Intersection is a methodological relationship, while interlacing is historical (Massimi, 2022, 340).

Attention to the historical and cultural context of scientific perspectives enables us to grapple with the tension between two ways of thinking about culture: *cosmopolitanism* and *multiculturalism*. Cosmopolitanism is a broad term referring to the general idea of "community among all human beings, regardless of social and political affiliation," with cultural cosmopolitans opposing "exclusive attachments to a particular culture" (Kleingeld & Brown, 2019). Multiculturalism, on the other hand, embraces the distinctness of cultures and opposes those forces that tend to erode their differences; it has been "advocated in political theory to reject assimilation or exclusion of . . . nondominant groups; to call out injustices done by policies of assimilation or exclusion; and to offer remedies for those injustices" (Massimi, 2022, 349).

In considering multiculturalism and cosmopolitanism in the context of science, Massimi examines two varieties of epistemic injustice that she labels *epistemic severing* and *epistemic trademarking* (Massimi, 2022, 349). Epistemic severing involves "cutting off specific historically and culturally situated communities to historically remove or blur their contributions to 'historical

[15] *Epistemic injustice* refers generally to forms of injustice in the treatment of people *as knowers or epistemic subjects*. See Fricker (2007) for a seminal treatment, and Bohman and McCollum (2012) for a collection of further developments and responses.

lineages' in the scientific knowledge production" (Massimi, 2022, 350). (Among several examples, Massimi cites "a narrative on the invention of Hindu–Arabic numerals that fails to mention the role played by the North African nomadic culture of Bedouins in the transmission of such knowledge to the Mediterranean communities" (Massimi, 2022, 351).) Epistemic trademarking constitutes a further infliction of injustice, subsequent to epistemic severing, that involves "the appropriation and branding of entire bodies of knowledge claims, with associated practices, as a 'trademark' of one particular epistemic community" (Massimi, 2022, 354). By epistemic trademarking, one epistemic community benefits at the expense of others, laying claim to the merchandising rights – in a broad sense – of knowledge that should rightly be viewed as the achievement of a larger complex of scientific perspectives.

Massimi illustrates her approach with a brief discussion of the multifarious manifestations of knowledge of terrestrial magnetism that predate the 1600 publication of William Gilbert's *De Magnete*, in which he attributes the invention of the mariner's compass to the sailors of the Italian town of Amalfi. Drawing on the work of historian Joseph Needham, Massimi notes the apparent beginnings of the study of magnetism in the Chinese Han dynasty, documented in a work by Wang Chung in 83 CE describing a 'south-controlling spoon' made from lodestone used for the purposes of divination by geomancers of that era. Subsequent Sung-dynasty (960–1279 CE) geomancers developed a 'wet' compass by placing a lodestone in a wooden fish floating in water. Needham connects this use of a compass by geomancers to the "Chinese symbolism of the emperor as the pole star facing south to his realm" and dates the use of such a wet compass for nautical navigation to the eleventh century CE or earlier. He also notes how the continued relevance of the phenomenon of terrestrial magnetism for subsequent Chinese dynasties, as manifested in the use of the magnetic compass, relates to "Chinese attachment to a doctrine of action at a distance, or wave-motion through a continuum, rather than direct mechanical impulsion of particles" (quoted in Massimi, 2022, 345).

This marks a significant difference between the perspective of Han and Sung dynasty geomancers and navigators and that of early modern European investigators working within the assumptions of a mechanistic point of view skeptical of the possibility that any body can act on something located where that body is not. When Isaac Newton asserted that a force of mutual attraction acts on every pair of material bodies in proportion to their masses, one of the most potent criticisms of his theory was that it made gravitational phenomena into effects of action at a distance. Subsequent development of European theories of electromagnetism involving fields came about through a process of separation from such mechanistic commitments.

There is more to the distinctness of the perspectives of the Chinese Han and Sung dynasties on terrestrial magnetism: "[T]he Han-Sung craftmanship of wet and later dry compasses was *situated in* a scientific practice which delivered knowledge of important phenomena about the Earth's magnetic field and that sits in a historical lineage alongside Gilbert's later research in *De Magnete* all the way up to" the electromagnetic investigations of nineteenth-century British investigators like Michael Faraday and J. J. Thomson. But this craft tradition was also "*situated for*" the purpose of divination and was "part of a complex network of cosmological beliefs with a clear sociopolitical undertone" (Massimi, 2022, 346, original emphasis).

For its Han–Sung dynasty users, the magnetic compass was a device for divination that also facilitated their encounter with the 'modally robust phenomenon of the Earth's magnetic field.' Massimi notes how a variety of other scientific perspectives on this phenomenon opened during the medieval period, as recorded in sources including those of the Benedictine monk Alexander Neckham, author of treatises appearing between 1175 and 1204 CE; reports about nautical compasses from French and Icelandic poets, from a Scottish astrologer, descriptions in Persian sources as early as 1232–1233 CE; and a thirteenth-century treatise by Al-Ashraf Umar ibn Yūsuf of Yemen (Massimi, 2022, 345–346).

The historical question of how these perspectives *interlace* is complicated and not easily resolved. Figuring it out involves answering specific historical questions for which clear evidence might be lacking: Did Chinese compasses and knowledge of their use make their way to Europe, or did Europeans develop this technology independently? The *intersection* of perspectives is a distinct methodological question with relevance, Massimi argues, for "refining the *reliability* of knowledge claims over time."

Massimi's version of perspectivism has a cosmopolitan aspect insofar as different perspectives can fruitfully intersect in this way, improving the reliability of knowledge claims made within distinct scientific perspectives. The phenomenon of terrestrial magnetism can be encountered in various ways. Those ways are not methodologically or historically isolated. They intersect and interlace over time, through exchanges and interactions that result in changes to the tools, techniques, and meanings particular perspectives use to engage with a given phenomenon. But Massimi's perspectivism retains a multicultural orientation insofar as she insists that this interlacing does not erase the historical and cultural situatedness of scientific perspectives. Recognizing this, Massimi concludes, is an important step toward a philosophy of science that recognizes and can respond to historical processes of interlacing that can produce varieties of epistemic injustice.

10.9 Alternate Realist Discourses

Several philosophers have sought to approach the question of scientific realism in ways that do not fit neatly into the framework in which the debates sketched above have been conducted. I mention here just a few of these alternate routes to indicate philosophers' varied approaches to the question of the aims and outcomes of scientific inquiry.

10.9.1 Fine's Natural Ontological Attitude

Arthur Fine's "Natural Ontological Attitude" (NOA) seeks to transcend the dichotomy between realism and antirealism. Writing in 1986, he notes that even as the successes of science accumulate, scientific realism as a philosophical position is in decline.[16] Since the ongoing success of the projects scientists undertake has not served to strengthen scientific realism (and has no prospects for strengthening antirealism), he concludes that the debate "does not concern issues that can be settled by developments in the sciences" (Fine, 1986, 150). He considers both realism and antirealism to be 'inflationary' philosophical adjuncts to science that seek to tie the legitimation of science to either metaphysics (realism) or epistemology (antirealism). By contrast, NOA includes a 'pro-attitude' towards science and even towards science yielding belief in the truth of claims about unobservables, but it avoids theorizing about truth itself or about the aims of science, the pursuit

[16] Interestingly, Fine takes one indicator of that decline to be a proliferation of different versions of scientific realism. He writes, with a tacit invocation of Imre Lakatos, that a token of the "degeneration" of "the realist programme . . . is that there are altogether too many realisms" (Fine, 1986, 149).

of which it regards as a "chimera" and an artifact of philosophical dispute (Fine, 1984, 1986).

According to Fine, scientific realism and antirealism (or instrumentalism) are both engaged in a misguided project of "making sense of science" as "part of a search for authority." Realism seeks that authority "outside" the discourse of science itself and is thus led to the philosophical concept of "the *World*" and the notion of "correspondence to the *World*" defining truth. By contrast, "instrumentalism looks inward" and, in its empiricist mode, delimits the range of knowledge by reference to what is observable, resulting in a "radical interpretation of scientific practice, surgically grafting altered significance on to the practice at precisely the point where science moves beyond the observable" (Fine, 1986, 171).

NOA abandons the search for the authority or authentication for science. "NOA urges not to undertake the construction of teleological frameworks in which to set science" (Fine, 1986, 172). The idea of NOA is not that scientists are aimless when they undertake inquiry but rather that it is a mistake to suppose there is something that we might call "the aim of science" (whether truth – as correspondence to "the *World*" – or empirical adequacy) that governs all such inquiry.

10.9.2 Stein's Realist/Instrumentalist Dialectic

Howard Stein argues for a subtle combination of realist and instrumentalist "attitudes" through a careful consideration of episodes from the history of physics.[17] Like Fine, he rejects the standard arguments for both scientific realism and antirealism. The explanationist defense of scientific realism, he argues, fails to illuminate just what the thesis of scientific realism asserts. The entities the realist insists "really exist" do their explanatory work only as "postulated within a natural-scientific theory" and not at the metascientific level of the advertised explanation for the success of science (Stein, 1989, 53).

[17] A caveat: It is difficult to do justice to Stein's argument without reproducing its entire dialectic, which I cannot do here. Arguably, the dialectic between instrumentalist and realist viewpoints is in fact the point of his piece (published in the journal *Dialectica*, no less). Here I intend not so much a summary of his view as an advertisement for reading the original (Stein, 1989).

This critique does not, however, lead Stein to embrace antirealism. On the contrary, his essay ends with the proclamation, "Realism – Yes, but ... instrumentalism – Yes also; No only to anti-realism" (Stein, 1989, 65). His "Yes" to realism consists in part in an acceptance of a point emphasized by structural realists: The "structural 'deepening'" that can be seen in the historical development of physical theory reveals a tendency of certain "abstract mathematical forms" to persist through changes in theory. The scientific realist has misplaced their attention on the referential semantics of theories.

What Stein adds to this historical observation, however, is an observation about the realist and instrumentalist attitudes of scientists themselves and the effect this has on their work. Referring specifically to James Clerk Maxwell, Isaac Newton, and Albert Einstein, Stein argues that a "dialectical tension ... between a realist and an instrumentalist attitude, existing together without contradiction, seems to me characteristic of the deepest scientists" (Stein, 1989, 64). By contrast, even otherwise great scientists can be misled by an imbalance between these two attitudes. Stein cites the example of Henri Poincaré, "whose work in physics suffered from an imbalance on the instrumentalist side." On the other hand, he notes how Christian Huygens and Lord Kelvin resisted the advances of Newton's theory of gravity and Maxwell's electromagnetic theory, respectively, on the basis of misplaced realist scruples.

10.9.3 Chang's Pragmatist Realism

For Hasok Chang, realism is a matter of "respect[ing] facts, not fantasies," being "open to learning from experience, instead of hiding behind the comforts of fixed opinions," and living "with the optimism that we *can* learn some truths about realities, rather than resign ourselves to living without improving our current state of ignorance" (Chang, 2022, 204). Chang's realism is also an "activist realism" that promotes the normative idea that "science should strive to maximize our contact with realities and our learning about them" (Chang, 2022, 209).

This activist realism lies at the heart of Chang's recent efforts at a realist philosophy of science that develops themes from the pragmatist philosophies of Charles Sanders Peirce and John Dewey, while also drawing upon the *operationalist* ideas of Percy Bridgman (see also Chang, 2017). Chang shares

Arthur Fine's resistance to supplementing scientific undertakings with abstract metaphysical or epistemological theorizing. His realism is "not a descriptive thesis, but a commitment to an ideal" (Chang, 2022, 208).[18] He describes that ideal as "an *activist ideal of inquiry*: a commitment to seek more and better knowledge about realities, along with a commitment to improve our epistemic practices to that end" (Chang, 2022, 208, emphasis in original).

Chang's pragmatist orientation treats knowledge broadly, without privileging the traditional philosophical orientation toward emphasizing propositional knowledge as a certain category of belief. Pragmatist views of knowledge emphasize the connection between knowledge and action or ability, and how activities of inquiry constitute ways of producing knowledge: "Pragmatism regards knowledge as an outcome of humble on-the-ground inquiry, and locates it in actual intelligent activities we carry out in life" (Chang, 2022, 5). Such an inquiry does not proceed from indubitable foundations, but begins in a state of uncertainty. As characterized by John Dewey, the inquirer begins from a kind of unstable relation to their environment. Chang emphasizes that scientific inquiry proceeds through a process of *epistemic iteration*: "[S]uccessive stages of knowledge are created, each building on the preceding one, in order to enhance the achievement of epistemic goals" (Chang, 2022, 208).

What Chang does *not* argue is that we have a reason, based on the evidence of the success of theoretical science, to believe that its theories are approximately true. He views attempts to argue philosophically that scientific theories describe the world "as it really is" as intellectual dead ends. A pragmatist orientation should instead orient us toward "a commitment to engage with, and learn from reality," and we should think of "(external) reality as whatever it is that is not subject to one's own will" (Chang, 2017, 33). We should engage with the world, and we can recognize the reality of the world in the resistance we sometimes encounter. Knowledge "is a state of our being in which we are able to engage in successful epistemic activities" (Chang, 2017, 33), which requires that such resistance not be so great as to prevent our success. But it is not enough, as a realist, to bask in the achievement of knowledge, which is always incomplete

[18] Chang also characterizes his realism as a *stance*, as has been articulated by van Fraassen (2002) and Kellert, Longino, and Waters (2006).

and always provisional. The activist realist seeks *progress* through the continuation of inquiry. In this way, Chang's philosophy extends the significance of what Charles Sanders Peirce declared in 1898 as a corollary to the "first rule of reason": "Do not block the way of inquiry."[19]

10.10 Conclusion

Scientists have developed remarkably sophisticated methods and technologies for learning about the world. It is perhaps shocking, therefore, that there is so little agreement about *what* we learn from their expert use of these methods and technologies. Imagine that you went to a restaurant run by a world-famous chef with a state-of-the-art kitchen, serving a spectacular menu of delicious selections, but none of the staff could agree whether any of the items on that menu would actually nourish you!

Considering the wide variety of 'realisms' we have encountered in this chapter, one might think that the situation is even worse: The restaurant workers hold a wide range of views about what counts as 'nourishing' and consider the arguments of others about this to constitute attempts to answer the entirely wrong question.

But perhaps we should not get too hung up on the different ways philosophers use words like 'realism,' 'antirealism,' or 'instrumentalism.' These differences in usage tend to track differences not only over the project of scientific inquiry but also over the goals of philosophy. When the differences are adequately explained they allow us to place the question of the significance of scientific thinking into the broader context of human thought in general. You might not take much comfort in seeing one large problem addressed by folding it into an even larger problem, but as the example of Socrates illustrates well, philosophy does not seek to provide comfort.

[19] Peirce articulated the first rule of reason itself as follows: "[I]n order to learn you must desire to learn, and in so desiring not be satisfied with what you already incline to think" (Peirce, 1931–1958, vol. 1, 135; see also Peirce, 1992; Haack, 2014; McLaughlin, 2009, 2011).

11 Explanation

11.1 Introduction

That we use scientific theories to *explain* things is a matter of broad (though incomplete) agreement.[1] But what is it to explain something? Intuitively, explanations enable us to *understand* the phenomena we observe, where understanding involves something more than merely knowing *that* something occurs. Intuitively, again, that *something more* seems to involve a knowing *why* or knowing *how* the phenomenon occurs. But none of these intuitive ideas helps very much with an analysis of what constitutes an explanation (or what constitutes a good explanation). For what constitutes understanding? And what does it take for an explanation to help us achieve it?

This is a very old question in philosophy. Aristotle insisted that in order to "grasp the 'why' of something" one should understand each of its four different kinds of *cause*: What it is made of, its form, its source of change or stability, and its end or purpose.

Aristotle's four causes are really four kinds of explanation, and they reflect his own approach to explaining natural phenomena. The discussion of explanation within contemporary philosophy of science likewise draws motivation from the explanatory practices of contemporary science; it really begins with Carl Hempel and Paul Oppenheim's Deductive-Nomological (D-N) model of explanation, which constitutes the core of the Covering Law model. We will therefore begin with the Covering Law account and objections to it before turning to several prominent alternative accounts.

[1] Pierre Duhem is among the dissenters. He rejected the view that the aim of a physical theory is to explain phenomena, holding that to explain "is to strip reality of the appearances covering it like a veil, in order to see the bare reality itself." This task belongs to metaphysics (Duhem, 1991, 7).

11.2 The Covering Law Model

We can start with a simple example:

> A mercury thermometer is rapidly immersed in hot water; there occurs a temporary drop of the mercury column, which is then followed by a swift rise. How is this phenomenon to be explained? The increase in temperature affects at first only the glass tube of the thermometer; it expands and thus provides a larger space for the mercury inside, whose surface therefore drops. As soon as by heat conduction the rise in temperature reaches the mercury, however, the latter expands, and as its coefficient of expansion is considerably larger than that of glass, a rise of the mercury level results. (Hempel & Oppenheim, 1948, 135–136)

In every explanation we can distinguish between the *explanandum*, a statement describing what is to be explained, and the *explanans*, the statements that tell us what explains the explanandum. In this example the explanandum has two parts: First, the height of the mercury column in the thermometer is briefly lowered; second, the height of the mercury column subsequently increases.

The explanation quoted above leaves out all mathematical details. It does not tell us how much the column of mercury falls or rises, or how long it takes to do these things. The laws that describe the processes of heat conduction and thermal expansion are invoked implicitly but not stated in their full quantitative form. But Hempel and Oppenheim, nonetheless, think that implicit references to these more detailed and quantitative expressions point us to the requirements of a 'sound' explanation, which they divide into *three logical requirements* and *one empirical requirement*:

11.2.1 Logical Requirements

*DNL*1 The explanans must deductively entail the explanandum. A D-N explanation is a deductive *argument* in which the explanans constitutes the premises and the explanandum constitutes the conclusion.

*DNL*2 The sentences that make up the explanans must include some sentences expressing general *laws*. (Typically, though not required, they will also include some sentences not expressing laws but describing so-called *initial conditions* that in conjunction with the laws make possible the derivation of the explanandum.)

*DNL*3 The explanans must have *empirical content,* meaning that "it must be capable, at least in principle, of test by experiment or observation" (Hempel & Oppenheim, 1948, 137).

11.2.2 Empirical Requirement

*DNE*1 The sentences constituting the explanans must be true.

Hempel and Oppenheim summarize these requirements through a general schema for D-N explanations:

$C_1, C_2, \ldots, C_n$ (statements of antecedent conditions)
$L_1, L_2, \ldots, L_m$ (general laws)

E (description of the phenomenon to be explained)

Here the explanans are given above the line, the explanandum below, and the former must deductively entail the latter.

According to the D-N model, an explanation is a certain kind of deductive *argument*. Hempel also argued that every explanation could function equally well as a *prediction* for someone who knows the explanans prior to the explanandum. Explanation and prediction share the same logical structure, and they do so because that logical structure enables explanations to satisfy the criterion of *nomic expectability*: An explanation must show that the phenomenon described by the explanandum *was to be expected* in virtue of the laws applicable to the situation in which that phenomenon occurred.

Hempel developed another model for explanations drawing upon laws that are statistical in nature, which he called the Inductive-Statistical (I-S) model. Here it will suffice to note that it resembles the D-N model, but with the universal laws of D-N replaced with *statistical laws*, from which (along with the statement of antecedent conditions) one can derive not that the explanandum must be true but that the explanandum has a very high probability, given the explanans. In this way, the I-S model also meets the nomic expectability requirement, though less perfectly. I will use the term Covering Law (CL) model to refer to the combination of D-N and I-S accounts.

As we noted, our thermometer example does not exactly match this schema as stated. It constitutes instead an *explanation sketch*, meaning that it does not state explicitly everything that an explanation must include, but gives enough of an outline that we can see what would need to be filled in to make a complete explanation.

The idea of an explanation sketch helps the CL model to accommodate our frequent use of explanations that do not explicitly include all of the elements the model requires. Especially in the social sciences, in history, and in our attempts to understand the purposive actions of human beings, we do not often find statements of general laws from which the description of the phenomena we seek to understand can be deduced, even with the help of additional background information or initial conditions. But Hempel and Oppenheim, nonetheless, contend that such explanations are really explanation sketches that, although incomplete, make "reference" to both specific antecedent conditions and general laws (Hempel & Oppenheim, 1948, 141).

11.2.3 An Example: The Hodgkin and Huxley Model of the Action Potential

To further explore the motivations and the limitations of the CL model, let us consider a somewhat extended example from the history of neuroscience.

Most cells in an organism can be thought of as batteries. That is, there is a voltage, or *electrical potential difference* (V_m) between the inside surface and the outside surface of a cell membrane, due to differences in the concentrations of ions (electrically charged atoms or molecules) on either side. In certain kinds of cells, that voltage sometimes changes significantly over very brief intervals following a characteristic pattern. Such changes are called *action potentials*, and they are due to changes in the permeability of the membrane that allow ions to diffuse across it. The movement of ions through the membrane constitutes an electrical current. Although action potentials can be found in a variety of cells, our present example concerns their role in neurons.

At the beginning of an action potential, the neuron is in a resting state in which the potential difference between the inside surface and the outside surface of the membrane is –60 to –70 mV (about 1/20th of the voltage of an AA battery). This is the *resting potential*, V_{rest}. Should V_m rise above a certain *threshold* value (~ –55 mV), it will then rapidly rise to a peak of +35 to +40 mV and then rapidly decline to a value a little below the resting potential, thus initiating the *refractory period* during which the cell is less excitable. Gradually V_m will then rise back to the level of V_{rest} (Figure 11.1).

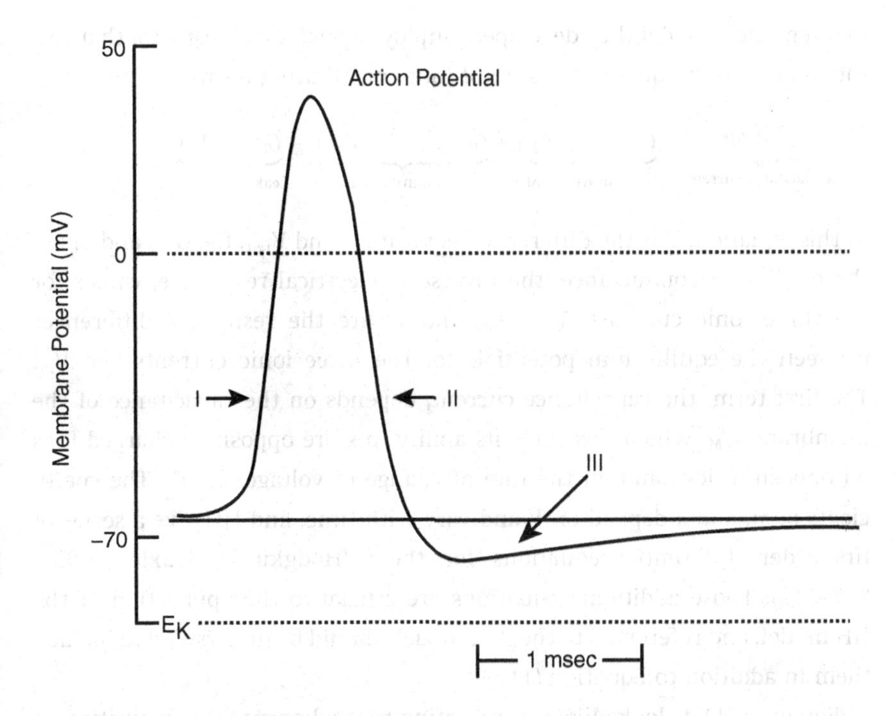

Figure 11.1 A schematic drawing of an action potential indicating the main features of changing voltage across the cell membrane: (1) a rapid rise of V_m, followed by (2) a steep decline to values below that of the resting potential and (3) an extended refractory period (Craver, 2007, 50). Reproduced with permission of Oxford University Press through PLSclear.

In neurons, electrical current produced in an action potential is chiefly due to the flow of sodium (Na^+) and potassium (K^+) ions across the membrane. Other ions contribute smaller amounts to the total current, and the total of these smaller contributions is called the *leakage* current. For each of the ions Na^+ and K^+, as well as for the leakage current, there is a distinct value of the potential difference at which that ion is in equilibrium, that is, there is no net current in either direction due to the movement of that ion.

Alan Hodgkin and Andrew Huxley (henceforth HH) developed a mathematical model of the action potential in neurons based on a series of experiments on an unusually large nerve fiber, or *axon*, found in squids. HH published the results of their experiments in a series of papers in 1952 in the *Journal of Physiology*, summarized in Hodgkin and Huxley (1952). The mathematical model they developed employs a *total current equation* that sets the total current equal to the sum of four contributing terms:[2]

$$I = \underbrace{C_M dVdt}_{\text{Capacitance current}} + \underbrace{G_K n^4 (V - V_K)}_{\text{Potassium current}} + \underbrace{G_{Na} m^3 h (V - V_{Na})}_{\text{Sodium current}} + \underbrace{G_l (V - V_l)}_{\text{Leakage current}}. \quad (11.1)$$

In this equation, V is the difference between V_m and V_{rest}. G_K, G_{Na}, and G_l are the maximum conductance (the inverse of electrical resistance) values for the three ionic currents. V_K, V_{Na}, and V_l are the respective differences between the equilibrium potentials for the three ionic currents and V_m. The first term, the capacitance current, depends on the capacitance of the membrane, C_M, which measures its ability to store oppositely charged ions on opposing sides and on the rate of change of voltage, dV/dt. The coefficients n, m, and h depend on V and vary with time, and HH give a series of first-order differential equations for them (Hodgkin & Huxley, 1952, 518–519). Those additional equations are crucial to the application of the HH model, and references to the 'HH model' should be understood to include them in addition to Equation (11.1).

Equation (11.1) looks like a law, stating that whenever the quantities on the right-hand side take on certain values, the quantity on the left-hand side – the total current across the membrane – will take on a value determined by the equation itself. Moreover, HH appear to use it to explain certain phenomena regarding action potentials. They summarize their paper by stating that Equation (11.1), along with associated differential equations for n, m, and h, "were used to predict the quantitative behavior of a model nerve under a variety of conditions which corresponded to those in actual experiments." They then enumerate eight properties of the action potential for which "good agreement was obtained," such as the "form, amplitude and threshold of an action potential under zero membrane current at two temperatures," the "form, amplitude, and velocity of a propagated action potential," and the "total inward movement of sodium ions and the total outward movement of potassium ions associated with an impulse" (Hodgkin & Huxley, 1952, 543–544).

According to Hempel, only the pragmatic distinction between deriving a statement prior to knowing it to be true and after knowing it distinguishes

[2] I follow Craver (2007, 51) in writing the equation as Hodgkin and Huxley wrote it, rather than as it will be found in contemporary textbooks.

prediction from explanation. So, if we view HH's summary through the lenses of the CL model, their claim that they used Equation (11.1) and its associated equations to *predict* these aspects of the action potential entails that we can likewise use those equations to *explain* those phenomena. Craver writes, "There is perhaps no better example of covering-law explanation in all of biology" than the HH model (Craver, 2008, 1022). Later in this chapter, we will consider reasons why we might, as Craver himself advocates, choose to revise this judgment.

11.2.4 Objections to the CL Model

Objections to the CL model have circulated long enough to form a set of criteria used to judge other models of explanation. Here we present four.

Explanations without Laws

The first objection holds that the requirements of the CL model are not *necessary* for explanation. We noted earlier that many examples of explanations even within science do not explicitly state laws and initial conditions that suffice for the deduction of the explanandum. The CL model holds that we should expect such efforts to *implicitly* refer to a more detailed explanation that does include such statements, in the way a sketch of a figure shows us where a more detailed rendering would provide more information about the appearance of the figure sketched. Can we explain something without either an explicit or implicit reference to a general law?

Michael Scriven gives the example of a person who, while reaching for a cigarette, knocks over an ink bottle, causing the ink to spill onto the carpet. The person can then give the following explanation for how the stain appeared on the carpet: She knocked the ink bottle over. "This is the explanation of the state of affairs in question, and there is no nonsense about it being in doubt because you cannot quote the laws that are involved, Newton's and all the others; in fact, it appears one cannot here quote any unambiguous true general statements, such as would meet the requirements of the deductive model" (Scriven, 1962, 198).

The ink bottle example reveals that we not only give and accept explanations that do not explicitly cite laws; we do so when we could not state any 'unambiguous true general statements' (let alone *laws*) that would enable us to deduce a description of the phenomenon to be explained (thus ruling out that we are somehow *implicitly* appealing to such a law). For example, the generalization 'Whenever an ink bottle is bumped, an ink stain appears on the carpet' might suffice for the required derivation, but is clearly not true. We might try to make the statement more specific: 'Whenever an open ink bottle is bumped hard and falls over, and there is ink in the bottle, and the bottle sits on a table, below which is a carpet, then an ink stain appears on the carpet.' But this is still not true, since one might intervene to set the ink bottle alright before enough ink spills to run off the table, or one might quickly cover the carpet with a towel, or the dog might lie down in the path of the spilling ink, or ... It is not obvious that there is any nontrivial way to build all of this into the needed generalization, and the more we build these conditions into the statement of the generalization, the less the resulting statement looks like a statement of *general law.* (Indeed, just what does constitute a law is a matter of ongoing debate, to which we could easily devote an entire chapter.[3])

Nomic Expectability

Another important criticism of the CL model strikes directly at one of its chief motivations: nomic expectability. Some seemingly satisfactory explanations do not show that the phenomenon to be explained was to be expected in light of the applicable laws, but only cite circumstances that are *causally relevant* to its occurrence. Consider another example from Michael Scriven (1959): Some people with syphilis develop a form of paralytic dementia known as paresis. Most people who contract syphilis do not develop paresis, but in those who do, having contracted syphilis explains why they developed paresis. At any rate, this is the best explanation we can give. Since having contracted syphilis explains – but does not render predictable – that a person develops paresis, nomic expectability is not a necessary condition for an explanation.

Advocates of a CL approach might of course object that there must be some difference between syphilis patients who develop paresis and those

[3] See Armstrong (1983), Cartwright (1983), and Lewis (1986) for some influential discussions. See Lange (2009) and Roberts (2008) for more recent proposals that have garnered attention.

who do not, and that a full explanation of an individual case of paresis would cite those additional distinguishing factors. We simply do not yet know enough to give that explanation.

The problem with this response, however, is that the CL model is committed not just to saying that it would be a *better explanation* to show that we should expect paresis in that individual but that citing the patient's prior condition of having syphilis does not explain his paresis *at all*. And in many cases – indeed in many fields of science – it simply is not obvious that we should expect to be able someday to give explanations that meet the criterion of nomic expectability; at any rate, whether we can *now* explain something ought to be independent of whether we will be able to explain it better in the future. Scriven gives his example in the course of arguing that Darwinian evolutionary explanations of survival by appealing to fitness exemplify the dispensability of the nomic expectability requirement.

Asymmetry

Many scientific laws express some kind of *equivalence* between quantities used to characterize a physical system. Such laws can be applied equally in either direction. For example, players of wind instruments create sound by vibrating, either with their lips (brass instruments) or with a reed (most woodwinds), the air in a tube. At resonant frequencies, this creates a standing wave in the horn. We can use a simple expression to approximate the frequency f of standing waves in a resonant cylindrical metal pipe that is closed at one end and has length L: $f = (nv)/(4L)$, where n is an odd number and v is the speed of sound in air. A D-N explanation of the frequency of sound waves coming from a given resonating horn would cite this expression as well as a description of the length of the horn.

The *asymmetry* problem for the CL account arises because we can run such derivations in reverse (Bromberger, 1966): A derivation of the length of the horn from sound waves it produces when resonating satisfies the requirements of the CL model of explanation just as well. But it seems that explanatory relationships do not share this symmetry. We appeal to the length of the pipe to answer the question 'why does it produce these particular sound waves when resonating?' (Call this the 'length-to-notes' explanation.) But we do not in the same way appeal to the sound waves produced to answer 'why does it have the length that it does?' (the 'notes-to-length' explanation).

I say 'in the same way' because we might well give a certain *kind* of notes-to-length explanation for the case of horns or other musical instruments using tubes. In those explanations, we note that the manufacturer gave the tube the length that it has so that it would produce the particular kind of sound waves that it does. But here the explanation invokes both the properties of sound waves and the intention of the manufacturer to produce an object with certain tonal properties, whereas no such considerations need enter into the length-to-notes explanation. The explanatory relationships are not symmetrical in the same way as the derivational relationships. Moreover the length-to-notes explanation applies to all tubes, but the notes-to-length explanation does not apply to natural or accidental resonant cavities (like conch shells or moonshine jugs).

Irrelevance

The final problem, like the asymmetry problem, challenges the *sufficiency* of the CL conditions for explanation. For any law-like generalization we might use in an explanation, such as 'every sample of sodium chloride (salt) crystals dissolves in water,' we can add an irrelevant condition to the antecedent, such as being hexed by a magician. This yields a new, seemingly law-like generalization: 'Every hexed sample of sodium chloride crystals dissolves in water.' We can use this new generalization, just like the original one, to derive an explanandum statement describing the dissolution of a particular sample of hexed sodium chloride in water, but the derivation that appeals to the law about hexed salt does not really seem to explain why the salt dissolved (Kyburg, 1965).

The problem of irrelevance also arises in cases allowing the derivation of explananda by appeal to generalizations that are 'accidentally' true. Suppose every member of a six-person band happens to read ancient Greek. This does not explain why a particular member of that band (Scott) reads ancient Greek, although we can derive the statement 'Scott reads ancient Greek' from this generalization along with the statement that Scott is a member of that band.

Of course, if we had settled criteria for deciding whether a given generalization is a law, we might be able to use those criteria to exclude these cases on the grounds that 'every hexed sample of sodium chloride dissolves in water' and 'every band member reads ancient Greek' are not really laws. But Hempel himself acknowledged that the determination of lawfulness remained something of an unsettled issue for his account.

11.2.5 The Hodgkin–Huxley Model as CL Explanation?

In discussing the Hodgkin–Huxley model of the action potential, we mentioned that it seems like a good candidate for a covering law explanation of the features of the action potential that HH discuss. After all, the equation appears to describe quantitatively the way total current across the membrane depends on the quantities on the right-hand side of Equation (11.1) and, by extension, on the quantities in the additional equations for the conductance factors in Equation (11.1). Moreover, HH state that they were able to use these equations to *predict* those features of the action potential in a way that yielded 'good agreement' with the results of their experiments.

In a careful study of this episode, however, Carl Craver points out that, although their mathematical model can do *some* explanatory work, HH themselves denied that every part of that model that could serve a *predictive* function was thereby explanatory: "The agreement [between model and data] must not be taken as evidence that our equations are anything more than an *empirical description of the time course* of the changes in permeability to sodium and potassium. *An equally satisfactory description of the voltage clamp data could no doubt have been achieved with equations of very different form, which would probably have been equally successful in predicting the electrical behaviour of the membrane.*" Consequently, "*the success of the equations is no evidence in favour of the mechanism of permeability change that we tentatively had in mind when formulating them*" (Hodgkin & Huxley, 1952, 541, quoted with emphasis added in Craver, 2007, 53–54).

More specifically, the HH model does, on Craver's account, explain the current involved in the action potential in terms of voltage and conductance changes. But in Equation (11.1), conductances for potassium and sodium ions are given by the products of maximum conductances and some function of time-dependent quantities n, m, and h. The additional partial differential equations that HH gave for the evolution of these latter quantities over time, according to them, give only an "empirical description of the time-course" of these sodium and potassium conductances. They predict the outcomes of measurements but do not give a *causal* story about how those quantities change.

Craver's point is that, although HH's conductance equations seem to meet the requirements of the CL model, they do not explain the changes in conductance. Of course, a defender of the CL model might simply insist that if those equations do not explain conductance changes, then they must not express laws.[4] Drawing upon the statements of HH and later neuroscientists who continued to study mechanisms of ion transfer between neurons, Craver imposes a causal requirement on explanations and concludes that what prevents the HH model from explaining the phenomena regarding action potentials is not that it does not express a law but that, at least for conductance changes, it does not show how the phenomena are "situated within the causal structure of the world" (Craver, 2007, 49).

Perhaps, then, we should consider another example in which we can see a phenomenon situated in the causal structure that brings it about.

11.2.6 Another Example: Curare Poisoning

Early nineteenth-century British explorers in South America reported how natives hunted using arrows dipped in a poisonous plant residue called wouralia.[5] The natives they encountered in Guiana and Brazil knew that this deadly substance would not only quickly kill monkeys they hunted but would bring certain death to an unfortunate hunter struck by an arrow gone astray.

There was no question that wouralia, which came to be known as *curare*, caused death, but how it did so posed a mystery. As lethal as curare was on the tip of an arrow, natives suffered no ill effects from eating curare-poisoned monkeys. Research published in the 1850s by physiologist Claude Bernard revealed that curare produces death only when it enters the bloodstream. Bernard determined that the structure of curare prevents it from being absorbed efficiently into the bloodstream through the stomach. Moreover, he established that the hearts of frogs poisoned with curare continued to beat for some time even after they stopped breathing. Further

[4] Marcel Weber argues that the explanations that make use of the HH model derive their explanatory force not from the HH model itself but from physical and chemical laws (describing the forces applied to charged particles, etc.) that underlie the HH model (Weber, 2005).

[5] See Craver and Darden (2013, ch. 1) for a very readable discussion of this episode. My own discussion follows theirs closely.

research revealed that curare stopped one from breathing by somehow breaking the *connection* between the motor neurons and muscles responsible for breathing. (Curare left unimpaired the functioning of both the motor neuron itself and the muscle.)

Bernard was able to take us beyond knowing that curare causes death to knowing that it does so by interfering with the connection between motor neurons and muscles. But how does it do that? Modern neuroscientists have worked out the answer. Roughly the explanation goes like this: Ordinarily, electrical stimulation of motor neurons triggers the release of the neurotransmitter acetylcholine (ACh). When ACh binds to ACh receptors in the muscle membrane, they open a pore through which charged ions can then pass, and it is this electrical signal – an ion current – that triggers muscular contraction. The active ingredient in curare (tubocurarine chloride) mimics the shape of ACh, binding to ACh receptors without opening pores through which an ion current can pass. Motor neurons can no longer activate muscle contraction because the curare blocks their activation signals from being received by the muscle.

Craver and Darden summarize the story nicely:

> How does curare kill its victims? It enters the bloodstream and makes its way to the neuromuscular junction. There, it blocks chemical transmission from motor neurons, effectively paralyzing the victim. When the diaphragm is paralyzed, the victim cannot breathe, and animals that cannot breathe do not get the oxygen required to maintain basic biological functions. Thus is hunter's lore about the irrevocable effects of curare transformed into scientific knowledge of mechanisms. (Craver & Darden, 2013, 2)

Later in this chapter we will consider perspectives on explanation that emphasize *causation* and the *mechanisms* that produce phenomena.

11.3 Achinstein's Pragmatic Theory

Thus far, we seem to have carelessly failed to distinguish between the *act* of explaining, the *explanation that is given* in that act, and the *evaluation* of that explanation. Peter Achinstein insists that one must begin an account of explanation with an analysis of the act of explaining as fundamental, and that "the concept of an explanation (as product), and that of a good explanation, must be understood, in important ways, by reference to" the concept of an explaining act (Achinstein, 1983, 4).

11.3.1 Illocutionary Acts

To explain, Achinstein tells us, is to perform an *illocutionary act.* Such acts are "typically performed by uttering words in certain contexts with appropriate intentions" (Achinstein, 1983, 16).[6] Other examples include promising, thanking, requesting, warning, and betting.

Consider a sentence of the form 'S explains *q* by uttering *u*,' where *S* is a person, *q* expresses an indirect question (such as 'why the mercury rose in the thermometer'), and *u* is a sentence. Achinstein proposes three conditions, individually necessary and jointly sufficient, on the truth of a sentence of this form:

1. *S* utters *u* with the intention that his utterance of *u* render *q* understandable.
2. *S* believes that *u* expresses a proposition that is a correct answer to *Q*, where *Q* is the direct question whose indirect form is *q* (e.g., *Q* might be 'why did the mercury rise in the thermometer?').
3. *S* utters *u* with the intention that his utterance of *u* render *q* understandable by producing the knowledge – of the proposition expressed by *u* – that it is a correct answer to *Q*.

So, for example, consider the indirect question *c*: 'Why there is an ink stain on the carpet,' the direct form of which is *C*: 'Why is there an ink stain on the carpet?' Uttering the sentence *IB*, 'There is an ink stain on the carpet because I knocked over the ink bottle while reaching for a cigarette,' might constitute an act whereby I explain why there is an ink stain on the carpet, provided these three conditions are met. I must utter *IB* with the intention that my doing so will make understandable why there is an ink stain on the carpet. If I address my utterance to the dog simply to see him wag his tail, which I know he does regardless of what I say, then I am not explaining *c*; I am teasing the dog. Achinstein's analysis also requires that I must believe that *IB* expresses a correct answer to *C*. If I believe that the ink stain was actually produced by the dog knocking over the ink bottle but I am protecting the dog from getting a scolding, then I am not explaining why there is an

[6] The term comes from the philosopher J. L. Austin (1962). For more on illocutionary acts, illocutionary force, and *speech acts* generally, see Searle (1969) and Searle and Vanderken (1985).

ink stain on the carpet; I am covering up why there is an ink stain on the carpet. Finally my intention in uttering *IB* must be not merely to render *c* understandable but to accomplish this by producing the knowledge that the proposition expressed by *IB* is a correct answer to *C*. This condition will not be met if, say, I know that although *IB* expresses a correct answer to *C*, you tend not to believe anything I say, but I also know that you compulsively look up scientific information on any topic that comes up in our conversation, and I intend, by uttering *IB*, to prompt you to go online to study the chemical processes whereby fabric becomes stained by ink.

These conditions thus allow Achinstein to avoid the *illocutionary force problem*. The utterance of any given sentence or group of sentences can be a product of quite distinct illocutionary acts, depending on the context and intentions of the speaker. 'I will pay for everyone's dinner' might be either a promise or a prediction. Achinstein regards this as a problem for any theory of explanation that specifies an ontology of explanations as certain kinds of sentences or arguments independently of the *act* of explaining.

We noted earlier in this chapter that the notion of understanding seems closely related to that of explanation. One might worry, then, that Achinstein's first requirement – that *S* must intend to make *q* understandable by uttering *u* – might make his analysis either circular (if our account of understanding invokes explanation) or unclear (if we leave understanding undefined). He avoids this charge by offering an analysis of understanding that does not in turn depend on the concept of explaining or explanation. I will forego detailing that analysis here, other than to note that it endorses the intuition that understanding consists in more than simply knowing that something happened or is happening. On Achinstein's account, understanding *q* is a certain *kind* of knowledge state with regard to *q*.[7]

[7] The philosophical significance of understanding and its relationship to explanation has become a topic of renewed interest among philosophers of science and epistemologists. Although Hempel regarded understanding as a subjective matter of psychological satisfaction unconnected to the logical nature of explanation (Hempel, 1966), recent work has attempted to separate the normative significance of understanding from its psychological components. The idea that the aim of explaining is to produce or contribute to understanding is taken up in this discussion (Humphreys, 2000; Potochnik, 2017, 127–134). On this topic see also the important contributions of Stephen Grimm (2014, 2019, 2021), Henk de Regt (2017), Catherin Elgin (2017), and Kareem Khalifa (2017).

11.3.2 The Ordered Pair View

Calling Achinstein's account a *pragmatic* theory of explanation emphasizes the fact that explanation for him must be understood primarily as something that people (including scientists) *do*, with certain intentions and in certain contexts. For reasons such as the illocutionary force problem, Achinstein argues that one cannot analyze what it is to be an explanation except by reference to an explaining act.

Achinstein gives an *ordered pair* view of explanations in which candidate explanations consist of a pair (*x*; *y*), where *x* is a certain kind of proposition and *y* is an action. Achinstein's conditions on (*x*; *y*) being an explanation of *q* (an indirect question) given by *S* (a person) include (1) that *y* must be the act of explaining *q*, (2) that *Q* (the direct question of which *q* is the indirect form) must be a *content question*, and (3) that *x* is a *complete content-giving proposition* with respect to *Q*. Rather than attempting to encapsulate the technical definitions of terms used in conditions (2) and (3), let us hope that some examples will suffice here. (See Achinstein, 1983, 23–53 for a more detailed account.)

'Cause,' 'effect,' 'excuse,' 'explanation,' 'function,' 'meaning,' 'purpose,' and 'reason' are all examples of *content nouns*. Abstract nouns earn this title in virtue of the role they play in *content-giving sentences* that express what their content is. Here are some examples of such sentences:

- The *cause* of death is that *the victim received a lethal dose of curare.*
- The *explanation* of the stain on the carpet is that *I was reaching for a cigarette and knocked over the ink bottle.*
- The *function* of the action potential is that *it makes communication between cells possible.*
- The *reason* you should turn off your phone in a concert hall is that *doing so will prevent people from considering you obnoxious.*

In each of these sentences a 'that . . .' expression *gives the content* of its content noun. By contrast, the following sentences tell us something about their respective content nouns, but do not give the content of those nouns:

- The cause of the victim's death was discovered by performing an autopsy.
- The explanation of the stain on the carpet is not very scientific.
- The function of the action potential was understood before the mechanism behind it was.

- The reason you should turn off your phone in a concert hall is apparently something you have failed to appreciate.

The sentences in the first list – but not those in the second – can be thought of as (possibly correct) answers to their respective *content questions*: What is the cause of the victim's death? What is the explanation of the stain on the carpet? And so on.

Returning to Achinstein's conditions, then, an explanation of indirect question *q* by person *S* is an ordered pair of a proposition *p* and an act of explaining. Requiring that the direct form *Q* of *q* must be a content question means that it must be a question asking for the content of a content noun. Requiring that the proposition *p* be a complete content-giving proposition with respect to *Q* means that *S* must use a sentence in the act of explaining that expresses a proposition giving the content of that noun ('completeness' involves additional technical requirements; Achinstein, 1983, 40).

Let us pause for a moment to compare this account with the CL model. The most obvious point of difference is that Achinstein's account does not require, but the CL model does, that an explanation must include a law. Achinstein's model is simply less demanding than the CL account in this respect. Of course, one would need to cite a law in order to give an explanation as an answer to a content question that asks for one, such as 'What are the laws from which it follows that voltage *V* in the final stage of an action potential will be lower than the resting potential?' Achinstein's account can thus handle cases like Scriven's ink stain.

In another respect, Achinstein's illocutionary account is *more* demanding. On the CL account, if an argument (a set of premises plus a conclusion) meets requirements *DNL*1–3 and *DNE*1, then it is an explanation, regardless of the context and intentions of anyone who might present it. But this is clearly not sufficient on Achinstein's account, since someone might utter such a set of sentences not as an act of explanation but as part of some other illocutionary act, such as reminding or lamenting.

However, one might object that the CL model is a model not of the act of explaining but of *scientific explanation*. Conceding Achinstein's point that what counts as an act of explaining depends on the intentions and context of the action, one might insist that the CL model tells us what must be included in any utterance offered as an explanation of some explanandum *q* for that utterance to count as an adequately scientific explanation of *q*. This move Achinstein regards as a change of question. Instead of asking what is an explanation, we must now consider how to evaluate an explanation.

11.3.3 Evaluating Explanations

Achinstein argues at length against the possibility of articulating universal rules for evaluating scientific explanations independently of the context in which the explanation is offered, aside from evaluating them as correct or incorrect. He gives the following analysis of the correctness requirement: "If (p; explaining q) is an explanation, then it is *correct* if and only if p is a correct answer to Q" (Achinstein, 1983, 106).

Beyond the requirement of correctness, what general instructions should explainers follow in order to give good scientific explanations? Achinstein articulates some conditions on the appropriateness of instructions that explainers might follow in the act of explaining q to a given audience, but these conditions crucially involve facts about the audience in question, such as what the audience knows, what would help the audience to understand q, the kind of understanding the audience is interested in having of q, what would be valuable for them to understand about q, and what it is reasonable for the explainer to believe about these facts about the audience. In application, then, one could not determine appropriateness without some substantive empirical knowledge about the context in which one offers an explanation. In this way, evaluating explanations requires attention to context.

11.4 Unificationist Theories

Many of the arguments against the CL model have targeted its treatment of explanations as a certain kind of argument. But Philip Kitcher claims that behind their 'official theory' of explanation, logical empiricists such as Hempel held an unofficial view about scientific explanation: Scientists seek to explain seemingly diverse phenomena by *unifying* them in a systematic way.

Moreover, according to Kitcher, one can reconcile this unification view of explanation with a rejection of the ontological thesis that explanations are arguments. Kitcher instead accepts Achinstein's ordered pair view of explanations (Kitcher, 1981, 509). Yet arguments remain important in Kitcher's account. Although an explanation is not simply a certain kind of argument, the act of giving a scientific explanation involves drawing upon arguments supplied by science. This leads Kitcher to a reformulation of the question about scientific explanation: "What features should a scientific argument have if it is to serve as the basis for an act of explanation?" (Kitcher, 1981, 510). Kitcher's broadly stated answer to this question is that it should be an instance of an argument pattern that scientists can use repeatedly to derive descriptions of a range of distinct phenomena, since "*[s]cience advances our understanding of nature by showing us how to derive descriptions of many phenomena, using the same patterns of derivation again and again, and in demonstrating this, it teaches us how to reduce the number of types of facts we have to accept as ultimate (or brute)*" (Kitcher, 1989, 432, emphasis in original).[8]

Citing the fact that you knocked over the ink bottle while reaching for a cigarette might well explain why there is a stain on the carpet, but stating that you did so does not show how the same kind of explanation could be given for stains from other kinds of liquids, or from other kinds of collisions.

Newton's theory of gravity and Darwin's theory of evolution by natural selection achieved tremendous unifications of phenomena. Showing that the same force – acting according to a single law – produces both the flow of ink downwards toward the carpet from the edge of the table and the motion of the Moon around the Earth contributed a new perspective that greatly increased our understanding of motions of both terrestrial and celestial objects. Newton thus completed the dissolution of Aristotle's boundary dividing the cosmos into sublunar and superlunar realms containing distinct physical elements following different principles of motion. As Kitcher points out, Newton in fact proposed – and his successors pursued – an even more ambitious program of unification seeking to understand all physical phenomena in terms of similar laws describing other attractive or repulsive forces (Kitcher, 1981, 512–514).

In a different way, Darwin's theory, too, unifies a wide variety of biological phenomena. Although he does not give detailed explanations of the characteristics of particular species in *On the Origin of Species*, he does provide a pattern of argumentation that can in principle be instantiated with details

[8] Kitcher echoes Michael Friedman's earlier account of explanation as unification (Friedman, 1974). But Kitcher seeks to repair technical problems that he pointed out in Friedman's account of unification (Kitcher, 1976).

for any given species. Instances of the pattern would cite facts about an ancestral population (its traits and their variability) and about the environment (determining what selective pressures were present) and would invoke laws of heredity and the principle of natural selection, thus showing how a particular trait would tend to emerge (Kitcher, 1981, 514–515; see Morrison, 2000 for a critical discussion of unification in both modern physics and Darwinian biology that challenges Kitcher's connection between unification and explanation).

11.4.1 Kitcher's Theory

Kitcher provides a detailed account of just what constitutes an argument pattern, as well as a theory of explanatory unification that in principle would allow one to determine which set of arguments achieves the greatest unification at a given time, earning them a place among those arguments that science provides for explanatory purposes, which Kitcher calls the *explanatory store*.

Roughly Kitcher thinks of argument patterns as involving schematic sentences (think of these as the skeletons of sentences, showing their logical structure but without any flesh on their bones displaying what they are about). Such a *schematic sentence* will use placeholder terms like *X* instead of substantive terms like *curare*. Not just any way of filling in these placeholder terms will yield an instance of a given argument pattern. *Filling instructions* indicate how to go about replacing placeholder terms with substantive ones. In a Darwinian argument pattern, a placeholder might be replaceable by terms denoting selective pressures like predation or drought, but not terms referring to divine intentions or vital forces. An argument pattern also needs to include a *classification* of sentences, telling which ones should serve as premises, which as conclusions, what inference rules one may use, and so on (Kitcher, 1981, 515–519).

Suppose that at a given point in a community's pursuit of inquiry they collectively accept a set of sentences *K*. Arguments allow one to derive some members of *K* from other members of *K*, and Kitcher refers to a set of arguments that does this for *K* as a *systematization of K*. The explanatory store relative to *K* is then the set of arguments that best systematizes or unifies *K*. Passing over the technical details, the gist of Kitcher's theory is that our assessment of how well a given set of arguments unifies *K* depends on the number of distinct argument patterns needed (fewer is better) to derive a number of accepted sentences (more is better). But these factors have to be evaluated in light of other relevant factors. Importantly, argument patterns must be *stringent*: The filling instructions must restrict the kinds of terms one may substitute for dummy terms in the pattern. Kitcher imposes this requirement to block cases of spurious unification by argument patterns, allowing one to derive large numbers of sentences using trivial argument patterns, such as 'God wishes it to be true that *p* (where *p* could be any proposition whatsoever). Whatever God wants to be true is true. Therefore *p* is true.'

Let's revisit the asymmetry problem to see how the unificationist account works. Recall our example: The law relating the dimensions of a cylinder to the frequencies of sound waves it produces in resonance allows one to derive the dimensions from frequencies no less than the frequencies from dimensions. Yet only the latter kind of derivation seems generally explanatory. This, according to Kitcher, is because deriving dimensions from frequencies employs an argument pattern that is much less unifying than a competing approach, which he calls 'origin and development derivations' (OD). These explain the dimensions of bodies by describing "the circumstances leading to the formation of the object in question" and "how it has since been modified" (Kitcher, 1981, 525).

The OD approach to explaining the dimensions of objects achieves greater unification than, say, deriving the dimensions of an object from its acoustic effects in resonance (let's use RF for such a 'resonant frequency derivation') because we can apply it so much more widely. Even if an object might resonate, it often does not and might never do so. Other objects have dimensions, but no known law relating resonating frequencies to dimensions would apply to them. In such cases, no appeal to an RF explanation is available, whereas an appeal to OD will be. Thus we may use OD to derive many more of our accepted sentences describing the dimensions of objects than we could by using RF.

11.4.2 An Objection to the Unificationist Account

Eric Barnes (1992) has objected to this treatment of the asymmetry problem by pointing to cases involving deterministic theories that describe the evolution of systems over time. *Deterministic* theories allow one – given a sufficiently precise description of the state of a system at a particular time as well as of forces acting on that system – to *predict* the future state of that system with a corresponding degree of accuracy, provided the system described is *closed.*[9] We may call a system closed if our description of forces acting on it includes all forces that will act on it over a relevant period of time that matter for the degree of accuracy we seek. Here is an example: We might apply Newton's celestial mechanics to the solar system to predict the positions and momenta of planets at some time t_1, based on a description of the positions and momenta of planets at an earlier time t_0 and the gravitational forces acting on them during period $[t_0, t_1]$, assuming no other forces will act on them during that time. We would thus use Newton's theory for a derivation using what Barnes calls a Predictive Pattern. Barnes's objection to unificationism notes that Newton's theory equally supports a Retrodictive Pattern allowing one to derive the state of planets at t_0 from the same information about forces and the description of their positions and momenta at t_1.

There is no reason to think that the Predictive Pattern is more unifying than the Retrodictive Pattern. Both patterns use the same kind of information as inputs and yield the same kind of information as outputs, with only the temporal direction reversed. Yet that shift in temporal direction is associated with a shift in explanatory direction. The positions and momenta of planets at a given time are *explained* by appeal to their positions and momenta at earlier times, not by their positions and momenta at later times. Since the two patterns unify equally well, the unificationist cannot account for this asymmetry.

11.5 Causal Explanation

This temporal asymmetry maps very nicely, however, onto an asymmetry in *causal* relations: Causes always precede their effects. Unificationists regard

[9] To state a satisfactory definition of determinism is a nontrivial philosophical problem (Earman, 1986). For example, my definition appeals vaguely to a degree of accuracy of predictions 'corresponding' to a particular degree of precision in describing the initial state. This *correspondence* will vary from one theoretical system to the next. In *chaotic* systems, for example, very small degrees of imprecision in that initial state description can result in very large divergences among the possible evolutions of the state in the future. Yet chaotic systems are deterministic. See Smith (2007) for a lucid and accessible introduction.

our practice of explaining effects in terms of causes as a byproduct of the unifying power of cause-to-effect derivations. An alternative view reverses the priority of these considerations, as when James Woodward states, "[T]he causal order ... is independent of and prior to our efforts to represent it in deductive schemes." Woodward cites Barnes's argument to support this claim (Woodward, 2003, 361). Although Woodward does not claim that all explanations or even all scientific explanations are causal, he does insist that providing such explanations is central to scientists' explanatory aspirations.

Clearly such a view demands an account of causal explanation. Woodward provides an account of his own, which we will consider after first discussing one of its leading competitors.

11.5.1 Salmon's Causal Mechanical Account

A common thread running through many accounts of causal explanation is that such explanations situate an explanandum in a network or nexus of causal relations.[10]

A particular version of this idea lies at the heart of Wesley Salmon's Causal Mechanical (CM) account of explanation (Salmon, 1984, 135–183). The central concepts in Salmon's theory are those of a causal process and of a causal interaction. *Causal processes* are physical processes capable of transmitting a mark in a continuous way, where a *mark* is a local modification of the structure of that physical process. Causal processes are extended in both space and time. A postcard traveling from St. Louis, Missouri, to Millersville, Pennsylvania, is a causal process. So is a baseball traveling from the pitcher's hand to the batter's bat, and then over the outfield wall to the windshield of a car in the parking lot. A shadow traveling along a wall is not a causal process. If I try to introduce the modification of shadow locally (i.e., doing something at the location of the shadow itself, rather than at the location of the object casting the shadow), such a modification will not be

[10] The regularity with which discussions of causation invoke the language of networks, webs, weaves, tangles, and the like is striking and indicates how naturally we think of causal relations as a kind of *connection* between things. David Hume, nonetheless, insists that regular association alone relates causes to effects. For Hume, when we speak of something as a cause, we simply refer to "an object, followed by another, and where all the objects similar to the first are followed by objects similar to the second" and "whose appearance always conveys the thought to that other" (Hume, [1748] 1993, 51).

transmitted as the shadow moves along the wall. The shadow exemplifies what Salmon calls a *pseudo-process.* A *causal interaction* is an intersection between two causal processes that modifies the structure of both, such as when I draw a pencil across the surface of a postcard, or when the baseball strikes the car windshield. A CM explanation of an event *E* then consists of a description of the causal processes and interactions that lead up to *E* (Salmon, 1984, 267–276; Woodward, 2003, 350–351). I explain the shattering of the car windshield by first describing the ball's departure from the pitcher's hand, noting how the interaction with the bat modified its trajectory, and then describing how it interacted with the windshield.[11]

The CM account seems well suited for many explanations in physics,[12] but, as Woodward has argued, many other kinds of causal explanations do not involve describing processes and interactions in the way the CM account requires (Woodward, 2003, 354–356).

Consider, for example, an explanation regarding the differences in unemployment levels between Spain and Portugal (Blanchard & Jimeno, 1995). Because of the significant and multifaceted similarities between the two countries, the fact that Spain had at the time the highest unemployment rate in the European Union while Portugal had the second lowest rate constituted, the authors say, "the biggest empirical challenge facing theories of structural unemployment." The authors tentatively endorse the following explanation: "In Spain, high unemployment protection and unemployment benefits have led to small effects of labor market conditions on wages. This led to large adverse effects of disinflation on unemployment in the first half of the 1980s. And high persistence since then explains why unemployment has remained high since." Lower unemployment benefits in Portugal have made wages more responsive to unemployment, resulting in "smaller adverse effects of disinflation on unemployment" and to less unemployment persistence (Blanchard & Jimeno, 1995, 216–217). Whether or not it is correct (see, e.g., Ball, Mankiw, & Nordhaus, 1999 for another perspective),

[11] Salmon later modified his view, replacing the CM model described here with a view of causal processes in terms of the transmission of conserved quantities (Salmon, 1994) that is a modified version of an account developed by Phil Dowe (1992, 1995).

[12] But see Hitchcock (1995) for a criticism that applies even to the use of CM in physics. Also, Robert Batterman (2001) explains how physicists often use *asymptotic explanation*, based on the analysis of what happens when some parameter in a theory approaches zero. Asymptotic explanations cannot be subsumed within the CM model.

this explanation employs a causal language, but without doing anything like what the CM model requires. The causes are policies regarding unemployment benefits. These policies make a difference to the decisions of individuals regarding the pursuit of employment, and the aggregation of many such individual decisions results in a particular pattern of economic activity. Although we might suppose that every interaction that contributes to these patterns involves physical human beings, the actions of whom can be traced down to the molecular level at which mark-transmitting causal processes and intersections between them can be identified, this explanation makes no reference to such processes. Moreover, we might not gain any additional insight into the phenomenon of disparate unemployment rates if we did take such processes into account.

11.5.2 Woodward's Manipulationist Account

If we want an account of explanation that applies not just to the processes described by physics but also to causal relations operating at a variety of scales (of size, organization, and complexity), then we will need a different approach. James Woodward's account of causal explanation emphasizes a feature of causal relations that applies across, he argues, all disciplines: When one quantity has a causal influence on another, we can in principle make a difference to the latter by *manipulating* the former (Woodward, 2003).

Woodward's account of causal explanation is linked to his account of the meaning of causal claims. Rather than provide a complete discussion, the following brief remarks I hope will illuminate why Woodward's account has generated significant discussion.

When we (rightly or wrongly) make a general causal claim such as 'smoking cigarettes causes heart disease,' 'taking birth control pills prevents pregnancy,' or 'listening to Mozart makes babies smarter,' what do we mean? For scientific purposes, we might sharpen the question: What meaning can we attribute to these statements that makes them amenable to empirical investigation through data collection?

One step toward answering this sharpened question would be to translate such claims into statements about the relationships between the values of *variables.* A reference to 'smoking' simpliciter conflates numerous dimensions of a complex behavior, different aspects of which can be represented by such different variables as number of cigarettes smoked per week, nicotine consumption per week, age at onset of smoking habit, duration of smoking habit, use or nonuse of filters, brand of cigarette, and so on. (We can represent 'all or nothing' states – such as being pregnant or not – using *dichotomous* variables that are capable of taking either of two values, such as 0 for the absence of a condition or 1 for its presence.) Making one's claim of causal relevance specific to a particular variable can help clarify the content of the causal claim more precisely (assuming the variable itself is well defined). In Woodward's account of the meaning of causal claims, the causal relation is a specific kind of dependence relation between variables.

When we give causal explanations, we appeal to a kind of dependence that in principle we can use to manipulate the explanandum variable. The qualifier 'in principle' needs a careful specification. We may not in fact be in a position to change the value of some causal variable. We regard the mass of Earth as relevant to the causal explanation of such features of our Moon's orbit as its period, but we cannot perform experiments in which we change the mass of Earth and observe changes in the period of the Moon's orbit. What makes such an explanation causal, nonetheless, according to Woodward, is that we provide information about the relationship between Earth mass and Moon period that allows us to answer *what-if-things-had-been-different* questions (Woodward, 2003, 11): If somehow we were to increase the mass of Earth by 10 percent, how would the period of the Moon's orbit change? What-if-things-had-been-different questions ask about *counterfactual* dependence relations.

Counterfactual conditionals are statements of the form 'If *p* were the case, then *q*' (Lewis, 1973). We often regard such claims as having clear meanings and readily determinable truth values. ('If you had called five minutes earlier, I would not have been able to answer the phone.') In other cases, it is not clear whether such a claim is true or false because it is not clear to just what kind of situation the antecedent refers. ('If the ski resorts of Colorado had the climate of the Florida beaches, you could ski in your bathing suit.' Is this false because having the climate of Florida would rule out the presence of snow? But the antecedent seems to require that under the proposed scenario there are still *ski resorts*, so perhaps we should suppose that the freezing temperature of water would be much higher? But that would seem to necessitate a completely altered human physiology – lest our blood freeze – among other things ...) The counterfactuals contemplated in what-if-things-had-been-different questions relating to causal claims need to be of the former kind if causal claims are to have clear meanings.

When we cite a relation of causal dependence to explain why some variable takes a certain value, the related counterfactuals describe how the value of that variable would be different were we to *intervene* on the causal variable so as to set it to any of a range of possible values. Suppose that we specify 'smoking causes heart disease' to mean that increasing the *number of cigarettes a person smokes per week* increases the *probability the person will at some point develop heart disease*. This claim entails that, were we to somehow increase cigarette consumption of a group of people, this would be an effective means to increase the rate of heart disease among them. This would be a terrible thing to do, and no one ought to carry out this manipulation. But this relationship between the manipulation and its terrible consequences is what the causal claim in question expresses.

Woodward's account thus draws a strong connection between causal explanation, experimentation, and control. As he states, "[O]ne ought to be able to associate with any successful explanation a hypothetical or counterfactual experiment that shows us that and how manipulation of the factors mentioned in the explanation . . . would be a way of manipulating or altering the phenomenon explained" (Woodward, 2003, 11). Such a manipulation takes place by way of *intervening* on the causal variable. The term 'intervention' has a technical meaning for Woodward. Through a detailed discussion he provides conditions intended to ensure that something counts as an intervention on variable *X* with respect to *Y* only if it changes *Y only* as a result of the causal influence of *X* on *Y* (Woodward, 2003, 94–104). This is necessary to avoid problems arising from cases where, for example, curare is administered by means of a curare-tipped bullet shot into a subject's heart (thus producing death, but not from curare).

Explanation then is "a matter of exhibiting systematic patterns of counterfactual dependence" (Woodward, 2003, 191). A causal explanation of the value of variable *Y* that cites the value of variable *X* in the explanans exploits a systematic relationship, expressible by a generalization *G*, between two variables *X* and *Y*. *G* must describe how changes to the value of *X* would make a difference to the value of *Y* (it must be a *change-relating* generalization). Woodward notes that this systematic relationship might take the form of a law, but need not. Rather than the CL model's requirement that explanations must cite laws, Woodward's account requires reference to a relationship that is *invariant* across some range of conditions. Specifically it must be invariant (i.e., hold true) across a range of possible interventions on *X* that, according to *G*, yield distinct values of *Y*.

Some change-relating generalizations are more invariant than others, though all must exceed some minimal threshold of invariance to count as explanatory. Suppose we determine the lethal dose of curare for a monkey of certain size. We can express the relationship between curare dosage and mortality with a change-relating generalization G_1 that could then be invoked to explain either the deaths of monkeys wounded by arrows delivering sufficiently strong doses of curare, or the survival of some monkeys struck with arrows not thus prepared. But the same intervention on curare dosage might not be sufficient to bring about death of an elephant. Presumably no dose of curare would suffice to kill a tree. Our generalization G_1 does exhibit some invariance. It would (we might suppose) apply equally well if the dose were administered with a syringe instead of an arrow, for example. A generalization that describes how curare interferes with the communication between motor neurons and muscles would have a greater degree of invariance. We would see this mechanism at work even in cases (like the elephant) where the dosage was insufficient to cause death. Woodward argues that the *depth* of a causal explanation corresponds to the degree of invariance of generalization that it cites. (See, however, Strevens, 2007 for a criticism of this claim, and Strevens, 2008 for an alternative account.)

Circularity?

According to Woodward, to say that variable *X* is a cause of variable *Y* is to say that intervening on *X* makes a difference to the value of *Y*. But to know whether the process that sets the value of *X* constitutes an intervention relative to *Y* requires sufficient causal knowledge to establish that the intervention changed the value of *Y* only through its effect on *X*, and not through other causal connections to *Y*. This raises the worry that Woodward's account of causation is circular: In explaining causation, Woodward uses the very notion he is trying to explain!

Woodward's reply emphasizes two points. First, the account of causation he is offering is intended not as a metaphysical theory that shows how causal relations can be reduced to more basic ontological categories but rather as a theory that clarifies the *meaning* of causal claims. Second, although determining whether a given means for fixing the value of *X* is an intervention with respect to *Y* does require some causal knowledge, it does not require the knowledge of causal relationship between *X* and *Y*. Thus, the circularity involved is not a vicious circularity. (Whether this response suffices is questioned in de Regt, 2004.)

11.5.3 Mechanistic Explanation

The advancement of our understanding from simply knowing that hunters kill monkeys with curare-poisoned arrows to understanding the mechanism by which curare induces death exemplifies a common pattern of progress in many areas of science, particularly life sciences. Let us finally consider an account of this kind of advance in our understanding that emphasizes the role of *mechanisms* in such explanations.

Previously we read a quotation from Carl Craver and Lindley Darden describing the mechanism whereby curare poisoning causes death. Craver and Darden, along with Peter Machamer (henceforth MDC), argue that a significant aspect of scientific practice, at least in neurobiology and molecular biology, consists of the search for mechanisms and the explanation of phenomena by appealing to mechanisms (Machamer, Darden, & Craver, 2000). (MDC leave as an open question how widely their account of mechanisms applies, although they suggest a broad range of disciplines exemplify the pattern.) Moreover, they hold, "To give a description of a mechanism for a phenomenon is to explain that phenomenon, i.e., to explain how it was produced" (MDC, 2000, 3).

MDC offer the following account of what such a description must include: "Mechanisms are entities and activities organized such that they are productive of regular changes from start or set-up to finish or termination conditions" (MDC, 2000, 3).

Note that an arrangement of entities, on the MDC account, does not suffice to compose a mechanism; one also needs *activities*. The mechanism whereby curare poisoning causes death includes things like tubocurarine chloride molecules and ACh receptors, but to achieve its final state the mechanism must include the *binding* of former to latter, which is how tubocurarine chloride molecules *block* the ACh neurotransmitter and *paralyze* the muscle, ultimately *asphyxiating* the organism. MDC write, "Activities are the producers of change. Entities are the things that engage in activities"

(MDC, 2000, 3). How the phenomenon is produced depends on how entities and activities are organized. A competing account of mechanisms by Stuart Glennan excludes activities from its ontology. On Glennan's account the mechanism for some behavior produces that behavior by "the interaction of a number of parts according to direct causal laws" (Glennan, 1996, 52).

MDC motivate the inclusion of activities in their ontology by appealing to the kinds of things scientists deploy in their explanations: "There are kinds of changing just as there are kinds of entities. These different kinds are recognized by science and are basic to the ways that things work" (MDC, 2000, 5). Moreover, describing how activities produce termination conditions is required for mechanistic explanations to elucidate or make intelligible the phenomenon described by the explanandum (MDC, 2000, 20–21). According to Craver, the reason that the HH equations for conductance changes in the action potential did not provide explanations of those changes is that HH did not have evidence that the terms in those equations correspond to the elements – both entities and activities – of the mechanism that produces those phenomena.

11.6 Conclusion

Scientists investigate systems, seeking to explain why they behave as they do. Many regard the understanding that such explanations yield as the principal benefit of the pursuit of scientific investigation. Thus, it is no surprise that philosophers of science have devoted so much attention to attempting to understand scientific explanation. That the CL account is not adequate seems to constitute the one point of consensus in this discussion. Above the grave of the CL model a profusion of alternatives have blossomed, however, and opinions remain divided over the merits of these contenders. That we should not be able to settle on a single account of what constitutes an acceptable scientific explanation is of course just what Achinstein's argument would seem to predict. If nothing else, however, the various accounts considered in this chapter have served to highlight aspects of scientific explanation – the unifying power of explanatory theories, the potential for manipulation that causal explanations illuminate, the elucidation gained from understanding the mechanism behind a phenomenon – that enable us to better appreciate the ways in which scientific knowledge enhances and deepens our experience of the world.

12 Values in Science and Science in Policymaking

12.1 Introduction

We have already discussed how logical empiricists sought to develop a *scientific philosophy* that would enable investigators to reason in a manner that would be independent of their commitments to ideologies or political causes.[1] Philosophers sometimes refer to this aspect of logical empiricists' project as the ideal of a *value-free* science. But what are values?

Humans spend a lot of effort *evaluating* things, whether they be sneakers, songs, or senators. We can evaluate how honest a person is, how tasty a doughnut is, or how probable it is that an asteroid will collide with the Earth. One way in which such evaluations differ concerns what we value and how we value it. We value both honesty and tastiness, but in different ways. Valuing the continued survival of ourselves and other species, we might prefer that an asteroid not hit the Earth, but it would be odd to say that we value the probability of this happening, although we may value knowing that probability. We might think of the category of things we value as those we take to be *good* and the features of good things that give us reasons to value them (i.e., to regard them as good) we might call values. We should

[1] This is not to say that the philosophers who pursued this project were not themselves motivated by political commitments of their own, for they were. Moreover, as the logical empiricist project developed, it became apparent that 'neutral frameworks' in which the degree of confirmation afforded particular hypotheses by particular bodies of evidence could not be determined by criteria that were in any absolute sense a priori. As we saw in Chapter 4, the most celebrated of logical empiricists working on theories of confirmation – Rudolf Carnap – acknowledged that only after a formal *language* had been chosen could one define an unambiguous measure of confirmation (Carnap, 1962), and that the choice of a language had to be made on the basis of either *pragmatic* criteria (Carnap, 1963, 980–983), or on the basis of what he later came to call *inductive intuitions* (Carnap, 1968).

perhaps reserve the label *values* for those more general and fundamental kinds of features that give us such reasons. Suppose a pop song has a particular hook that you think helps make it a good pop song. *Having that particular hook* might not merit inclusion on a list of musical values for pop songs, but *being catchy* – the general attribute the song achieves by having that particular hook – might well constitute such a value.

We can take things – including scientific theories, hypotheses, and methods – to be good in different *ways*. One distinction in particular is important here. We have already encountered the idea that scientists evaluate theories in terms of such characteristics as simplicity, accuracy, fruitfulness, and so on. Kuhn emphasized this point in his account of how theory choice could remain subject to incommensurability-driven underdetermination in spite of agreement among scientists about the importance of these values. It has become common, since Kuhn's essay was published, to distinguish *epistemic* from *nonepistemic* values (McMullin, 1982). The motivation for the distinction lies in a widely shared hunch: Certain kinds of values have a legitimate role to play in scientific thinking, while others do not. A scientist may, for example, quite legitimately prefer one theory over another because it is more accurate. The scientist may not, however, legitimately prefer one theory over another because it reinforces their political opinions. We will see that this hunch must be qualified and clarified if it is to have any chance of being upheld.

Values then play a role in guiding decisions. What roles should values play in decisions made in the course of scientific inquiry?

We may start with some useful distinctions between different kinds of value freedom that science might exhibit. Hugh Lacey distinguishes the following claims (Lacey, 1999):[2]

1 *Impartiality:* The only values that figure among the grounds for accepting or rejecting theories in science are epistemic[3] ones.
2. *Neutrality:* Accepting a scientific theory implies no commitments regarding nonepistemic values; neither does such acceptance undermine or support the holding of particular nonepistemic values.

[2] These approximate Lacey's formulations. See Lacey (1999, ch. 10) for his more detailed statements.

[3] Lacey uses the term 'cognitive' here. Douglas distinguishes cognitive values from epistemic criteria of adequacy (Douglas, 2009, 93–94).

3 *Autonomy:* Scientific inquiry is best carried out in a way that is free from 'outside interference' by social or political values or forces.

12.2 Decisions about Research Directions and Methods

Not much reflection is needed to see that nonepistemic values must play *some* role in science. For example, the question of what kinds of scientific investigations are worth conducting depends significantly on value judgments about ethics and about what is socially beneficial. Given the finite resources available for the pursuit of scientific investigation, decisions about what kinds of scientific projects deserve support depend on judgments about the value of the knowledge those projects might yield. To some extent, funding agencies providing support for scientific research depend on expert judgments of scientific peers to judge not just the likelihood a specific proposed project will succeed but also that the aims of the project are worth pursuing. At a more coarse-grained level of judgment, decisions about whether and how much to support broad categories of knowledge pursuit – whether it is the treatment of juvenile diabetes, or the study of conditions in the early universe – are made by funding agencies themselves. For governmental agencies, some choices may even be directed or constrained by an elected legislature, for example, by the differential allocation of funds to agencies pursuing varied research aims.

The use of public funds to pay for the costs of scientific research began in the nineteenth century.[4] Before then, investigators had either to spend their own money or convince other individuals to spend theirs. The increasing involvement of government in supporting science has coincided with the development of increasingly costly scientific undertakings, some of which – such as the Large Hadron Collider at the CERN laboratory in Geneva, Switzerland – are too expensive for any single nation to fund. Of course, funding for scientific research continues to come from private individuals or organizations, but public funding remains dominant, in spite of recent declines. According to the 2022 Science and Engineering Indicators report by the US National Science Board, research and development performed in

[4] Laura Snyder tells the history of this development in England, and the philosophical issues that surrounded it, in her engaging and illuminating book, *The Philosophical Breakfast Club* (Snyder, 2011).

the US in 2019 was valued at $656 billion. The percentage of funding for science and engineering research and development provided by the US federal government in 2019 was 21 percent, down from 31 percent in 2010. When research carried out by the business sector is set aside and one looks at research in higher education and nonprofits, the share of funding from government sources becomes considerably more significant (Burke et al., 2022).

The involvement of legislative bodies in determining the levels of support for scientific research entangles the scientific community with broader political negotiations. Although scientists have never operated in complete isolation from their political and social contexts, the era of government dominance in research funding brings with it increased public attention to and involvement in the directions pursued by researchers. The results may yield occasions for either celebration or condemnation depending on your own political convictions. The US Congress in 1996 removed from the budget of the Centers for Disease Control and Prevention (CDC) exactly the amount of money (US$2.6 million) that the CDC's National Center for Injury Prevention and Control spent on research into injuries caused by firearms, while adding to the CDC's appropriation a prohibition on the use of its funding "to advocate or promote gun control." This restriction was later extended to all agencies within the US Department of Health and Human Services, and the effect has been to greatly diminish the amount of research carried out on the patterns and causes of gun-related injuries and deaths and on the effectiveness of measures to reduce them. The National Rifle Association, an organization that lobbies against restrictions on firearms, criticized federally funded scientists who presented the results of their research at meetings where others advocated gun control measures, and eventually the research stopped (Davidson, 2013; Kellermann & Rivara, 2013). Resumption of funding for such research by the CDC only began in 2020, seven years after US president Barack Obama called for its resumption in the aftermath of the killing of schoolchildren at Sandy Hook Elementary School in Connecticut.

Finally, citizens have influenced the ways in which research is carried out based on their ethical values by acting to prohibit the use of research methods they consider unethical. One point of ongoing debate concerns the use of nonhuman primates as model organisms in biomedical research. Researchers who use them note that their genetic and physiological similarity to human beings makes nonhuman primates ideal for studies in which it would be impractical or unethical to use human subjects. Critics of this practice point to the same resemblances as a reason why we should regard such uses as morally problematic or even repugnant. A number of European countries, as well as New Zealand, have banned the use of great apes in biomedical research, and in the US, the director of the National Institutes of Health (NIH) announced in December 2011 that the NIH would suspend new grants for biomedical and behavioral research on chimpanzees (Gorman, 2011). Subsequently the NIH has undertaken to 'retire' most of the chimpanzees it holds for research purposes to a sanctuary where they would be ineligible for use as research subjects and to both reduce the number of studies using chimps and raise the standards of chimpanzees' living conditions in laboratories where they are still used (Gorman, 2013). A report commissioned by the Institute of Medicine that found that "most current use of chimpanzees for biomedical research is unnecessary" provided the basis for these changes (Committee on the Use of Chimpanzees in Biomedical and Behavioral Research, 2011, 4).

Clearly, in such cases, scientists and other interested people act to influence the direction of research as well as the methods used to pursue that research based on their ethical and social values. Scientists and trained ethicists may influence government agencies through participation in a committee that either a government body or an independent organization has called upon to prepare an advisory report. Citizens who are simply interested in an issue, whether pursuing scientific research as a profession or not, may also exert pressure on policymakers to restrict or encourage research on a particular topic and to change or restrict the methods researchers may use. It seems entirely reasonable in a democracy that scientific research be subject to such influences, casting considerable doubt upon the *autonomy* version of the value-free ideal.

12.3 The Argument from Inductive Risk

Regardless of one's stand regarding questions about the direction of science or the moral acceptability of certain research techniques, that nonepistemic values should play a role in these debates seems appropriate and does not raise any serious concerns about the objectivity or rationality of scientific inquiry or its results. Whatever considerations might have affected the selection of the research question, and whatever the source and nature of the constraints imposed on permissible methods in pursuit of an answer to that question, once the scientist is engaged in the collection and analysis of data, nonepistemic values have no further role to play. Determining an answer to the research question that is best supported by the data generated using the selected method requires no further consideration of what is permissible or desirable, morally or otherwise.

Or so it might seem.

Even when they apply them correctly, the methods scientists use to make inferences from data may lead them to draw conclusions *in error*. One argument for the relevance of nonepistemic values to the core tasks of scientific reasoning points to this feature and to the fact that not all errors a scientist might make are equally serious. Insofar as nonepistemic values figure into our differential attitudes toward such errors, they affect our choice of methods for drawing conclusions from data, and hence the conclusions themselves.

Somewhat more systematically, the *argument from inductive risk*, as it is sometimes called, may be reconstructed as follows:

1. If scientists accept or reject hypotheses on the basis of data, then they must choose a method that uses the data to determine whether to accept or reject a particular hypothesis.
2. If a scientist chooses a method that uses data to determine whether to accept or reject a particular hypothesis, then they must choose among methods with varying probabilities of producing different kinds of possible errors (accepting the hypothesis although it is false, or rejecting it when true).
3. If a scientist must choose among methods with varying probabilities of producing different kinds of possible errors, then they may consider preferences regarding those possible errors, including what kinds of consequences they may yield, when making that choice.
4. Preferences regarding the consequences of errors are appropriately influenced by nonepistemic as well as epistemic values.
5. If a scientist may consider their preferences among the consequences of different possible errors when choosing a method for deciding whether to accept or reject a hypothesis, then they may legitimately consider nonepistemic values in that choice.

6. If a scientist may legitimately consider nonepistemic values when choosing a method to draw a conclusion from data, then those values may legitimately make a difference to the conclusion the scientist draws from data.
7. Scientists accept or reject hypotheses on the basis of data.
8. Therefore, nonepistemic values may legitimately make a difference to the conclusion the scientist draws from data.

12.3.1 Example: Formaldehyde Again

The argument from inductive risk may be best appreciated if we consider a concrete example of statistical hypothesis testing in which nonepistemic values are clearly at stake in some way. In Chapter 9 we discussed how scientists used statistical tests to investigate the carcinogenicity of formaldehyde.

We can start with a simplified reconstruction of the inferences and decisions involved in the formaldehyde case so that we may consider how the argument from inductive risk might apply in this case. (We will subsequently note the problems when relying on this simplification.)

The tests performed on rats and mice in order to evaluate carcinogenicity amounted to tests of the *null* hypothesis that the inhalation of formaldehyde does not increase the risk of cancer. (Recall our discussion of such tests in Section 9.3.) The *alternative* hypothesis then is that formaldehyde does increase the risk of cancer by some amount. In order to specify a Neyman–Pearson test, we need to specify the *critical region*, that is, those values of the test statistic for which the test rejects the null and accepts the alternative hypothesis. Assuming the test statistic has been determined, the choice of critical region determines the Type I error probability for the test, that is, the probability that the test will reject the null hypothesis, assuming it is true. So in order to choose the critical region, we need to decide what Type I error rate we wish to achieve, that is, what is the maximum probability of rejecting the null hypothesis when the null is true that we will tolerate? In answering this question, of course, we must keep in mind that the lower we set the Type I error rate (the significance level) of the test, the higher the Type II error rate for any given member of a set of possibilities that constitutes the alternative hypothesis. Recall that a Type II error consists of not rejecting the null when the alternative hypothesis is true.

In the formaldehyde case, then, a Type I error would be to declare formaldehyde as increasing the risk of cancer when it does not, while a Type II error would be to fail to declare formaldehyde as increasing the risk of cancer when it does. The argument from inductive risk would draw our attention to the fact that our decision regarding the appropriate method of testing for these hypotheses will involve a choice about how to balance these error probabilities. That decision in turn may quite appropriately draw upon our differential attitudes regarding the consequences of one error as opposed to another. Since what is at stake is the health of human beings (especially those employed in certain industries), we are well justified if we design the test to have a significance level that is relatively high (i.e., there is a non-negligible chance that the test leads us to incorrectly reject the claim that formaldehyde does not increase the risk of cancer). However, we must also consider the negative economic impacts on certain businesses and their employees that might result from such a Type I error.

We certainly would not want to demand a Type I error probability as low as that invoked by particle physicists in announcing the discovery of the Higgs boson, that is, about 3×10^{-7}. Particle physicists can afford to be very conservative about Type I errors, because the consequences of failing to report the discovery of a new particle when one is present would not be so serious – the experiment would simply have to run a bit longer to have sufficient data to meet the statistical threshold. (By contrast, erroneously announcing a discovery could be disastrously embarrassing for particle physicists and undermine public support for their quite expensive, publicly funded research.) But, if formaldehyde exposure really does increase the risk of cancer and scientific investigations fail to detect it, then dire consequences may result. Government agencies responsible for regulating workplace safety standards would not have a basis for eliminating or limiting exposure to a carcinogenic substance, leaving people working in industries in which prolonged exposure to formaldehyde is prevalent vulnerable to high rates of potentially lethal cancers that could be prevented by appropriate regulation.

We should pause to note an important complication, however. Testing for the carcinogenicity of formaldehyde follows the typical procedure of setting a point null hypothesis (the increased risk of cancer equals zero) against a compound alternative (the increased risk is greater than zero). This means that although a Type I error can be described simply, Type II errors vary widely in their consequences. Failing to reject the null even though formaldehyde exposure increases the risk of cancer by 1,000 percent commits a Type II error. So does failing to reject the null when the increase is 0.01 percent. The severity of the consequences of these two errors on human health differs greatly.

From the perspective of the argument from inductive risk, this complication – that different Type II errors will differ in their potential for causing harm – simply reinforces a crucial premise: That the investigator may, on nonepistemic grounds, judge some possible errors as more undesirable than others. Choosing a significance level for a test that would have a good chance of detecting even as small an increase in risk as 0.01 percent may well, given the limitations on affordable sample size, turn out to be achievable only by having a test that will quite frequently reject the null, even when it is true. A rejection of the null based on such a test would not constitute strong evidence that the null is false, according to the severity principle of Section 9.4.1. Allowing people to be exposed to substances that increase the risk of cancer always (all else being equal) increases the probability of a terrible outcome for someone. Choosing instead to test the null in a way that has a lower Type I error rate at the expense of being insensitive to such very small increases in risk *might* indicate that the investigator has made a decision that the importance of avoiding such a relatively low-probability outcome is trumped by the epistemic value of conducting a test that has the potential to produce strong evidence either for or against the hypotheses under consideration.

In his version of the argument from inductive risk, Richard Rudner identifies the crucial premise of the argument as the claim that scientists *accept* hypotheses on the basis of tests they perform (premise 7 in the reconstruction above). Since those tests do not in general yield a conclusive proof, the decision to accept a hypothesis depends on a judgment that the evidence supporting that hypothesis is *sufficient* to make such acceptance a reasonable choice. As long as there are practical *consequences* of such acceptance, value judgments regarding those consequences will be relevant for determining how good the evidence must be in order to be sufficient in this sense (Rudner, 1953).

12.3.2 Skeptical Responses to the Argument from Inductive Risk

To this, however, one might object that the scientist in doing so assumes a task that is not properly theirs. As a scientist, they ought to report the strength of evidence regarding relevant hypotheses, and not presume to decide whether such evidence is sufficient to accept any particular hypothesis for the purposes of practical decision-making.

Richard Jeffrey argues this point in a direct response to Rudner's essay. The scientist ought not to accept hypotheses, according to Jeffrey, because they could not make a decision about acceptance in a way that would be optimal for each problem for which that hypothesis would be relevant (Jeffrey, 1956). Jeffrey offers an alternative view of the task of the scientist, according to which they ought only to report a probability for whichever hypothesis is under investigation, leaving it to others to use that probability assessment – in combination with their own values – as a basis for practical decisions. (Such a probability would have to be some sort of Bayesian probability, of course, since frequentist probabilities do not apply to general hypotheses.)

As Jeffrey notes, Rudner had already anticipated such a response and argued that such a move would not excuse the scientist from the task of accepting hypotheses because the claim that the probability of a given hypothesis is p is itself a hypothesis. But, replies Jeffrey, the scientist should not *accept* this kind of hypothesis either! (Jeffrey, 1956, 246). Jeffrey does not explain how reporting a probability statement to be used in subsequent practical decisions differs from accepting that same probability statement.

Moreover, even if one supposes that the scientist can in this way avoid accepting the hypotheses on which they report, they must, as Duhem emphasized, *rely* on hypotheses in the design of experiments, operation of instruments, and analysis of data, which surely counts as a kind of acceptance. Thus, even were it possible to decline accepting hypotheses that are under investigation, one must accept other hypotheses in order to carry out an experiment. Considerations that enter into such decisions may rightly include not only the sufficiency of experimental arrangement for the intended investigation but also the safety, affordability, or ethical acceptability of the proposed procedure. We have been supposing that we can neatly segregate the entry of nonepistemic values into considerations of the moral acceptability of experimental techniques from questions regarding the analysis and interpretation of data, but the dependence of the latter on accepting previously tested or at least testable scientific hypotheses casts doubt on how well we can maintain that separation.

Another response to Rudner's argument acknowledges that scientists do accept hypotheses but denies Rudner's conclusion that scientists must make nonepistemic value judgments in doing so. According to Isaac Levi, Rudner's argument assumes a behavioristic understanding of statistical inference that attempts to reduce scientific reasoning to "technology and policy making" (Levi, 1962, 47). Levi offers an alternative account according to which the goal of hypothesis testing is sometimes to replace doubt with belief, and in some such cases the scientist seeks "the truth and nothing but the truth" (Levi, 1962, 49). In such cases, the investigator must not differentiate between the seriousness of different errors but must treat all errors as equally serious.

Levi directs his arguments primarily not at the argument from inductive risk but at the *behavioristic* interpretation of hypothesis testing in science, according to which the act of accepting a hypothesis must be given an interpretation as a decision to take some practical action. Yet he clearly considers the two issues to go hand in hand. Nonepistemic values are necessarily involved in the inferences scientists draw from data if and only if all such inferences must be interpreted in behavioristic terms (Levi, 1962, 50).

Such a linkage between the role of nonepistemic values in scientific reasoning and the behavioristic interpretation of hypothesis testing becomes understandable in the context of the literature of the mid twentieth century to which Levi was responding. C. West Churchman had linked scientific hypothesis testing with practical action in his version of the argument from inductive risk (Churchman, 1948). Churchman insisted that "every scientific hypothesis is considered to be a possible course of action for accomplishing a certain end, or set of ends ... Statements like 'we merely want to find out so-and-so' represent pragmatically incomplete formulations" (Churchman, 1948, 259).

Critics of the argument from inductive risk apparently concede that within the behavioristic interpretation of hypothesis testing, the argument from inductive risk validly supports the conclusion that nonepistemic values may legitimately influence scientific reasoning. In Chapter 9 we discussed an alternative *evidential* approach that regards statistical hypothesis testing as a framework for the evaluation of the evidence supporting inferences from data to hypotheses under test. Is the argument from inductive risk still valid if we employ this evidential interpretation?

To exclude nonepistemic values from playing a role in scientific reasoning is, in practice, to suppose that we could impose a division of labor between scientists – responsible for evaluating evidence regarding some scientific hypotheses – and policymakers – responsible for weighing the potential costs and benefits of various policies in light of the evidence relevant to the suitability of those policies for attaining particular public goods at issue. As long as we stick to our simplistic conception of how testing works in the formaldehyde and similar cases, such a division of labor might seem possible, but actual hypothesis testing involves numerous kinds of judgments that complicate this simple picture.

In a discussion of the very same formaldehyde example, Deborah Mayo articulates a number of questions that must be answered in the course of utilizing scientific data from both epidemiological studies and animal model experimental studies to inform policy decisions. These questions concern "inference options" and have "no unequivocal scientific answers," but because "these choices have policy implications, they are intertwined with policy" (Mayo, 1991, 257). Such questions include what weight to give to studies yielding different results, what significance level to consider as indicating a positive result, how to extrapolate from doses used in experimental studies to doses encountered in policy-relevant situations, and how to extrapolate from the dose response of model organisms to the response of humans (Mayo, 1991, 258). Making appropriate decisions on these matters requires significant scientific expertise and judgment, and thus cannot be left to policymakers. At the same time, different choices have clear policy implications insofar as they raise or lower the likelihood that, for example, a substance such as formaldehyde will be found to pose a significant risk. Similarly, Manuela Fernandez Pinto argues that the methodological trustworthiness of scientific results in a "broad" sense depends upon scientists "weighing their microdecisions during the research process against the broad epistemic and social goals of research as understood by their scientific community, as well as against standards for scientific integrity" (Fernandez Pinto, 2020b, 1010). Thus, at least in contexts in which scientists assess evidence that is relevant for policymaking, nonepistemic values ought to inform decisions concerning inference options. That entanglement of nonepistemic values in scientific reasoning does not depend on a behavioristic interpretation of hypothesis testing. We may take the scientists studying the carcinogenicity of formaldehyde to be using statistical methods to evaluate the evidence supporting a variety of possible inferences from data, as the evidential interpretation requires. The argument from inductive risk requires only that scientists' choices regarding inference options make a difference to their conclusions regarding evidence, and that those choices are based in part on the potential influence such conclusions may have on policymakers' decisions. Even if there are contexts in which scientists do, as Levi puts it, seek "the truth and nothing but the truth" (a premise pragmatically oriented philosophers such as Churchman would deny), it does not follow from this that we should as a general rule regard the entry of nonepistemic values into scientific reasoning as *illegitimate* or *unscientific*.[5]

12.3.3 Direct and Indirect Roles for Values

You would be right to be dissatisfied with leaving the matter there, however. Supposing we accept that nonepistemic values *may* play a legitimate role in scientific reasoning, we still face the question of when they *ought* and *ought not* do so. The idea that scientists should seek to understand how the world really is in a way that is not influenced by how they wish the world to be retains its intuitive appeal.

Moreover, allowing values to enter into scientific reasoning willy-nilly might exacerbate current threats to the integrity of scientific research. To cite just one example, the role of pharmaceutical companies in funding research into the safety and effectiveness of their own products might lead to bias in the results of such research (Fernandez Pinto, 2020b; Lemmens & Waring, 2006). Researchers have found a statistically significant association between industry funding and pro-industry results (Bekelman, Li, & Gross, 2003; Lundh et al., 2017). Although investigators may certainly maintain the independence of their results from the source of their funding, the potential for bias is significant (Wilholt, 2009). Investigators clearly should not allow the potential for profit to influence their judgments regarding the effectiveness and safety of a treatment. Given that we allow some role for nonepistemic values in science, on what basis do we exclude cases such as this?

To answer this question, Heather Douglas argues, we need to distinguish distinct *roles* values might play in science, which she labels *direct* and *indirect*. In their direct role, values "act as reasons in themselves to accept a claim,

[5] For an argument that seeks to show how the argument from inductive risk applies to a scientific episode that has no direct bearing on policy, see Staley (2017b). In that essay, I also suggest an alternate reading of Levi's position regarding the legitimacy of allowing values to influence scientific reasoning.

providing direct motivation for the adoption of a theory" (Douglas, 2009, 96). When acting in their indirect role, values "weigh the importance of uncertainty about the claim, helping us to decide what should count as *sufficient* evidence for the claim" (Douglas, 2009, 96, emphasis in original). She goes on to observe that in their direct role "values determine our decisions in and of themselves, acting as stand-alone reasons to motivate our choices . . . In this direct role, uncertainty is irrelevant to the importance of the value in the judgment" (Douglas, 2009, 96).

In characterizing Douglas's position, note that I have referred to these as roles for values in general and not nonepistemic values alone. Douglas rejects the distinction between epistemic and nonepistemic values. According to her, what are typically considered epistemic values are really *criteria* of adequacy that scientific claims must satisfy, rather than values. For her, not only is the distinction between different roles for values in science more important than the distinction between epistemic and nonepistemic values; it completely replaces that distinction. Daniel Steel argues that Douglas's distinction between roles does not succeed in supplanting this distinction between types of values. He defends the view that epistemic values are those that promote the attainment of truth (Steel, 2010). We will not pursue this aspect of the debate further; we may read Douglas's references to 'values' as indicating 'nonepistemic values' in order to focus on the question of what roles such values may legitimately play in science.

Values may play indirect roles throughout science without necessarily threatening scientific integrity, according to Douglas, but their direct roles must be limited if we are to maintain "the integrity of the scientific process" (Douglas, 2009, 96). We have already discussed some legitimate direct roles values may play, such as in decisions concerning what questions to investigate or what research methods to use. Even here, Douglas notes, we need to scrutinize how values influence decisions, since they could, for example, lead one to choose a method of inquiry that "predetermines (or substantially restricts) the outcome of a study" (Douglas, 2009, 100). For one's study of the effects of formaldehyde on rats, one could select only rats that are already in extremely frail health, for example. Or, underlying values-based beliefs might bias an individual or an entire community toward considering only certain kinds of questions. Some critics, for example, have alleged that modern medical science's emphasis on intervention and control – what Hugh Lacey (2005) has in another context called the "modern valuation of control" – biases it toward research approaches to problems, such as cancer, that emphasize pharmacological and surgical treatments at the expense of investigations into environmental factors that might be causing increased rates of cancer (Epstein, 1998; Kourany, 2010, 122–125).

Douglas insists, however, that any direct role for values in science must be restricted to these 'early stages' of research (choosing a question to investigate, choosing a method of investigation). "Once the study is underway," values should play no direct role, except in "unusual circumstances when the scientist suddenly realizes that additional direct value considerations need to be addressed" (Douglas, 2009, 101). In a clinical study of some potential medication, for example, one might find that patients in the treatment group are suffering unexpected adverse effects, requiring one to terminate or modify the study for the good of the subjects. Alternatively, one might find that patients in the treatment group are so obviously benefiting in a way patients in the control group are not that moral demands compel the assignment of all patients to the treatment group (Lilford & Jackson, 1995; Worrall, 2008). Aside from such situations (which to some extent may be accounted for in advance through appropriate research plans), once the investigation is under way, "values must be restricted to an indirect role" (Douglas, 2009, 102).

The argument against allowing a direct role for values in the 'later stages of science' is that doing so would allow inferences from scientific data to be influenced or even determined by the values of the investigator. Scientists could "reject data if they did not like it, if, for example, the data went against a favorite theory," or they "could select an interpretation of the evidence because they preferred it cognitively or socially, even if the evidence did not support such an interpretation," or they could reject a theory they dislike "regardless of the evidence supporting it." By doing so, "we would be causing harm to the very value we place in science," that is, its ability to provide us with a "more reliable, more true understanding of the world" (Douglas, 2009, 102).

It would not harm this value, however, for scientists to allow values an *indirect* role in the later stages of inquiry, such as determining the level of significance that must be achieved to reject a null hypothesis. Douglas thus upholds the argument from inductive risk (though with a narrowed scope for its conclusion): Scientists may legitimately consider the different values attached to different possible errors in deciding what counts as sufficient evidence for accepting a hypothesis. Allowing a direct role for values in the drawing of conclusions regarding the questions under investigation, however, would allow human preferences to supplant the role of empirical data in scientific inquiry.

Douglas's distinction between direct and indirect roles marks a noteworthy contribution to our examination of this issue, yet it does not seem to address fully the question of when nonepistemic values may or may not enter into scientific reasoning. As Kevin Elliott has pointed out, the terms in which Douglas draws the distinction are somewhat unclear. What does it mean for something to function, for example, as a "stand-alone reason" (Elliott, 2011, 306–310)? Moreover, it seems that the problems that arise from values entering into a direct role in drawing conclusions from an investigation can result from indirect roles as well. The investigator who wishes either to cling to a theory she does not want to give up, or reject a theory she loathes, may simply let those preferences serve as a basis for the criteria of sufficiency that, by being permissive or restrictive, make her preferred outcome more likely (Wilholt, 2009).

12.4 Values and the Objectivity of Science

Our discussion thus far has concerned the negative or positive roles values might play in influencing the decisions of scientists engaged in research. A somewhat different set of questions concerns how values shape the conduct of scientific work. Does the achievement of the aims of science require commitment to certain values, and if so, which ones? Should scientific research be organized and institutionalized according to certain values? How might broader social values contribute to or erode the objectivity of science?

12.4.1 Merton on the Ethos of Science

The sociologist of science Robert K. Merton,[6] writing in 1942, sought to describe what he called "the ethos of science": "that affectively toned

[6] When Merton published his doctoral dissertation 'Social and Cultural Contexts of Science' in 1938 (Merton, 1938), the sociology of science did not really exist as an established discipline within anglophone sociology. A lively field bristling with

complex of values and norms which is held to be binding on the man of science" (Merton, 1973, 268–269). Merton posits the *institutional goal* of science to be "the extension of certified knowledge," an aim to be achieved through technical methods adapted to the attainment of "empirically confirmed and logically consistent statements of regularities" (Merton, 1973, 270). From the institutional goal and technical methods, Merton claims, derive four sets of *institutional imperatives* that function as "moral as well as technical prescriptions in science": universalism, communism, disinterestedness, and organized skepticism (Merton, 1973, 270). Merton does not conclude that these norms function as imperatives within science from having observed how scientists always adhere to them – on the contrary, scientists rather frequently or even systematically violate all four of them – but from observing the ways in which violations are rationalized by the violators and condemned by defenders of these norms.

Universalism, according to Merton, finds expression in the requirement that scientific claims should be evaluated according to "preestablished impersonal criteria" rather than by consideration of the "personal or social attributes of their protagonist" (Merton, 1973, 270). Moreover, this norm requires that one's prospects for a career in science should be determined by one's abilities and accomplishments rather than by one's station in society. "To restrict scientific careers on grounds other than lack of competence is to prejudice the furtherance of knowledge" (Merton, 1973, 272). Merton had, at the time of his writing, ample opportunity to observe widespread and systematic violations of this norm. German national socialists had insisted, for example, that only Aryans could produce 'good' science and as a consequence had expelled Jews from their academic institutions. (Merton did not at the time discuss the widespread exclusion of African American or Black people from science in the US and elsewhere, or the obstacles women faced in the pursuit of scientific careers.) Merton argued that systematic violations of universalism were bound to arise because "science is part of a larger social structure with which it is not always integrated," and when the broader

theoretical and empirical debates, contemporary sociology of science has developed in ways that the erudite Merton could not have anticipated in those early years when, as he recalled many years later, "[a]n abundance of [sociological] monographs dealt with the juvenile delinquent, the hobo and saleslady, the professional thief and the professional beggar, but not one dealt with the professional scientist" (Merton, 1973, 173).

culture surrounding science develops anti-universalistic tendencies, this will put the ethos of science under considerable strain, leading to rationalizations seeking to show that the members of some group despised in the broader culture are incapable or ill disposed toward producing worthwhile scientific contributions (Merton, 1973, 271).

The norm of *communism* holds that the results of scientific investigation "are a product of social collaboration and are assigned to the community ... Property rights in science are whittled down to a bare minimum by the rationale of the scientific ethic" (Merton, 1973, 273). We might reward discovery by attaching a scientist's name to a result, but the estate of James Clerk Maxwell does not get to charge royalties whenever 'Maxwell's laws' of electromagnetism are invoked. The increasing role of private capital in the pursuit of potentially profitable scientific research has exerted considerable pressure on this norm in recent years, particularly in the US, where laws protecting intellectual property are especially robust.[7]

In his discussion of *disinterestedness* as a norm in science, Merton emphasizes that this is to be interpreted as a norm operating at the institutional level rather than as a description of the motivations of the individual scientist. However biased or partisan the individual scientist might be in his thinking about the subject of his investigations, the demand that his results be both testable and publicly available subjects the investigator to "the exacting scrutiny of fellow experts" (Merton, 1973, 276). This explains what Merton describes as the "virtual absence of fraud in the annals of science" (Merton, 1973, 276). (It is not clear how well this assessment stands up in the present day.[8]) As a kind of corollary to the self-enforcement of the norm of

[7] But not unlimited. The US Supreme Court ruled in 2013 that Myriad Genetics could not patent a DNA sequence targeted by its test for variants of BRCA genes associated with an increased risk for breast cancer. The court found that "naturally occurring" DNA is a "product of nature" and as such not eligible for patenting. The court's reasoning distinguished between isolating the gene and an "act of invention." The court excluded from this decision complementary DNA, or cDNA, which is made in the laboratory, rejecting claims of the plaintiffs in the case the cDNA is also not eligible for patenting (Simoncelli & Park, 2015).

[8] There have been some scandals involving scientific fraud recently. The story of social psychologist Diederik Stapel at the University of Tilburg, who admitted to fabricating data in a significant number of publications, is both dramatic and troubling (Bhattacharjee, 2013). Still contested is the case of behavioral scientist Francesca Gino, accused by the authors of the Data Colada blog of including fraudulent data in research

disinterestedness *among* scientists, Merton notes in contrast that in relations between scientific experts and laypersons, the authority of science "can be and is appropriated for interested purposes, precisely because the laity is often in no position to distinguish spurious from genuine claims to such authority" (Merton, 1973, 277).

Merton describes *organized skepticism* as "both a methodological and an institutional mandate" (Merton, 1973, 277) that periodically puts science into conflict with other institutions. The willingness of scientists to subject to critical examination beliefs that religious or political authorities regard as inviolable "appears to be the source of revolts against the so-called intrusion of science into other spheres." Such conflicts arise from the application of scientific inquiry to questions regarding which "institutionalized attitudes" have been established or when other institutions "extend their control over science" (Merton, 1973, 278).

Merton's analysis of these institutional norms makes clear the *sociological* function of these imperatives without delving into their *epistemological* aspects. More recent discussions among the philosophers of science have sought to illuminate how these and other values might *contribute* to the objectivity of scientific knowledge.

12.4.2 Longino on Contextual Values

Helen Longino seeks to reconcile an active role for contextual values in science with scientific objectivity. No one doubts the importance of *constitutive values* in science, meaning values "generated from an understanding of the goals of science" (these might include what we have been calling "epistemic values"; they may also encompass Merton's norms). But calls for a "value free" science presumably mean to exclude "those group or individual preferences about what ought to be" that Longino calls *contextual values* (Longino, 1990, 4).

on dishonesty, a charge that she rejects (Scheiber, 2023). Yet, in the spirit of Merton's emphasis on the self-policing of scientists to enforce disinterestedness, one can also visit the blog Retraction Watch, which tracks retracted publications "as a window into the scientific process." Although most of the retractions reported there are not instances of fraud, Retraction Watch does display the wide variety of ways in which scientists can – intentionally or unintentionally – wind up in a pickle.

Longino's reconciliation begins with a claim about evidence: No fact or sentence describing a factual state of affairs serves as evidence for a particular hypothesis all on its own. Considered in isolation, facts are evidentially meaningless; they serve as evidence for particular hypotheses (meaning – for Longino – that scientists can *use* them to argue for such hypotheses) only in the context of particular sets of background beliefs. A crucial problem about evidence arises from this claim: To the extent that background beliefs used in evidential arguments are themselves susceptible to influence from contextual values, how can the objectivity of scientific inferences from evidence be secured?

Longino discusses a broader range of value influences in science than that which emerges in a discussion of inductive risk. The latter involves value judgments concerning the costs and benefits of erroneous or correct inferences. Longino's discussion encompasses a range of ways in which contextual values "shape the knowledge" that emerges from research programs in science, sometimes involving subtle entanglements among moral and epistemological considerations. We will here focus on Longino's analysis of cases in which values influence judgments about what kinds of explanations are best for certain kinds of phenomena. Such background assumptions often take the form of what Longino calls "explanatory models," which describe "the sorts of items that can figure in explanations of a given sort of phenomenon and of the relationships those items can be said to bear to the phenomena being explained" (Longino, 1990, 134).

In the sciences that deal with human behavior, such value judgments have long vexed investigators, the more so because they often remain unstated. In her 1990 book, Longino describes what she then took to constitute the dominant explanatory model assumed in investigations into gender role behavior, the development of homosexuality, and sex differences in cognitive performance: The *linear hormonal model* posits certain types of behaviors as the outcome of a process in which exposure of the fetal organism to testosterone, estrogen, or their metabolites plays a causal role in the determination of the brain's structure and functioning, so that animals wind up with "male brains, programmed for characteristically male behavior, or female brains, programmed for characteristically female behavior" (Longino, 1990, 137). Longino argues that only in the context of this linear hormonal model do the kind of data gathered in studies of hormonal exposure and behavior provide evidence for causal hypotheses linking behavior to hormonal exposure.

Longino goes on to show how a different model might lead to a significantly different understanding of the behaviors being investigated. According to what Longino refers to as the *selectionist model*, the functions of the brain are carried out by groups of about 50–10,000 neurons, and connections between them can be strengthened or eliminated in the process of brain development. Longino postulates that such a model allows for a much greater potential role for the experience of a developing individual (including experience of culturally influenced responses of others to physical appearance) in the explanation of patterns of behavior.

Neither the linear hormonal model nor the selectionist model, Longino explains, enjoys definitive direct empirical support. Both can be given plausible but inconclusive empirical and conceptual motivations. Neither does either model obviously suffer fatal empirical defects. Moreover, such scientific research into the nexus of gender and behavior enters a broader culture of laypersons and policymakers eager to learn 'what science has to say' about such issues as child-raising, gender differentials in academic fields and professions, gender relationships in the family, and sexual orientation. Some are even more eager to exploit the latest results to defend their own value judgments and policy proposals. In such circumstances, where constitutive scientific values fail to single out an explanatory model that will enable the determination of causal explanations from otherwise ambiguous data, implicit appeals to contextual values can be expected to play an important role in filling this gap between data and evidence.

This returns us to our previous question, however. With contextual values playing such an important role in how we draw conclusions, it seems that the very possibility of achieving objectivity lies in peril. Advocates of a value-neutral ideal for science will see in this situation the conditions that give rise to 'bad science' and demand that we strip away any influence on our conclusions from contextual values. Longino agrees both that objectivity is an important aim of scientific inquiry and that some ways in which contextual values influence scientific reasoning threaten that objectivity. But because she does not think that contextual values can be eliminated from influencing how the gap between data and evidence gets filled, she offers a different vision of how objectivity is to be secured: Objectivity results from the *interaction* of investigators holding different, competing contextual values through social mechanisms that, in virtue of satisfying specifiable criteria, enable transformative criticism.

Longino's response deploys a certain notion of objectivity. One might think of the objectivity of science in terms of an "accurate description of the facts of the natural world as they are." To conceive of the objectivity of scientific knowledge in these terms is to embrace some version of realism about science, in the sense of making the truth of descriptions derived from scientific theory central to the aims of science. Longino's account of the objectivity of science employs a different notion, according to which "objectivity has to do with modes of inquiry" (Longino, 1990, 62).[9] To describe science as objective in this methodological sense is to regard the development, acceptance, and rejection of scientific theories and hypotheses as guided by criteria that are "nonarbitrary and nonsubjective" (Longino, 1990, 62).

To see the extent to which scientific inquiry achieves methodological objectivity, according to Longino, we must consider not only the logical forms of argumentation scientists instantiate or the rules they follow but also the *practices* of science, conceived of as implemented by social groups. Longino here draws on work by Marjorie Grene (1966, 1985) in the philosophy of science and Alasdair MacIntyre (1981) in ethics. The production of scientific knowledge requires not only cooperative and coordinated action of groups of researchers as a practical matter but also, as Grene emphasizes, skill in the practices that constitute the methods in any given discipline. Such skill requires a kind of initiation into "the traditions, questions, mathematical and observational techniques, 'the sense of what to do next,'" and this initiation must take place at the hands of someone who has already been through it and practiced the acquired skill within the community (Longino, 1990, 67). With Grene, Longino concludes that the activities of a scientific community constitute something akin to what MacIntyre describes as *practices*. This term, MacIntyre explains, refers to "any coherent and complex form of socially established cooperative human activity through which goods internal to that form of activity are realized in the course of trying to achieve

[9] As emphasized by K. Brad Wray, Longino rejects the traditional view of knowledge as requiring truth. She wishes to retain the normative function of the philosophically traditional concept of knowledge, and the distinction between knowledge and mere belief, but extend the category of knowledge to include representations (including non-linguistic models of the sort discussed in Chapter 7) that are empirically adequate within some domain and sufficient to support both practical and investigative projects (Longino, 1994; Wray, 1999).

those standards of excellence which are appropriate to, and partially definitive of, that form of activity" (MacIntyre, 1981, 175).

Precisely what constitute the distinctive internal goods of science Longino declines to specify, considering such specification to be a part of the evolving self-conception of science itself, an outcome of a socially complex and historically contextualized process (Longino, 1990, 19). Instead, Longino offers criteria by which to judge the social organization of science with regard to its ability to achieve methodological objectivity. It is precisely the social character of science that enables the mitigation of subjective preferences, but not just any social organization will do (the pop music industry is also socially organized, after its own fashion). Science can only achieve objectivity if it is organized to promote "criticism from alternative points of view" and "the subjection of hypotheses and evidential reasoning to critical scrutiny ... A method of inquiry is objective to the degree that it permits *transformative* criticism" (Longino, 1990, 76, emphasis in original).

Longino specifies four criteria "necessary for achieving the transformative dimension of critical discourse" (Longino, 1990, 76). A scientific community must have:

1 "recognized avenues" for criticism that may be directed at all aspects of scientific reasoning (including methods and background assumptions). Scientific journals and professional meetings obviously fulfill this function.[10]
2 "shared standards" by which to evaluate both the arguments advanced by investigators and the criticisms directed at those arguments. Longino notes the similarity of her view of such standards and their importance with that articulated by Kuhn in his essay "Objectivity, Value Judgment, and Theory Choice," but emphasizes that "the open-ended and non consistent nature of these standards ... allows for pluralism and for the continued presence, however subdued, of minority voices" (Longino, 1990, 77–78).
3 a capacity and habit of responding to critical discussion by changing its beliefs. Longino does not take this to require that investigators "recant"

[10] The preeminence of the scientific journal was more secure when Longino wrote her 1990 book than it is today, when scientists post prepublication drafts and other documents to preprint archives such as arXiv.

whenever criticism is directed at their data or assumptions. "Indeed, understanding is enhanced if they can defend their work against criticism" (Longino, 1990, 78). But it does require that "the assumptions that govern their group activities" remain "logically sensitive" to such criticism.

4 an equal distribution of intellectual authority "among qualified practitioners" (Longino, 1990, 76). Rather like Merton's norm of universalism, this criterion prohibits the scientific community from privileging the arguments of one investigator over another on the basis of attributes such as gender, race, or religious affiliation. Longino notes some historical and ongoing threats to or outright violations of this criterion, such as the "bureaucratization of science" and the "exclusion, whether overt or more subtle, of women and members of certain racial minorities from scientific education and the scientific professions" (Longino, 1990, 78).

12.5 Values and Scientific Evidence in Policy Contexts

Scientific objectivity, in this methodological sense, is important simply on the basis that scientific inquiry aims to produce knowledge. The results of scientific inquiry are also relied upon in making decisions, whether at an individual level or at the level of public policy. That scientific conclusions may yield such practical consequences is an assumption of the argument from inductive risk previously discussed. These consequences typically have the most wide-ranging, significant impacts in the context of policy decisions. It is not surprising, then, that the recently growing discussion of the roles played by value judgments in science has been accompanied by increased attention to the roles of science and scientists in the process of making public policy. The COVID-19 pandemic that began as an outbreak in late 2019 provides an especially vivid example of both the complexities and importance of this issue. The pandemic exhibited every feature that makes scientifically informed policymaking practically and epistemically difficult. The stakes were very high. The disease itself spread rapidly, with staggering (but hard-to-ascertain) numbers of deaths and hospitalizations worldwide. The economic consequences were likewise enormous, both directly, from the impacts of so many illnesses and deaths, and indirectly, from the responses to the disease. The psychological impacts ran wide and deep, in terms of grief and loss, but also stemming from abrupt disruptions of normal life that interrupted ordinary activities and relationships, resulting in widespread feelings of isolation. (These impacts were not experienced equally by everyone, though, and those with access to resources were better able to shield themselves from the harms being produced.) Both governments and citizens looked to scientists to provide guidance in how to respond. The situation was novel in many respects, however.[11] The SARS-CoV-2 coronavirus that causes COVID-19 was new to scientists. It would take time to understand its structure, transmission, and the mechanisms through which it causes disease. Ideas about how to limit the spread of disease (mask wearing, maintaining distance between individuals) were for the most part not well tested, and the reasons for adopting or not adopting them changed as the dynamics of the crisis unfolded. Perceptions of scientists varied widely among the public. Many people praised medical professionals and other service providers deemed essential as heroes for their courage and persistence in the face of risks of infection and surging numbers of cases at overwhelmed medical facilities. Scientists succeeded in quite quickly sequencing the genome of SARS-CoV-2, and a vaccine was quickly developed. But public health responses became entangled with deep political divisions in many countries, exacerbated by pervasive and persistent uncertainties regarding the medical, economic, and psychological consequences of many of the remedies being proposed. Promoters of doubt about vaccination suddenly gained publicity and converts. The situation in general contained a complex, even bewildering, mix of disagreements over both factual claims and value judgments, not easily separated.

As argued by Inmaculada de Melo-Martin and Kristen Intemann in a book published shortly before the pandemic, dissent plays important positive roles in science, but can also be "normatively inappropriate" (de Melo-Martin & Intemann, 2018). They argue that combating normatively inappropriate dissent directly is likely to be counterproductive and that a better approach is to address the ways in which "scientific institutions and practices fail to facilitate and sustain warranted public trust" and propose strategies for facilitating trust (de Melo-Martin & Intemann, 2018, 5). It is hard not to

[11] This is not to say that the features of the pandemic were unanticipated. On the contrary, public health researchers had been warning for many years of the potential for an outbreak and had predicted in particular the inequitable distribution of harm such an event would cause (Valles, 2020).

conclude, in the aftermath of the pandemic, that this project of building trust has been made more difficult, but also even more important.

The value-free ideal yields an attractively simple account of the role of science in policymaking, with a clear division of labor: Scientists do their best to present objective and unbiased reports of facts that are relevant to the decisions policymakers face. Policymakers apply appropriate values to arrive at a decision that, in light of scientific facts obtained from scientific experts, will maximize the promotion of the public good.[12] Scientists are qualified for their contribution to this process by their scientific expertise, not by the values they hold or their ability to make judgments in light of those values. Policymakers are qualified for their role by their selection through a political process to which they are responsible, and that ideally links their decision-making authority with the aim of promoting the public good. Put into the terms of risk analysis, it is the scientist's job to *assess* risk (determine the probabilities of various costs and benefits of policy options under consideration), while it is the policymaker's job to *manage* risk, that is, to make normative, political judgments about the acceptability of risks thus assessed (Douglas, 2009, ch. 7).

Scientists who are concerned about maintaining public trust in science's authority to inform policymaking may find the value-free ideal appealing as a way to "distinguish the point where science ends and policy-making begins" (Rosenstock & Lee, 2002, 17). According to the argument from inductive risk, however, this neat picture constitutes a fiction. Assuming that we accept the conclusion of that argument, we need a new, more nuanced picture that we can draw upon to delineate the roles and responsibilities of scientists in the policymaking process.

12.5.1 Scientists' Moral Responsibilities and the Science Advisor Role

The distinction Heather Douglas draws between the direct and indirect roles for value judgments in scientific reasoning, discussed in Section 12.3, is

[12] This description can be mapped onto the framework of decision theory by attributing to scientists the role of determining the probabilities – for each option under consideration in the decision – that certain outcomes result from choosing that option. The policymaker's task is to determine the goodness or badness (i.e., *utility*) of each such outcome. In this idealized account, the optimal choice emerges as that which maximizes *expected utility*, calculated by multiplying, for each outcome, the probability of that outcome by its utility and then adding all such products for each option under consideration.

central to her response to this problem. According to Douglas, the value-free ideal is not merely unachievable, as shown by the argument from inductive risk, but also ethically inappropriate. Scientists have a moral responsibility, she argues, to consider the risks posed by the research they undertake, and in doing so they must rely on value judgments (Douglas, 2009, ch. 4). What safeguards the objectivity of science is not the exclusion of value judgments from the scientific process but their *confinement* to specific roles at various stages of that process. Specifically, value judgments (including judgments of social or political value that are specific to the case at hand) have an appropriate *indirect* role in drawing conclusions from empirical data, meaning that they should be considered in deciding how much weight should be given to the uncertainty of a given conclusion. Douglas writes that such a confinement allows for "holding to detached objectivity" (Douglas, 2009, 155). Value judgments must be excluded, however, from playing a direct role in the drawing of conclusions, that is, they must not function as reasons to accept or reject any particular conclusion. That role is appropriate only in making choices pertaining to earlier stages of the research process, such as the acceptability of using certain kinds of test subjects or test procedures.

Granting all this, a cluster of difficult problems remains: Policymakers must rely on inputs from scientific advisors in the process of making policy decisions. These policymakers typically do not themselves have the scientific expertise to judge directly either technical competence of the scientists who advise them or the ways in which they have deployed value judgments in arriving at their conclusions. How, then, can the public – to whom policymakers are answerable – be confident that policy choices have been based on sound scientific conclusions arrived at via an inquiry process drawing on values that are conducive to the public good? What should policymakers expect from the scientists providing them advisory assistance? And what should scientists consider to be an appropriate mode of engagement in the policymaking process? (See Douglas, 2021 for a clear articulation of this problem and a broader consideration of the relationship between science and democratic governance.)

One basis for responding to this problem constitutes the "linear model" of scientifically informed policymaking, as described by Roger Pielke, Jr. According to this model, "achieving agreement on scientific knowledge is a prerequisite for a political consensus to be reached, and then policy action can occur" (Pielke, 2007, 13). The linear model has a clear affinity with the value-free ideal in the way it rests on a demarcation between the scientist's role (working to achieve consensus in the scientific realm) and the policymaker's role (relying on that scientific consensus to identify a policy option around which a political consensus may form). According to Pielke, maintaining this model forms an "iron triangle" of narrow self-interest: It serves policymakers' interest in avoiding having to make hard decisions; it serves scientists' interest in having their work seen as important and deserving of support; and it serves the interests of issue advocates who can invoke "following the science" as a sufficient basis for preferring their own policy proposals (Pielke, 2007, 143). There are good reasons to regard the linear model as inadequate, however. Some of the contexts in which policy decisions are needed most urgently also involve scientific questions that are subject to significant uncertainty. The linear model lacks a mechanism for scientists to communicate uncertainty to policymakers in a manner that can help with vital decisions. The linear model also tends to obscure the role of values not only in science but also in policy decisions that invoke scientific findings as justification. (This is particularly so when the value judgments involved are widely shared among policymakers.) Finally, the linear model suggests that the role of the scientist in policymaking is a feed-forward mechanism: Scientists decide facts, and policymakers use those facts to make policy. Pielke, drawing upon the work of various authors in science policy and philosophy and social studies of science, proposes an alternative "stakeholder" model, the core idea of which is that there is a "mutual education process" (Pielke, 2007, 14, quoting Harvey Brooks) between scientists and policymakers.

Within the stakeholder model, the scientist may choose to be an *issue advocate*, explicitly taking on board some policy proposal and using their scientific credentials to advance policy aims (as exemplified by nuclear physicists advocating nuclear disarmament). Pielke proposes, as an alternative, the role of *honest broker* of policy alternatives. Both the issue advocate and the honest broker take consideration of policy options and outcomes to be important for the responsible fulfillment of their roles as science advisors. But whereas issue advocates seek to reduce the range of policy options that are considered by policymakers (to favor those preferred by the issue advocate), the honest broker seeks to expand that range by helping policymakers think about what options are compatible with the current state of scientific findings and their attendant uncertainties (Pielke, 2007, 21).

Issue advocates may operate in different ways, and it is important to draw some distinctions among these. Physicists involved in the Manhattan Project that produced the world's first nuclear weapons provided crucial scientific knowledge that made the use of such weapons possible. Some of them also sought to provide advice to influence how, when, and whether they would be used. Leo Szilard was instrumental to the creation of the Manhattan Project. He had achieved some of the earliest insights into the possibilities of uranium as a source of nuclear fission leading to a chain reaction.[13] Convinced of the potential significance of nuclear fission for a potential weapon, Szilard conceived of a plan for a letter,[14] which he helped draft, to be signed by Albert Einstein. That letter was ultimately sent to President Franklin Roosevelt and occasioned Roosevelt becoming aware of the possibility of nuclear energy and nuclear weapons (Rhodes, 1986, 303–315). The Einstein letter was motivated by a concern that Germany would develop a nuclear weapon, a concern shared among Szilard, Einstein, and other physicists who had emigrated from Central Europe around the time fascists were coming into power.

After Germany surrendered, but the war with Japan continued, Szilard became convinced that the actual use of a nuclear weapon in an attack on civilians would be wrong for both moral and geopolitical reasons. He drafted a memorandum presenting those reasons for President Harry Truman, but Truman directed Szilard to meet instead with James F. Byrnes, soon to be Truman's secretary of state. Szilard's argument rested in part on the question of who had the right kind of expertise to be able to anticipate the future directions research on nuclear weapons might take, both in the US and other countries, once the possibilities of such weapons became widely known. Szilard concluded that "this situation can be evaluated only by men who have first-hand knowledge of the facts involved, that is, by the small group of scientists who are actively engaged in this work" (quoted in Rhodes, 1986, 637). Byrnes did not respond positively to Szilard's argument: "[Szilard] felt that scientists, including himself, should discuss the matter with the

[13] In his authoritative history of the Manhattan Project, Richard Rhodes describes Szilard as "the man who thought longer and harder than anyone else about the consequences of the chain reaction" (Rhodes, 1986, 635).

[14] Szilard received encouragement and support from his fellow Hungarian émigrés Eugene Wigner and Edward Teller.

Cabinet, which I did not feel desirable. His general demeanor and his desire to participate in policy making made an unfavorable impression on me" (Rhodes, 1986, 637). Byrnes was not persuaded by Szilard that the deployment of the bomb against Japan – or even its demonstration in public – would initiate an arms race with the Soviet Union after the conclusion of the war. To Byrnes, Szilard was politically naïve for not realizing that the bomb had to be demonstrated to convince Congress to continue funding research and development, and that it would intimidate the Soviet Union into a less aggressive stance. To Szilard, Byrnes was scientifically ill informed. For example, Byrnes mistakenly thought that nuclear weapons required a kind of high-grade uranium ore that was not to be found in the Soviet Union, whereas Szilard knew that abundant low-grade ores would suffice and were readily available to the Soviets (Rhodes, 1986, 637–638). Certainly, Byrnes had little interest in the opportunities for "mutual education" that a closer involvement of physicists in policymaking might afford.

The example of Szilard and Byrnes has some peculiar features. The very nature of the decision facing President Truman and the knowledge held by Szilard entailed that this episode unfolded in tremendous secrecy. The question of public confidence in the decision process could not even arise. Secrecy and concerns about threats to it also complicated or even ruled out possibilities for mutual trust and transparency between scientists and political actors. Szilard himself was being shadowed by security agents as he traveled to and from South Carolina for his meeting with Byrnes. General Leslie Groves, who directed the Manhattan Project, considered Szilard to be one of a number of scientists working on the program "of doubtful discretion and uncertain loyalty" (Rhodes, 1986, 649). The example illustrates, though, how scientists and policymakers may have not only different kinds of knowledge but also different views about what kind of knowledge is important and relevant for the decision that needs to be made. It also illustrates the complicated and varied motivations a scientist may have for becoming an issue advocate. Szilard, like many physicists who became involved in developing the bomb, had been motivated by worries about the grave threat posed by the possibility that Hitler's Nazi government would develop a nuclear weapon first. When it became apparent that this would not happen, the assessment of risks changed, as did Szilard's views about the moral and geopolitical dangers of the bomb. Szilard also worried that an *adequately informed* decision about the use of the bomb could not happen without a direct involvement of the scientists "actively engaged in this work." Finally, it would make sense that a scientist who had participated in the development of a weapon with so much lethal capacity would have a personal sense of moral responsibility leading them to feel called to act. Many other participants in the Manhattan Project besides Szilard felt such a responsibility, but the ways in which they responded varied significantly.

A very different mode of scientist involvement in shaping policy involves the use of scientific authority to amplify doubt in the minds of the public and policymakers about scientific claims that could be invoked to make policy changes that one opposes. This strategy is especially effective when an issue advocate seeks to limit policy options that would constitute significant departures from the status quo. Its use in an organized and well-funded manner has been well documented, particularly in the case of groups seeking to oppose policy responses to the warming of Earth's climate by increased carbon in the atmosphere (Brulle, 2014; McCright & Dunlap, 2010). As documented by Naomi Oreskes and Erik Conway, this response to the climate issue grew out of the efforts of a small but influential group of scientists working with industry groups, a partnership that began with the mid twentieth-century discovery of the severe health threat posed by cigarette smoke (Oreskes & Conway, 2010).

The strategy described by Oreskes and Conway spans decades and industries, but for our purposes, its crucial element is the way in which industries – seeking to avoid government regulations and controls, working with the help of public relations firms – enlisted the assistance of some prominent scientists. Those scientists were inclined to cooperate on the basis of their political beliefs, including a strong opposition to communism or anything they judged to have affinities with it. They included Frederick Seitz, who was trained by Szilard's colleague Eugene Wigner in the 1930s and had done important work in solid state physics. Important appointments such as president of the Rockefeller University (where he initiated new programs in the life sciences) and president of the US National Academy of Sciences ensured Seitz's status as an important figure among research scientists in the US.

Because the strategy pursued by Seitz and others aimed to prevent the implementation of new policies and preserve the status quo, it involved emphasizing uncertainty about scientific claims that provide an apparent basis for the implementation of new policies or regulations. In the case of the dangers of cigarette smoke, scientists (whose research was funded either directly or indirectly by tobacco companies) would testify, for example, that because other things besides cigarette smoke can result in cancer, one could not say with certainty in any individual case that even a three-pack-a-day habit would be a contributing cause to that individual's lung cancer (Oreskes & Conway, 2010, 30–31).[15]

12.5.2 Uncertainty, Doubt, and Decisions

Scientists use the term 'uncertainty' in a variety of ways. Not all of these correspond to its use in common speech, which typically treats uncertainty as simply another term for doubt. To say that someone doubts something is to describe their state of mind or attitude toward a proposition, an idea, or a choice. But when scientists speak of uncertainty – of a measurement result, or a hypothesis, or an experimental finding – they typically are trying not to describe anyone's state of mind but instead characterize the potential for some kind of discrepancy. Often, this is approached as a quantitative matter, such as when a measurement result or statistical conclusion is quoted with 'error bars' or a 'margin of error.' One source of such uncertainty may be that the result depends on a finite amount of data, so that the number being reported is subject to fluctuations due to variability in ways data are produced. This is often described as 'sampling error' or 'statistical uncertainty.' But measurement or experimental results may also depend on assumptions or procedures that have their own possibilities of being erroneous or inadequate for the purposes of the inquiry being conducted. Often scientists attempt to quantify the possibilities that their results may be subject to discrepancy as a result of these uncertain assumptions or procedures, which

[15] In 2006, Judge Gladys Kessler of the US District Court for the District of Columbia found nine tobacco companies and two tobacco industry trade associations guilty of charges brought under the Racketeer Influenced and Corrupt Organizations (RICO) Act. Tobacco companies' own research departments had established that cigarette smoke is harmful to human health in the 1950s, even as they publicly denied this claim and sought to avoid restrictions on the sale and marketing of tobacco products. Judge Kessler found that the defendants had "devised and executed a scheme to defraud consumers and potential consumers" regarding these hazards. What is remarkable is how many decades passed before they were held legally responsible (Oreskes & Conway, 2010, 31–32).

may be reported as "systematic uncertainty" (de Courtenay & Grégis, 2017; Grégis, 2019; Staley, 2020; Willink, 2013).

In other cases, uncertainty receives a qualitative assessment. For example, the Intergovernmental Panel on Climate Change (IPCC) uses a calibrated set of qualifiers of "confidence" to characterize its "evaluation of underlying evidence and agreement" for the claims that it puts forth (IPCC, 2005). The qualifiers range from 'very low' to 'very high.' Just as with the term 'uncertainty,' statements involving terms like 'confidence' and 'agreement' might be understood as statements about psychological states, but here are better understood as a way of characterizing an evaluation of the support or evidence for a given statement (Rehg & Staley, 2017).

However evaluated, and however reported, uncertainty in this scientific usage is not equivalent to doubt. Indeed, when a measurement report includes a well-conducted evaluation of uncertainty in its report, this provides us with a statement that is *less* doubtful ('wind speeds on Neptune at the equator are 180^{+70}_{-60} m/s') because it includes a range of possibilities compatible with the underlying evidence, as compared to a statement lacking such an evaluation ('wind speeds on Neptune at the equator are 180 m/s'). The latter kind of statement, all else being equal, has much greater possibilities for being wrong. (Are the winds really blowing exactly that speed? Not a little faster or a little slower? See Carrión-González et al., 2023 for the basis of this example, in which the authors attribute uncertainties to all their measurement results.)

Or, consider the following statement from the IPCC's 2023 "Summary for Policymakers," included in their *Synthesis Report* from their sixth assessment report on climate change: "Without a strengthening of policies, global warming of 3.2 [2.2 to 3.5]°C is projected by 2100 (medium confidence)" (IPCC, 2023, 11). This statement incorporates both a quantitative and a qualitative dimension of its expression of uncertainty. The quantitative part indicates a favored value for the projected warming of 3.2°C, but with a range of values also judged to be compatible with the underlying evidence from 2.2°C to 3.5°C. The qualitative part tells us that evidence supporting this projection is not as strong as it is for some other claims (such as "The 10% of households with the highest per capita emissions contribute 34–45% of global consumption-based household [greenhouse gas] emissions, while the bottom 50% contribute 13–15% (high confidence)," IPCC, 2023, 5). We might, on the basis of this qualitative statement about "confidence," adopt a more doubtful attitude toward the former projection than we do to the latter attribution, but in either case, the statement *qualified with respect to uncertainty* makes for a more trustworthy expression to the extent that it includes both an assertion and a unbiased characterization of the support for that statement.

With uncertainty thus distinguished from doubt, the question naturally arises as to how scientific communication of uncertainty can be used and misused in the context of policymaking, drawing on our already introduced context of climate science and policies regarding climate change.

The scientific basis for assessing changes in the Earth's climate, and for attributing such changes to increased concentrations of gases produced by carbon-releasing human activities, is inherently complex. In Chapter 7 we briefly touched on the complexities involved in the simulations provided by climate models, for example. The challenges facing human beings around the world as the climate warms, therefore, provide a clear case where: (1) scientific research is essential to understanding the nature of the problem that needs to be addressed, (2) the results produced by that research rests on a complex research process attended by a variety of sources of uncertainty, and (3) the problem is of such a magnitude that some of the relevant policy options for responding to it constitute significant changes to current social, economic, and political conditions. These three features provide a clear opportunity for scientists who, like Frederick Seitz, wish to employ the strategy of opposing new policies by emphasizing uncertainty. Oreskes and Conway document at length the ways in which scientists associated with groups like the George C. Marshall Institute (founded in 1984 to support Ronald Reagan's Strategic Defense Initiative, with Frederick Seitz as it first chair) grabbed hold of statements by climate scientists about the uncertainty of their conclusions and promoted these as reasons to doubt that the planet is getting warmer at all, much less that the warming has been caused by human activities. Much of the communication from such efforts has taken place not in recognized scientific venues but in non–peer-reviewed publications or publications devoted to policy advocacy, such as newspaper opinion pages like that of the *Wall Street Journal* (Oreskes & Conway, 2010, 208–210).[16]

[16] This would be an apt moment to recall that one of Longino's four criteria for achieving a "transformative dimension of critical discourse" is that there be "recognized avenues" for criticism directed at all aspects of scientific reasoning. The cited passage from

The conditions that enable and encourage a strategy of exaggerating uncertainty to generate doubt also present a dilemma for scientists who seek to communicate the results of their research to policymakers in a manner that can inform the pursuit of policies that would help prevent or mitigate harms that might be produced by climate change. Scientific standards of clear and unbiased communication of scientific results require that those results include statements that characterize their uncertainty. But doing so provides opportunities for strategies that seek to distort policymaking by amplifying such uncertainties. On the other hand, not communicating such uncertainties confuses crucial distinctions between results that have quite different degrees of support or variability and can ultimately lead to charges of dishonest reporting of scientific findings. The fact that scientific reports on climate change, like the assessment reports published by the IPCC, include lengthy and detailed discussions of the uncertainty of each finding and the manner in which these uncertainties are evaluated is a testament to the strength of commitment of scientists to the principle that clear and honest communication of scientific findings requires responsible treatment of reports of uncertainty. The questions of how to assess, communicate, and employ uncertainty in these contexts remain subject to ongoing inquiry (e.g., Drouet et al., 2015; IPCC, 2005; Rehg & Staley, 2017).

12.6 Conclusion: Underdetermination and Honesty

The ideal of a value-free science faces challenges to its achievability as well as its usefulness as an ideal.[17] Douglas has argued persuasively that, taken in too strong a sense, this ideal is incompatible with an acknowledgment of the moral responsibilities of scientists (Douglas, 2009). But just where and how can values enter into scientific reasoning without compromising the objectivity of scientific inquiry?[18]

Oreskes and Conway's book provides ample reason to doubt that the editorial page of the *Wall Street Journal* should qualify as such. It is debatable whether the editorial pages of newspapers in general can fulfill such a role.

[17] Menon and Stegenga (2023) argue that the value-free ideal is worthy of pursuit even if it is unachievable.

[18] For a thoughtful and even-handed discussion of the issues taken up in this chapter, see Elliott (2017).

Longino's argument that values enter into the use of evidence in support of scientific theories relies on an appeal to a version of the underdetermination thesis: Treating data as evidence relevant to a given hypothesis requires the use of an explanatory model, which may lack independent warrant. Contextual values, for Longino, contribute to a "rich pool of varied resources, constraints, and incentives to help close the gap left by logic" (Longino, 2002, 128).

As Kristen Intemann has argued, this appeal to an underdetermination gap does not suffice to show that contextual values have a legitimate role to play in determining evidential relationships between data and hypotheses (Intemann, 2005). After all, where such gaps do arise, it may be possible to fill them by relying on purely descriptive claims, and where that is not possible, it may be that the epistemically responsible choice would be to suspend judgment regarding the hypotheses at issue (see also Menon & Stegenga, 2023). Moreover, Deborah Mayo, operating within an error-statistical framework, has argued that assumptions that are needed to close the logical gap between data and hypothesis can often be provided with their own evidential warrant (Mayo, 1996, ch. 6, 1997a).

These comments do not, however, lead directly to a vindication of the value-free ideal of science, since, as also argued by Intemann, contextual values may, nonetheless, play an important role to the extent that scientific hypotheses themselves have content that implies or supports certain contextual values (in opposition to what Lacey calls the thesis of *neutrality*), or to the extent that the aims of science implicate certain contextual values. The former claim has been defended by numerous scholars (e.g., Anderson, 1995; Fausto-Sterling, 2000). Philip Kitcher defends the latter claim (2001). Moreover, the claim that allowing value judgments into evidential reasoning will leave conclusions dependent on arbitrary or unreasoned judgments assumes that value judgments themselves cannot be provided with evidential warrant. One might reject that assumption and argue that value judgments supporting conclusions drawn from scientific data can enjoy independent support of their own (Anderson, 2004).

The value-free ideal, as articulated in Lacey's impartiality thesis especially, enjoys a strong intuitive appeal. "Only the data matter," scientists will sometimes claim. We might imagine that data are unambiguous, quantitative, and not up for debate, while values are fuzzy, qualitative, and subject to disagreement. By now it should be evident that this contrast at best constitutes a drastic overstatement and at worst is dangerously misleading. Longino is surely correct (and in agreement with most other philosophers of science) in denying that data display their evidential significance *self-evidently*. Whether we describe it in terms of models, assumptions, or auxiliary hypotheses, data need to inhabit some richer structure in order to become evidentially *useful*. (See Boyd, 2017 for a noteworthy articulation of just how to conceive of this richer structure in terms of lines of evidence enriched with metadata.) The important questions regarding the possibilities of impartiality and neutrality in science concern the roles of value judgments in undergirding such structures and the epistemic status of such judgments.

To the extent values play an *unavoidable* role in some areas of science, as Longino might argue for the study of human behavior, the value-free ideal of science poses a potential threat to science. If the conception of 'good science' requires that one exclude contextual values from playing a role in the conclusions one draws from data, yet it is impossible in a given field to meet that requirement, then we may expect investigators to seek public acceptance of their work by obscuring the role values play. Better, under such circumstances, for scientists to feel they are free to discuss openly the ways in which values influence their interpretations and inferences from the data and the kind of reasoning that leads them to rely on those values and not others.

When scientists make explicit the supporting structures they use to turn data into evidence, they can then turn to the crucial task of asking how those structures might fail to be adequate for the aims of their inquiry.[19] I have elsewhere referred to this task as the *securing* of evidence. To secure an evidential inference from a given body of data is, roughly, to determine the various ways the world might be, given the relevant state of knowledge, such that the data in question would fail to be good evidence for the conclusion drawn (Staley, 2012, 2020; Staley & Cobb, 2011).[20]

[19] I state the target in terms of adequacy rather than truth because sometimes the auxiliary assumptions scientists rely upon are sufficient for this task without being true, perhaps because a simplified model is sufficient for the level of accuracy one desires, or perhaps because one makes overly 'conservative' assumptions that one knows to be false but that reduce the risk of committing some particularly undesirable error.

[20] Nancy Cartwright has likewise called for making assumptions explicit that are necessary for extrapolating from the conditions under which data has been collected to contexts in which conclusions from those data are to be applied. She has particularly called

Attending to the security of evidence requires an explicit consideration of the kinds of knowledge one brings to bear on a given body of data, as well as a comprehensive consideration of the possibilities for error at each stage of the scientific process. In this way it constitutes an extension of the error-statistical orientation discussed in Chapter 9: Rather than asking how probable or well confirmed a scientific claim is, we might ask about the ways in which it might be in error, and what steps investigators have taken to eliminate or otherwise take into account such errors. If we learned that scientists reached a conclusion in part by relying on value-laden assumptions on no stronger basis rather than that such assumptions reinforce their own subjective preferences, then we should certainly consider the inference to suffer from a significant potential source of error that had not been secured. On the other hand, should we find that reliance on a contextual value in a given case was warranted with further evidence, or that it promoted the aim for which the inference was drawn, this could contribute to the securing of the inference at hand.

We began this book by noting the tremendous impact and authority of scientists and the work they do. Our discussion of the ways scientific evidence and scientists themselves contribute to policymaking underscores this point. That great responsibility comes along with this power is a truism, but we have in a sense been discussing through all of these chapters the nature of that responsibility. What is the undertaking of science? To what standards should we hold scientists beyond the norms that we consider applicable to all human behavior generally? One point that I hope has emerged in this chapter is that science cannot be regarded as an entirely detached, disinterested enterprise. To undertake scientific inquiry itself reflects a certain kind of commitment and a certain judgment about what is worth doing. Deciding what kind of scientific inquiry to conduct and how to carry it out requires further evaluative deliberation. Whatever further roles contextual values might play, the securing of evidence and the susceptibility of conclusions to error-probing and transformative criticism demand that those roles be laid bare. For such an undertaking, a partnership of philosophers and scientists may prove fruitful, or even essential.

attention to this issue in the context of "evidence-based policy making" where the dangers of mistakenly inferring from the effectiveness of a policy intervention under one set of conditions to its effectiveness under other circumstances are especially grave (Cartwright, 2012).

13 Feminist Philosophies of Science

13.1 Introduction

Among the ways in which value judgments and other normative claims figure in the conduct of science, some of the most significant and most discussed are those that relate to sex and gender. As a matter of historical record, the pattern of Western science has been clearly documented: As the institutions and practices of 'modern science' first took shape, some women found paths that enabled them to engage in a wide variety of scientific inquiries. As scientific institutions became more powerful and professionalized, they developed increasingly rigid ways to close off the paths to scientific participation that had been open to women, leading to widespread patterns of exclusion of women from scientific communities and the devaluing of their contributions.

For a sampling of women's accomplishments in the beginnings of modern science, consider the following:[1] Margaret Cavendish (1623–1673) wrote extensively on natural philosophy and developed a distinctive and nuanced view of matter as imbued with intelligence while tackling a wide range of political and philosophical issues and writing speculative fiction, poetry, and plays (Cavendish, 2019; Lascano, 2021). Gabrielle-Émilie Le Tonnelier de Breteuil, Marquise du Châtelet (1706–1749, usually referred to simply as Émilie du Châtelet) became an important figure among Newtonians in France during the early eighteenth century. She surpassed her collaborator Voltaire in mathematics and in time developed her own natural philosophy in her *Institutions de physique* (*Foundations of Physics*), published in 1740.

[1] Londa Schiebinger's *The Mind Has No Sex?* (1989) remains an excellent introduction to women's involvement in the sciences in the early modern period and the ways in which scientific communities excluded them. Schiebinger provides helpful summaries of the lives and accomplishments of the women in the examples given here.

Du Châtelet's approach resembles that of Descartes's *Principles of Philosophy* in its scope and structure while attempting to provide a metaphysical foundation for a Newtonian physics. Her translation and commentary of Newton's *Principia* remains an influential and authoritative resource for studying Newton's work in French. Both Cavendish and du Châtelet drew upon privileges of aristocracy in pursuit of their scientific careers (Detlefsen, 2021; Detlefsen & Janiak, 2018). The path followed by Maria Sibylla Merian (1647–1717) illustrates quite a different route through the "craft tradition" (Schiebinger, 1989, ch. 3).[2] Her father was a successful artist and engraver in Frankfurt am Main, and it was in this context that Merian acquired the skills of observation and detailed illustration that she would put to scientific use as an entomologist and botanist. Merian drew also upon an unusual reservoir of resourcefulness and independence of mind. After enjoying success in the publication of important and painstakingly illustrated studies of caterpillar metamorphosis (established as a subject of potential economic significance by the importance of the silkworm) and a subsequent study of flowers produced in color, Merian sailed to the Dutch colony of Surinam in 1699 at the age of fifty-two, accompanied by her daughter Dorothea. Struck by malaria and offended by the cruel treatment by colonists of the Indigenous population, Merian managed to conduct sufficient research in two years in Surinam to complete her major work *Metamorphosis insectorum Surinamensium*, which included not only careful accounts of the life cycles of a variety of insects not previously studied, with sixty detailed illustrations, but also information about the practical uses of such plants, drawing from knowledge gained from the Indigenous people of Surinam and using the names they used for the plants in question.

As scientific academies – like the Royal Society of London, the Académie des Sciences in France, and the German Akademie der Wissenschaften – became more established and powerful, they faced a choice: Should their scientific communities consist only of men? The conditions in which they faced this question were not neutral; the societies of England, France, and Germany, which were homes to the most prestigious of the scientific academies, all featured strongly patriarchal power structures with strictly enforced gender roles. Yet the answer was not predetermined or beyond

[2] Women also followed the path of the craft tradition to make important contributions in astronomy (Schiebinger, 1989, ch. 3).

dispute. The role of the empirical scientist in its modern guise was itself an innovation, and the features of that role were contested even as it formed. Perhaps that role could be open to women as well as men? As it happened, the very process by which scientific societies became allied with political and financial power led to the same result in every case: Membership in the scientific elite was denied, even to women with all the advantages of class and wealth, such as Maria Cavendish. Although there continued to be women scientists, the barriers to their participation and their flourishing were significant, and there is no doubt that the barriers discouraged many women from pursuing scientific inquiry as a calling or career (Schiebinger, 1989).

That the science produced by the overwhelmingly male (white, heterosexual, cisgendered, socially advantaged) scientific community very often served to promote the interests of that very same demographic has also been well documented (see, e.g., Kourany, 2010, 4–12). The history of medical science provides some especially vivid illustrations of this. The displacement of the mostly female practice of midwifery by the male professional obstetrician/gynecologist in the nineteenth century was not the result of any advantage held by the medical profession in the safe management of pregnancy or women's health in general. On the contrary, the 'heroic' measures of nineteenth-century modern medicine posed a much greater threat to women's health than anything in the practices of midwives. The latter may not have been subjected to a controlled clinical trial, but they tended at least to be mild in effect. The main beneficiaries of this shift from midwives to "regular" doctors at the time were the men who made up the latter profession (Ehrenreich & English, 2010).[3]

[3] Barbara Ehrenreich and Deirdre English's *Witches, Midwives, & Nurses* (2010) constitutes a noteworthy development in its own right in the history of feminist writing about the sciences. Published in 1973 by Feminist Press, Ehrenreich and English's book – not much more than a pamphlet in length – was written for readers like their students at a branch of the State University of New York (College of Old Westbury). The college served primarily a population of "nontraditional" students, "for the most part black and Hispanic" (Ehrenreich & English, 2010, 10), many of them "practical nurses seeking RN degrees, who often brought with them memories and experiences of female healing traditions" (Ehrenreich & English, 2010, 10). Ehrenreich and English were also part of, and inspired by, the "women's health movement," members of which had also produced the book that would become known as *Our Bodies, Ourselves* and sell millions of copies around the world (Boston Women's Health Collective, 1973).

That science has served to promote the interests of the mostly male and otherwise privileged demographic that has historically dominated its institutions is just one part of the picture. The capacity of scientific inquiry to deliver surprising, innovative, disruptive findings has also on occasion yielded new sources of independence and other benefits for women, or at least knowledge about the problems women face. It is true that economics, for example, has often failed even to recognize the economic relevance of much of the labor women (and others) perform outside of the usual conceptions of "the market," such as care for children or elders (Ferber & Nelson, 2003). But it is also true that economic research provides the evidential basis for our knowledge of the existence of a gender gap in wages and other economic measures (including research by Claudia Goldin that led to her 2023 Nobel Prize in Economics; e.g., Goldin, 1990). As Janet Kourany writes, science "has done much to perpetuate and add to the problems women confront rather than solve them, But, of course, science has also produced much of the available information regarding these problems, and scientists have also provided at least some of the wherewithal for solving them" (Kourany, 2010, 11).

The other complication of the historical narrative of the exclusion or marginalization of women is that it is not only the oppression of women that has been facilitated by scientific institutions and empowered by scientific research but also oppression on the basis of many other considerations. These include race, ethnicity, religion, economic status, sexual orientation, being transgender or nonbinary in one's gender identification, not conforming to standard conceptions of ability, and others. The history of using science to justify and propagate oppression in all these respects is long and deep, including such things as 'race science,' eugenics, studies of the biological basis of intelligence, and much more. In this chapter I focus on feminist philosophy of science as a mode of engagement with the problem of oppression in the context of science that has a significant and well-established literature. Authors writing about feminist philosophy of science are also concerned about broader issues of intersectionality and other bases of oppression. Consideration of these issues is important and growing among philosophers of science.

13.2 The Terrain of Feminist Studies of Science

Although this book focuses on the *philosophy of science* as a distinct discipline, we have all along seen how the achievement of insights within this discipline has relied on, or at least gained from interacting with, disciplinary approaches to science other than philosophy (including history, sociology, and anthropology), as well as areas of philosophy other than the philosophy of science (including epistemology, metaphysics, ethics, and logic). Feminist philosophy of science displays this interdisciplinarity especially vividly and may be regarded as just one element of the interdisciplinary field of feminist critical studies of science (Crasnow, 2023), which seek to scrutinize the roles gender and gender norms have played and continue to play in science. Feminist philosophy of science also contributes to, and draws contributions from, *feminist epistemology* (Anderson, 2020).

Feminist philosophers of science also seek to contribute to the articulation and application of *feminism*, the characterization of which is itself the subject of extensive and consequential debate. Feminism can be approached both as a theory and as a practical, political project, but in neither respect can one point to a consensus definition. Nonetheless, one might consider feminists to be united in their opposition to *patriarchy*, which may be understood as "a system in which men rule or have power over or oppress women, deriving benefit from doing so, at women's expense" (Finlayson, 2016, 6). Positive formulations of feminist theory vary significantly insofar as they draw upon distinct political theories (Marxism or liberalism, for example) and insofar as they make stronger or weaker – or descriptive or normative – claims (Finlayson, 2016, 6–14). A 'classically' liberal orientation toward the feminist project might prioritize the importance of gains for women in personal autonomy and equality of opportunity for economic gains. In this respect, liberal feminism emphasizes a positive orientation that posits certain ideals or goals of justice (Finlayson, 2016, 83–88). Other feminists reject such formulations of ideals derived from liberal political philosophies. For bell hooks, the project of feminism should be thought of not in terms of the achievement of "equality of opportunity" in relation to men. After all, "men are not equals in white supremacist, capitalist, patriarchal class structure," so "which men do women want to be equal to?" (hooks, 2000, 19). hooks instead speaks of feminism in terms of the "struggle to end sexist oppression," entailing that it must also be a "struggle to eradicate the ideology of domination that permeates Western culture on various levels" (hooks, 2000, 26). The more radical and revolutionary feminism of hooks also has positive aspirations, but articulates explicitly a project of dismantling or eradicating an existing power structure as a means toward those positive aims.

Given the heterogeneous nature of feminism as such, it should come as no surprise that feminist philosophies of science vary in their aspirations and in their relationships to other strains of thought in the philosophy of science. In an influential 1986 book, Sandra Harding laid out an influential map of the terrain of feminist thinking about science, identifying three broad orientations: feminist empiricism, feminist standpoint theory, and feminist postmodernism.[4] In the decades since, feminist thinking (including Harding's own) has developed in ways that have tended to complicate Harding's classification, at least in the rough terms in which it is typically reproduced.[5]

13.2.1 Feminist Empiricism

Harding describes feminist empiricism as arguing that "sexism and androcentrism are social biases correctable by stricter adherence to the existing methodological norms of scientific inquiry" (Harding, 1986, 24). But feminist empiricism is not only a thesis about bias and scientific method; it also constitutes a contribution to a movement for greater opportunities for women in science, thus encouraging "an enlarged perspective" and "more women scientists" who are "more likely than men to notice androcentric bias" (Harding, 1986, 25).

When we examine current feminist philosophy of science and seek to locate the feminist empiricist tradition, we face two problems. What is feminism? What is empiricism? Since these are both contested terms, we

[4] I will not attempt in this chapter to provide a summary of feminist postmodernist approaches to science; they do not lend themselves to easy summary. Postmodernism has exerted influence on philosophical thinking about science, at least through a kind of dialectical engagement. For example, Sandra Harding acknowledges the importance of critical exchanges with Donna Haraway in her own work (Harding, 1986, 164n2). See Haraway (1988) for an engaging and thought-provoking postmodern reflection on feminism and science.

[5] Summary discussions, such as this one, of Harding's classification sometimes omit its nuances, which include important indications of the kinds of instabilities and tensions in the views described that have led to subsequent developments. Harding described these feminist epistemologies in 1986 as "transitional meditations upon the substance of feminist claims and practices" and that "we should expect, and perhaps even cherish" the "ambivalences and contradictions" she discussed as sources of creative efforts toward a "feminist society" (Harding, 1986, 141).

cannot expect the criteria we seek to be both sharply drawn and uncontroversial. A recent survey by Kirstin Borgerson explicitly sacrifices the former attribute in favor of the latter, adopting the "broadest and most inclusive" definition of the core feminist project as "the identification and elimination of all forms of oppression" and of the core empiricist project as "creating knowledge through methods that draw on experience" (Borgerson, 2021, 80). One might take these definitions to be so broad that any feminist (and maybe some nonfeminist) philosophy of science would fit within its boundaries. Yet Borgerson's characterization reflects the way positions and arguments of feminist philosophers of science have developed. The overall effect of such developments has been to render Harding's classification as originally devised largely inapplicable as changes in, particularly, feminist empiricism and feminist standpoint epistemology have complicated or diminished disagreements between these two approaches.

We could start by articulating, as a kind of foil, a position that probably has no defenders among those who have engaged in a careful philosophical reflection on science, which we might call 'naïve feminist empiricism.' Philosophers use the term 'naïve empiricism' to refer, pejoratively, to a view that holds scientific knowledge to be best pursued by means of 'experience alone' while failing to acknowledge the significant obstacles to defining what 'experience' might mean in this context and in what ways it is supposed to suffice for producing knowledge. A naïve *feminist* empiricist might hold that all problems connected with gender bias in science would be eliminated if we simply stopped allowing anything other than raw experiences or raw data (whatever those might be) to play a role (somehow) in our scientific conclusions. I hope the previous twelve chapters of this book have provided ample resources for understanding why this cannot count as a serious proposal for good science. Accordingly, Harding's characterization of feminist empiricism begins not with this view, nor indeed with anything explicitly formulated in terms of experience, but instead in terms of "existing methodological norms of scientific inquiry."[6] But what are those?

[6] Note that Borgerson's characterization of the empiricist component of feminist empiricism *does* explicitly refer to experience, though as something to be drawn upon in the pursuit of knowledge, rather than to the exclusive source of such knowledge. In Harding's earlier formulation, the relevance of experience is presumably implicit, as partly constitutive of the methodological norms in place.

Among the most significant questions here is that highlighted in Chapter 12: What, if any, role should *values*, and especially such gender-related values as *androcentrism* or *gynocentrism*, play in the conduct of scientific inquiry? In particular, how should a feminist philosophy of science respond to a problem sometimes referred to as the 'paradox of bias': If feminist or gynocentric values are permitted to play a substantive role (even a *direct* role, in Douglas's sense) in pursuit of scientific knowledge, then on what grounds can the feminist justify the exclusion of sexist or androcentric values from the same? (Antony, 2022; Rolin, 2006).

Some critics (including some feminists) argue that responding to the paradox of bias calls for a return to the ideal of value-free science that prohibits scientists from allowing values to influence their interpretations of their data and their conclusions from those data. According to this line of thought, the best philosophical response to the problem of gender bias in scientific theorizing (and the best strategy for achieving feminist goals) is the removal of all bias from science (Pinnick, Koertge, & Almeder, 2003), and the way to remove bias is to prevent value judgments from influencing the conclusions scientists draw. These critics interpret the feminist philosophies of science discussed below as introducing inappropriate political considerations into decisions about what theories are best supported by the evidence (Koertge, 2004). Most feminist philosophies of science, including the feminist empiricist approach, attribute significant roles to value judgments in the shaping of scientific knowledge, although there are differences in the characterization of those roles and their relationship to the advancement of the goals of feminists. Particularly prominent among such approaches has been the work of Helen Longino, introduced in Chapter 12. (See Oreskes, 2021 for a recent deployment of Longino's model.) Recall that Longino's model of scientific objectivity does not attempt to exclude 'contextual values' from shaping scientific knowledge. On the contrary, since facts for her become useful as evidence only in the context of explanatory models and background beliefs, and those models and beliefs can be influenced by judgments about contextual values, some involvement of such values is to be expected.

To see how Longino's account can be brought to bear on the problem of gender bias in scientific theorizing, consider her treatment of how gender-related values have directly influenced the study of human evolution. She draws attention specifically to: *androcentrism* (the "perception of social life from a male point of view"); *sexism* ("statements, attitudes, practices, behavior, or theories presupposing, asserting, or implying the inferiority of women, the legitimacy of their subordination, or the legitimacy of sex-based prescriptions of social roles and behaviors"); and *heterosexism* (in part "the idea that 'they' are made for and hence complementary to 'us'") (Longino, 1990, 128–132). Numerous studies have documented such biases in various aspects of the field of human evolutionary studies (e.g., Dahlberg, 1981; Lloyd, 2005).

For example, critics have noted the androcentric character of the *man-the-hunter* approach to the interpretation of archaeological evidence regarding early hominids. Researchers operating within this framework assumed that tools were developed for the purposes of hunting and that the hunting activities of males drove the process of human evolution. In response, feminist archaeologists developed the *woman-the-gatherer* framework, regarding the activities of women as the main evolutionary force (Dahlberg, 1981). From the standpoint of the advocates of value-free science, both frameworks would seem to constitute failures of objectivity, in which researchers have allowed their personal biases to influence their handling of data.

Longino's response acknowledges that the woman-the-gatherer framework "is as gynocentric as its rival is androcentric," but notes that "its great value from a logical point of view is its revelation of the epistemologically arbitrary character of the man-the-hunter framework. As long as both frameworks offer coherent and comprehensive accounts of the relevant data, neither can displace the other" (Longino, 1990, 130). According to Longino's analysis of evidential argumentation, the data offered by archaeological evidence cannot be used to draw conclusions about human evolution without invoking some kind of explanatory model, and the needed models themselves lack sufficient evidential backing to remove a degree of arbitrariness in the choice between them. This, for Longino, makes the value-free approach unavailable. A proliferation of distinct value-inflected approaches enables us to draw explicit contrasts and critically examine the values on which these theories depend. In this way, Longino's approach resembles Feyerabend's advocacy of theory proliferation. Whereas Feyerabend's pluralism was part of his thoroughgoing skepticism about the value of scientific ideals like objectivity, however, Longino eschews epistemological anarchism in favor of a contextualist and pluralist account of objectivity.

As Longino acknowledges, such an account generates a dilemma. When we seek to understand the scientifically supported answer to any particular question, we look to see where the consensus of the relevant scientific community lies, that is, a kind of *uniformity*. But a pluralist looks to *diversity* within that community as an indication of objectivity. "How is scientific knowledge possible while pursuing socially constituted objectivity?" The dilemma presented by this problem takes the following form: "[I]f objectivity requires pluralism in the community, then scientific knowledge becomes elusive, but if consensus is pursued, it will be at the cost of quieting critical oppositional positions" (Longino, 1996, 274). Longino's strategy for avoiding the dilemma involves two moves that 'detach' scientific knowledge from ideas commonly associated with it. The first move is to deny that scientific knowledge is a matter of consensus, understood as "agreement of the entire scientific community regarding the truth or acceptability of a given theory" (Longino, 1996, 274). The second move is implicated in the first and involves "detaching knowledge from an ideal of absolute and unitary truth" and replacing it with the idea that the aims of at least some scientific inquiries are "satisfied by embracing multiple and, in some cases, incompatible theories that satisfy local standards" (Longino, 1996, 274).

This response to the dilemma is rooted in two central features of Longino's pluralist contextualism. The first is that, as noted in Chapter 12's discussion of Longino's views, she treats science as a matter of practice. "If we understand science as a practice, then we understand inquiry as ongoing, that is, we give up the idea that there is a terminus of inquiry that just is the set of truths about the world," and we should then view scientific knowledge not as "the static end point of inquiry but a cognitive or intellectual expression of an ongoing interaction with our natural and social environments" (Longino, 1996, 276). Viewed as such, knowledge becomes compatible with a plurality of views. It does not require *con*sensus but can embrace a degree of *dis*sensus. The second feature of Longino's philosophy implicated here is her acceptance of a model-theoretic or semantic view of theories, as opposed to a syntactic view. As discussed in Chapter 7 and in the discussion of perspectival realism in Chapter 10, a model-based view of theories lends itself to pluralism insofar as one can acknowledge the abilities and relevance of multiple models in treating a particular target system.[7]

[7] As noted in Chapter 10's discussion of Massimi's perspectival realism, a *realist* philosophy of science must show how to reconcile realism with such a pluralism of model-based

Longino notes that for "feminists and other oppositional scientists" a consequence of such an approach is that their aim "does not consist in finding the one or best correct feminist model" that can be the object of "a general and universal consensus." Instead, dialogue within the community should aim "to make possible the refinement, correction, rejection, and sharing of models. Alliances, mergers, and revision of standards as well as of models are all possible consequences of this dialogic interaction" (Longino, 1996, 277).

A case study from Elizabeth Anderson (2004) provides a good example of feminist values in application to a specific research problem. Anderson bases her analysis on a premise that the relationship between value judgments and empirical evidence goes in both directions. What makes it possible for responsible and trustworthy science to include reliance on value judgments is that value judgments are themselves subject to scientific appraisal on the basis of evidence. She argues that the "orthodox case" for the claim that science should be value-free "depends on the claim that value judgments are science free" (Anderson, 2004, 3). On the contrary, she argues that value judgments can be subjected to scrutiny on empirical grounds. Emotional experiences, for example, can serve, under certain conditions, as relevant evidence for judging whether we would do well or poorly to pursue a particular career path. "If we condition our acceptance of value judgments on evidence, we will not hold our values dogmatically, and they can be integrated into scientific theorizing without making it dogmatic" (Anderson, 2004, 3).

To demonstrate how reframing the discussion of values in science in terms of a two-way relationship to empirical evidence can yield research outcomes that are value-influenced but not biased, Anderson compares two studies on the effects of divorce. One group approached the research from a traditionalist value orientation that regards "a model of the family in which the husband and wife are married for life, live in same household, and raise their biological children" as an ideal. For these researchers, divorce is necessarily a harm and a departure from what is best. Anderson reports the other feminist researchers as holding ambivalent attitudes toward divorce and as being "unsure how to assess divorce from the standpoint of opposition to

representations of a target system. Longino does not attempt to reconcile her view with realism.

sexism" (Anderson, 2004, 3, 12). This is just one aspect of the differences in value orientations of the two groups of researchers, which Anderson analyzes across multiple stages of the research process. To distinguish legitimate from illegitimate ways in which values influence that research process, she relies on the criterion that value judgments must not "drive inquiry to a predetermined conclusion" (Anderson, 2004, 3, 11).

For example, Anderson notes how different value judgments play into the conception of the object of inquiry for the two groups of researchers. The traditionalist conception of divorce in terms of trauma or loss is not inherently flawed and may come naturally to researchers whose experience with divorce comes from a clinical setting in which clients seek help in coping with the negative consequences of a disruptive event they experience in those terms. Feminist researchers bringing their own evaluative conception of divorce might see the object of research in terms of a process rather than an event. This conception might make more sense for someone who places divorce in the context of a failing or failed marriage and as affording opportunities for growth as well as for loss. Both of these conceptions of the object of research are value-laden, and they both are legitimate within the research process. Neither guarantees that evidence will be found to support the judgments underlying those conceptions.

As Anderson reviews the various stages of research, she shows how value judgments can influence research outcomes in illegitimate ways, but also how they can be shielded from doing so. A clinical orientation tends to lend plausibility to the traditional conception of divorce as trauma or loss. However, relying on a sample of people seeking clinical help when studying the effects of divorce introduces bias in the results. "A sample drawn from psychological clinics will be biased toward those experiencing great difficulties coping with divorce, or misattributing their difficulties to divorce, and against those who find divorce liberating" (Anderson, 2004, 16). Feminist researchers sought to avoid bias by sampling cases based on divorce dockets instead. A value judgment may legitimately inform researchers' conception of the object of study, but this becomes problematic if it leads to a sampling procedure that introduces bias.[8]

[8] A biased sampling procedure may not guarantee that a particular outcome is reached, but only make it more probable that a certain conclusion is drawn, even if it is incorrect (i.e., it makes the test of that conclusion less *severe*, as discussed in Chapter 9). Hence,

Judgments about data analysis may also depend on value judgments in ways that either generate or prevent bias in research outcomes. One decision concerns whether to report only main effects of an independent variable or to also consider interaction effects. For example, Anderson reports how feminist researchers studied children's response to divorce with respect to their ability to grasp events from another's perspective ('mature perspective taking'). They found *no main effect*, meaning that when they compared the level of adjustment to divorce between children who were capable of mature perspective taking and those who were not, they did not see a significant difference. But they went further and looked for interaction effects: Did mature perspective taking make a difference when conjoined with other factors? They found that when parents had high levels of conflict, mature perspective taking was associated with better adjustment, while for children of parents with low levels of conflict, it was associated with worse adjustment.[9] The relationship of this choice to value judgments, according to Anderson, is that a study that relies only on a main effects analysis does not consider individual variation. "This makes sense if one believes that a single way of life is best for everyone." But for researchers who think that "[w]ays of life should be tailored to individual differences," it makes sense to consider what interaction effects reveal that may be relevant for different kinds of families and individuals.

13.2.2 Feminist Standpoint Theories

Feminist empiricists such as Longino and Anderson (see also Antony, 2022; Campbell, 1998; Nelson, 1990; Okruhlik, 1994; Solomon, 2001) approach the paradox of bias by seeking to reconcile a robust role for feminist values in scientific inquiry with such epistemic aims as absence of bias, methodological objectivity, and reliability. Feminist standpoint theorists argue instead that feminist inquirers themselves enjoy certain advantages over masculinists when it comes to achieving reliable knowledge about certain

Anderson's criterion that value judgments should not "drive inquiry to a predetermined conclusion" should be understood in a probabilistic sense.

[9] Anderson notes, "On reflection, this makes sense. Mature perspective taking enables children to come to terms with their parents' fighting. But when they don't see their parents fighting, it leads to confusion, as the perceptive children try to make sense of their parents' divorce with inadequate information" (Anderson, 2004, 16).

kinds of phenomena.[10] The concept of *standpoint* is central to a theory that intends to capture how, within a given social order, individuals may vary – as a consequence of how that order works – in their ability to gain cognitive access to certain kinds of facts. Marxist standpoint theory, for example, holds that members of the proletariat, by organizing and becoming conscious of their role in the capitalist system, have "epistemic privilege over fundamental questions of economics, sociology, and history" (Anderson, 2012). Nancy Hartsock proposes that feminists appropriate this Marxian approach: Because "women's lives differ structurally from those of men" they "make available a particular and privileged vantage point on male supremacy … which can ground a powerful critique of the phallocratic institutions and ideology which constitute the capitalist form of patriarchy" (Hartsock, 1983, 36).

More generally, standpoint theory can be understood as comprising two theses in which the concept of standpoint figures strongly. According to the *situated knowledge* thesis, "Social location systematically influences our experiences, shaping and limiting what we know, such that knowledge is achieved from a particular standpoint." The thesis of *epistemic advantage* states that "[s]ome standpoints, specifically the standpoints of marginalized or oppressed groups, are epistemically advantaged (at least in some contexts)" (Intemann, 2010, 783; Wylie, 2003).

Harding's 1986 discussion of feminist epistemologies explored what she regarded as incoherences in both feminist empiricism and feminist standpoint epistemology. For our purposes, her critique of incoherence in feminist standpoint views is relevant because Harding's own attempts to articulate a standpoint approach in response to this critique have figured significantly in subsequent discussions of feminist philosophies of science. In particular, she raises the problem of "fractured identities" for feminism. When women embrace identities "as Black women, Asian women, Native American women, working-class women, lesbian women," the notion of *a* feminist standpoint comes under strain. Where might we find the "common

[10] Janet Kourany offers an interesting alternative classification of feminist responses to the value-free ideal of science that distinguishes Longino's approach as "social value management" from naturalized empiricist philosophies (which includes both standpoint approaches and naturalized views of the advantages of feminist values, like Anderson's). Kourany then offers a more explicitly political alternative of "socially responsible science" (Kourany, 2010, 58–77).

experiences as women" that "create identities capable of providing the grounds for a distinctive epistemology and politics?" (Harding, 1986, 163).

Harding's subsequent work aims to address these issues by developing the idea of a standpoint as part of a *methodology* for doing science that aims at "strong" objectivity. Her argument is that both feminist empiricism and feminist standpoint theory originated in the 1970s and 1980s as a response to a need for an improved standard for objectivity. "This standard had to be stronger than the prevailing ones since the latter had permitted sexist and androcentric assumptions and practices to shape some of the very best research in biology and the social sciences" (Harding, 2015, 26). A point regarding objectivity that standpoint approaches and feminist empiricism share is that research that seeks to test or replicate previous findings can reveal how results might depend on researchers' values and assumptions, but that this becomes much less effective when those are shared among all the members of a scientific community. We have already seen how Longino proposes a diversity of value judgments within the community as a way of responding to this issue. Harding proposes instead that achieving stronger objectivity requires that one "start research from outside one's discipline." This is not merely a matter of cultivating diversity within the discipline because "[m]ere diversity doesn't have the theoretical and analytic resources to capture what is so valuable about 'missing perspectives'" (Harding, 2015, 35). Harding points instead to examples of "alternative research communities" growing out of social justice movements as an example of what it looks like to begin research from outside of a discipline. "The standpoint of poor people, of racial and ethnic 'minorities,' of people in other cultures, of women, of sexual minorities, and of disabled people are perhaps the most widely used diversity standpoints from which dominant knowledge claims in every discipline have begun to be reevaluated" (Harding, 2015, 36).

How exactly such research can be pursued, and what kinds of researchers play which roles in its pursuit, remain contested issues. As an example of what a standpoint methodology might involve, consider the approach, or family of approaches, within collaborative archaeology. Collaborative archaeology is pursued through the cooperative efforts of academic archaeologists alongside people who identify historically with an archaeological site. The Eastern Pequot Tribal Nation began a collaborative relationship with the University of Massachusetts Boston's Anthropology Department in 2003. The purpose was to conduct "archaeological research and training as part of initial cultural and historical preservation efforts" on the reservation lands of the Eastern Pequot Tribal Nation. An article authored by four Eastern Pequot community members (including the lead author) and one member of the Anthropology Department at the University of Massachusetts Boston discusses how this collaboration worked to navigate the terrain of academic research expectations and the interests of the Eastern Pequot community. For example, to build "trust and familiarity" in the early years of the collaboration, the research group "moved forward slowly," with graduate student theses and publications vetted by the Tribal Council and "focusing the first publications on the process of doing collaborative archaeology rather than on sensitive archaeological information that Eastern Pequot community members might not be ready to release to the broader public" (Dring et al., 2019, 359). The authors, writing a decade and a half into the collaboration, note the need to shift from the published products being written by "university academics and students" to "more community voices" so the project's output will include "the input of all participants" (Dring et al., 2019, 360). They describe examples of how the collaboration can both "give more voice to the Eastern Pequot who participate in it regularly" and "make its products more useful and accessible to tribal members" (Dring et al., 2019, 360). These include the transfer of Eastern Pequot archaeological collections from the University of Massachusetts Boston (where they had been fully catalogued) back to Pequot land and the authorship of the article itself, which "marks the first time that an Eastern Pequot community member has taken the lead author position" (Dring et al., 2019, 361–363). A further goal is to produce "heritage products of use to the community" such as an already completed commemorative volume documenting the first decade of the collaboration's archaeological field school. The political context of the project, which concerns efforts of the Eastern Pequot Tribal Nation to receive recognition by the US federal government's Bureau of Indian Affairs, is explicitly invoked by the authors.[11] The authors make clear that they do not proclaim "mission accomplished" for their collaborative project, but instead call for more inclusion of and service to the Eastern Pequot community.

[11] An initial granting of federal recognition in 2002 was appealed by the state of Connecticut and twenty-nine Connecticut towns and withdrawn in 2005. To date the Eastern Pequot Tribal Nation has not been recognized by the federal US government. See www.easternpequottribalnation.org/history.

The paper just discussed makes no reference to standpoint epistemology. The authors and other participants in this collaborative archaeological research project may not have any intention of pursuing a project based on standpoint methodology. Yet, it is clear that to the extent that this project has relied from the outset on investigators pursuing an inquiry in part "from the outside" of academic archaeology, and in particular as it aims to make contributions not only to academic archaeology but also to the very community whose heritage is being inquired into, it provides an example of the kind of research Harding considers a contribution to the achievement of "strong objectivity."[12]

Philosopher of science Alison Wylie (in the same special issue as the Dring et al. paper) notes how collaborative archaeological projects can indeed be defended as a way of "mobilizing diverse epistemic resources" to achieve "better outcomes" in the form of what archaeologist Bonnie Clark describes as "better science" in the form of "more accurate reconstructions of a richer, embodied past" (Wylie, 2019, 579, quoting Clark, 2019, 469). But Wylie also notes a theme running through examples of collaborative archaeological research that emphasizes "ways in which the process of working in partnership can be transformative and should be valued in its own right" even as it is "constrained by conventional disciplinary measures of success and productivity" (Wylie, 2019, 580).

In the case of collaborative archaeology just described, the relevance of feminism per se may be less evident than in Anderson's example of divorce research, or Longino's discussion of androcentric assumptions in the study of human evolution. Yet, the example shows how the approach of a standpoint methodology can be brought to bear on scientific research in a way that is adapted to the varied manifestations of oppression or marginalization. For those feminist philosophers of science sympathetic to the more revolutionary aims of the feminist movement, this *adaptability* of the standpoint methodology could serve as a basis for understanding how certain forms of scientific research might contribute to the "struggle to eradicate the ideology of domination that permeates Western culture" (hooks, 2000, 26).

[12] Harding acknowledges the link between collaborative archaeological research (as an example of a broader category of *participatory action research*) and standpoint epistemology (Harding, 2015, 167).

13.2.3 Assessment of Feminist Empiricism and Standpoint Methodology

As noted in a 2010 assessment of feminist empiricism and standpoint theory by Kristen Intemann, both approaches have developed beyond the characterizations Harding gave them in 1986 (as she then suggested they would need to). Indeed, those developments have tended to diminish disagreements between the two approaches.

Feminist empiricism developed a critical stance toward existing methodological norms of science, taking shape as a contextualist epistemology "with respect to the aims, cognitive values, and methods that govern particular research contexts." It also adopted a normative stance acknowledging that "aims, cognitive values, methods, and other background assumptions are not always independent of social, ethical, and political values." Feminist empiricism also became a social epistemology, in the sense of locating "objectivity and justification [in] scientific communities rather than individual scientists," based on organizing scientific communities "in ways that minimize the negative influence of individual biases" (Intemann, 2010, 782).

Standpoint theorists meanwhile have developed responses to interpretations of their theses that dismiss them as either obviously false or true but trivial. For example, it is obviously false that simply being a member of an oppressed group "is sufficient for having a less distorted view of the world and that this epistemic advantage would be present in any epistemological context," since an individual member of an oppressed group may have internalized the biases of their oppressor, or may lack the background to hold certain kinds of knowledge (Intemann, 2010, 783–784). On another interpretation, one might take standpoint theorists to be asserting the obvious point that there are kinds of knowledge that are gained by experiences that are specific to particular groups of people. One can know a lot about racial prejudice, for example, but a direct knowledge of subjective experience of being a target of racial hatred is not accessible from "outside" that experience. At the same time, most of us will never know the subjective experience of having a billion US dollars.

Intemann notes how standpoint theorists, to avoid these problems, have emphasized an empiricist element in the situated knowledge thesis: Rather than basing the differences in situated knowledge on *essential* differences between people, standpoint theorists base those differences in knowledge on differences in experience.[13] The questions of which social locations produce what differences in experience, and how they produce those differences, are empirical questions about contingent matters (Intemann, 2010, 785). Intemann emphasizes that standpoints themselves are not to be understood simply as socially located perspectives but as, in Alison Wylie's words, "a critical consciousness about the nature of our social location and the difference it makes epistemically" (Wylie, 2003, 31). A standpoint is thus an *achievement* that depends not only upon membership in a social group but on gaining "sufficient scrutiny and critical awareness of how power structures shape or limit knowledge in a particular context" (Intemann, 2010, 785). To achieve this, moreover, requires a community effort (Intemann, 2010, 786). Finally, standpoint feminists articulate a normative approach that "takes certain ethical and political values to be central to inquiry and rejects the view of science and objectivity as 'value-free'" (Intemann, 2010, 786).

These features of the situated knowledge claim in standpoint theory have important implications for understanding the thesis of epistemic advantage. Research communities that include members of marginalized communities have an automatic advantage based not on social location alone but also on the opportunities afforded by such location for achieving relevant critical consciousness. The latter is needed to understand how the social locations of researchers can be relevant to the questions they ask, the methodological choices they make, the assumptions they rely on, and the way they collect and interpret data. Intemann considers the example of a "female biomedical researcher who grew up in sub-Saharan Africa and is now working in the United States on a new HIV vaccine." Her training, like that of her fellow researchers, will have equipped her with the requisite understanding of the biological mechanisms relevant to the research. But perhaps her experience of growing up in an area where the vaccine is intended to be deployed also allows her to question some assumptions of other team members about the conditions in which the vaccine would be put into use, such as the availability of refrigeration or other economic factors. Such criticisms might in turn be relevant to inferences about the usefulness of the vaccine in practice that are based on the results of clinically controlled trials. Her understanding of

[13] This mode of argument can be found much earlier, in a 1913 broadsheet by Jane Addams (Haslanger, 2017).

culture and regional economics may form the basis for improved articulations of, and strategies for achieving, the research team's goals. In this way, the epistemic advantage is gained by the research group – and the broader research community to which they contribute – as a consequence of including researchers who have the ability to increase "the rigor of critical scrutiny" needed to arrive at better justified "assumptions, methods, models, and explanations" (Intemann, 2010, 788).[14]

Given that feminist empiricism and standpoint theory have a shared emphasis on the empirical basis for their endorsement of greater diversity as a path to better science,[15] on the normativity of their proposals, on the contextual nature of scientific knowledge, and on the social nature of scientific inquiry, what remains to distinguish the two approaches? Intemann locates the crucial remaining disagreement in their respective explanations of the epistemic value of diversity in research communities, which rest on different conceptions of diversity itself. Feminist empiricists such as Longino have tended to focus on the *diversity of values and interests* in scientific communities as a way of achieving greater objectivity through "checks and balances" that "ensure that the idiosyncratic values or interests of scientists do not inappropriately influence scientific reasoning" (Intemann, 2010, 790). For standpoint theorists, the *diversity of social position* is key to greater objectivity. Given the historical legacy of structured power relations and the oppression these have enabled, "individuals from diverse social positions and backgrounds are likely to have had different experiences" that make them more likely to bring certain kinds of relevant evidence to bear on the plausibility of assumptions, methods, and explanations used in the research process. They are "more likely to identify limitations or problems with background assumptions that have gone systematically unnoticed" (Intemann, 2010, 791).

Intemann's assessment of this difference is that the standpoint account is more plausible and feminist empiricists should adopt the account of

[14] For an argument that epistemic advantage in the standpoint framework applies in particular to the study of social power relations, see Rolin (2009).

[15] See Wylie and Hankinson Nelson (2007) for an empirical assessment of feminist interventions in science through a discussion of case studies. Londa Schiebinger's "Gendered Innovations" project provides more detailed case studies: http://genderedinnovations.stanford.edu.

diversity and its benefits offered by standpoint theory.[16] Given that standpoint feminism has already incorporated important elements of feminist empiricism, the resulting view could appropriately be called *feminist standpoint empiricism*.

13.2.4 Gender and Scientists' Relations to Objects of Inquiry

We have focused in this chapter on feminist approaches to the methodology of science in its relationship to the social organization and identities of scientists. Another important discourse regarding gender and science concerns the role of gender concepts in the historical formation of scientific aims and norms, and the complex and differential effects this history can have on researchers. The writings of Evelyn Fox Keller have played an important role in encouraging a critical reflection on these issues and deserve a longer treatment than I will be able to afford them here.

Keller began her academic training in physics (see Keller, 1977 for reflections on this experience). She went on to explore how sex and gender have figured into the conceptions and pursuit of knowledge in the writings of a variety of philosophers and scientists, including Plato, Francis Bacon, and Robert Boyle, who have helped shape subsequent ideas about what it is to be a scientist and engage in scientific inquiry. Her approach employs tools of literary analysis and psychoanalytic theory, as well as many ideas and concepts drawn from the philosophy of science. Her investigations return often to the sometimes subtle (sometimes not) ways gender figures in descriptions of the relationship between the scientist as investigator and the subject of that investigation, "nature" (or "Nature").

Francis Bacon (1561–1626) set forth a programmatic vision for the pursuit of modern science as an empirical quest to empower human beings for the improvement of human life. In setting forth this vision, Bacon's writings appeal in several instances to vivid metaphors suggestive of sexual domination.[17] For example, Keller quotes Bacon writing, "For you have but to

[16] Intemann also argues that feminist standpoint theory provides a better solution to the "bias paradox" mentioned previously (Intemann, 2010, 791–792).

[17] The scholarship on this aspect of Bacon is a field of significant contention. Carolyn Merchant's 1980 book *The Death of Nature* ignited some of this controversy; some elements of her treatment were taken up by Sandra Harding (Merchant, 1980; Harding, 1991). See Landau (1998) for an argument against this reading of Bacon. Vickers (2008) argues

follow and as it were hound nature in her wanderings, and you will be able, when you like, to lead and drive her afterwards to the same place again" (quoted in Keller, 1985, 36). But Keller finds in Bacon's writings "traces of a dialectic that is far more complicated and hence richer in meaning than either the critics or the defenders of science tend to assume" (Keller, 1985, 34). Keller's interpretation of the dual aggressive and defensive elements of Bacon's imagery – by way of a psychoanalytic analogy to paternal and maternal relations – leads to her conclusion that "the aggressively male stance of Bacon's scientist could, and perhaps now should, be seen as driven by the need to deny what all scientists, including Bacon, privately have known, namely, that the scientific mind must be, on some level a hermaphroditic mind" (Keller, 1985, 42). The scientist who always dominates nature will miss opportunities for knowledge that arise only by yielding to or joining with nature. For all its nuance, though, Bacon's way of talking about science "provided the language from which subsequent generations of scientists extracted a more consistent metaphor of lawful sexual domination" (Keller, 1985, 34).

To gain some appreciation for what Keller might mean by a "hermaphroditic mind," one might consider her notion of "dynamic objectivity" and her study of the biologist Barbara McClintock. Keller articulates dynamic objectivity in contrast with a 'static' conception of objectivity as encoded in the idea of the scientist as a disinterested, even disconnected, observer of an ontologically distinct nature over which the scientist seeks to wield control. Scientific practice already departs, she argues, from the ideology of a static conception of objectivity. Keller relates dynamic objectivity to ideas in developmental psychology about "allocentric" or other-centered perception that is related to such cognitive activities as attention, focus, or "absorption in the object before one" (Keller, 1985, 119). Keller's book on Barbara McClintock (Keller, 1993) emphasizes the ways in which the Nobel Prize-winning biologist conducted her investigations into the genetics of maize employing an exceptionally powerful ability to perceive the organisms she studied "from *many sides*, as fully as possible" (Keller, 1993, 165, quoting developmental psychologist Ernest Schachtel). In Keller's interpretation, McClintock, as investigator, related to the maize plant primarily through not *division* but

against Merchant and Harding, but does not address the more nuanced reading offered by Keller; see Park (2008) and Merchant (2008) for responses to Vickers.

difference. Where division "severs connection and imposes distance," difference and the recognition of it invite the investigator to look for relatedness. "It serves both as a clue to new modes of connectedness in nature, and as an invitation to engagement with nature" (Keller, 1993, 163). Keller describes the resulting set of relations: "Self and other, mind and nature survive not in mutual alienation, or in symbiotic fusion, but in structural integrity" (Keller, 1993, 165).

13.3 Conclusion

Feminist philosophies of science encompass numerous philosophical orientations and are the subject of a lively and wide-ranging discourse. In this chapter I can only hope to have provided enough hints of some of the ongoing debates to encourage readers to delve more deeply.

One conclusion that deserves to be made explicit is that feminist engagements discussed here do not occupy some separate realm of the broader field of discussion in the philosophy of science. On the contrary, feminist arguments engage many of the same issues taken up in the previous chapters of this book: What methodology has the descriptive and normative resources to make sense of and guide scientific research? How should we think about scientific theories and the aims they promote? What constitutes scientific knowledge? What and how do scientific models or theories represent? How do social forces and relations figure in the production and status of scientific knowledge? What is the role of value judgments in the process of scientific inquiry? Not only does the feminist scholarship discussed in this chapter implicate arguments more generally on all these topics, but each of them to some extent implicates all the others!

To separate the questions that animate this field into separate chapters of a textbook is, consequently, an exercise in imposing an artificial but needed order on a naturally entangled and, frankly, rather messy network of implications and applications. Experiencing that messiness more directly will naturally come with delving deeper. I find that experience to be a source of great happiness that I hope you will share.

References

Aad, G., Abajyan, T., Abbott, J., et al. (2012). Observation of a new particle in the search for the Standard Model Higgs boson with the ATLAS detector at the LHC. *Physics Letters, B716*, 1–29.

Achinstein, P. (1968). *Concepts of science: A philosophical analysis.* Baltimore, MD: The Johns Hopkins University Press.

(1983). *The nature of explanation.* New York: Oxford University Press.

(2000). Proliferation: Is it a good thing? In J. Preston, G. Munévar, & D. Lamb (eds.), *The worst enemy of science? Essays in memory of Paul Feyerabend* (pp. 37–46). New York: Oxford University Press.

(2001). *The book of evidence.* New York: Oxford University Press.

(2002). Is there a valid experimental argument for scientific realism? *Journal of Philosophy, 99*, 470–495.

(2010). Mill's sins or Mayo's errors? In D. G. Mayo & A. Spanos (eds.), *Error and inference: Recent exchanges on experimental reasoning, reliability, objectivity, and rationality* (pp. 170–188). New York: Cambridge University Press.

(2020). Scientific realism: What's all the fuss? In W. J. Gonzalez (ed.), *New approaches to scientific realism* (pp. 27–47). Berlin: De Gruyter.

Ackermann, R. (1985). *Data, instruments, and theory.* Princeton, NJ: Princeton University Press.

(1989). The new experimentalism. *British Journal for the Philosophy of Science, 40* (2), 185–190.

Anderson, C. (1933). The positive electron. *Physical Review, 43*, 491–494.

(1999). *The discovery of anti-matter: The autobiography of Carl David Anderson, the youngest man to win the Nobel Prize* (R. J. Weiss, ed.). Singapore: World Scientific.

Anderson, E. (1995). Knowledge, human interests, and objectivity in feminist epistemology. *Philosophical Topics, 23*, 27–57.

(2004). Uses of value judgments in science: A general argument, with lessons from a case study of feminist research on divorce. *Hypatia, 19*(1), 1–24.

(2020). Feminist epistemology and philosophy of science. In E. N. Zalta (ed.), *The Stanford encyclopedia of philosophy* (Fall 2012 edition). http://plato.stanford.edu/archives/spr2020/entries/feminism-epistemology.

Antoniou, A. (2021). What is a data model? *European Journal for Philosophy of Science, 11*, 101.

Antony, L. (2022). *Only natural: Gender, knowledge, and humankind.* New York: Oxford University Press.

Armstrong, D. M. (1983). *What is a law of nature?* Cambridge: Cambridge University Press.

Ashford, N., Ryan, C. W., & Caldart, C. C. (1983). A hard look at federal regulation of formaldehyde: A departure from reasoned decision making. *Harvard Environmental Law Review, 7*, 297–370.

Austin, J. L. (1962). *How to do things with words.* Oxford: Oxford University Press.

Ball, L., Mankiw, N. G., & Nordhaus, W. D. (1999). Aggregate demand and long-run unemployment. *Brookings Papers on Economic Activity, 1999*(2), 189–251.

Barnes, E. (1992). Explanatory unification and the problem of asymmetry. *Philosophy of Science, 59*(4), 558–571.

Batterman, R. W. (2001). *The devil in the details: Asymptotic reasoning in explanation, reduction, and emergence.* New York: Oxford University Press.

Batterman, R. W., & Rice, C. C. (2014). Minimal model explanations. *Philosophy of Science, 81*, 349–376.

Baum, R., & Sheehan, W. (2003). *In search of planet Vulcan: The ghost in Newton's clockwork universe.* New York: Basic Books.

Bayes, T. (1763). An essay towards solving a problem in the doctrine of chances. *Philosophical Transactions of the Royal Society of London*, 53, 370–418.

Beatty, J. (1981). What's wrong with the received view of evolutionary theory? *PSA: Proceedings of the Biennial Meeting of the Philosophy of Science Association, 2*, 397–426.

Beauchemin, P.-H. (2017). Autopsy of measurements with the ATLAS detector at the LHC. *Synthese, 194*(2), 275–312.

Bekelman, J. E., Li, Y., & Gross, C. P. (2003). Scope and impact of financial conflicts of interest in biomedical research: A systematic review. *JAMA*, 289(4), 454–465.

Berger, J., & Bernardo, J. (1992). On the development of reference priors. In J. M. Bernardo, J. O. Berger, P. Dawid, & A. Smith (eds.), *Bayesian statistics* 4 (pp. 35–60). New York: Oxford University Press.

Bernardini, C. (2004). AdA: The first electron–positron collider. *Physics in Perspective, 6*, 156–183.

Bernardo, J. M. (1979). Reference posterior distributions for Bayesian inference. *Journal of the Royal Statistical Society, B41*, 113–147.

Bhattacharjee, Y. (2013, April 26). The mind of a con man. *New York Times*.

Bird, A. (2000). *Thomas Kuhn*. Princeton, NJ: Princeton University Press.

Birnbaum, A. (1977). The Neyman–Pearson theory as decision theory, and as inference theory; with a criticism of the Lindley–Savage argument for Bayesian theory. *Synthese, 36*(1), 19–49.

Bissell, C. (2007). Historical perspectives – The Moniac a hydromechanical analog computer of the 1950s. *IEEE Control Systems Magazine*, 27, 69–74.

Blackett, P., & Occhialini, G. (1933). Some photographs of the tracks of penetrating radiation. *Proceedings of the Royal Society of London, 139A*, 699–726.

Blackwell, R. J. (2006). *Behind the scenes at Galileo's trial*. Notre Dame, IN: University of Notre Dame Press.

Blanchard, O., & Jimeno, J. F. (1995). Structural unemployment: Spain versus Portugal. *American Economic Review, 85*(2), 212–218.

Bloor, D. (1976). *Knowledge and social imagery*. Chicago: University of Chicago Press.

Boge, F. (Forthcoming). Why trust a simulation? Models, parameters, and robustness in simulation-infected experiments. *British Journal for the Philosophy of Science*.

Bogen, J., & Woodward, J. (1988). Saving the phenomena. *Philosophical Review*, 97 (3), 303–352.

Bohman, J., & McCollum, J. (eds.). (2012). Special issue: Epistemic injustice. *Social Epistemology: A Journal of Knowledge, Culture, and Policy, 26*(2).

Bokulich, A. (2008). *Reexamining the quantum-classical relation: Beyond reductionism and pluralism*. Cambridge: Cambridge University Press.

Bokulich, A., & Parker, W. (2021). Data models, representation, and adequacy-for-purpose. *European Journal for Philosophy of Science, 11*, 31.

Borgerson, K. (2021). Feminist empiricism. In C. Crasnow & K. Intemann (eds.), *The Routledge handbook of feminist philosophy of science* (pp. 79–88). New York: Routledge.

Born, M., Heisenberg, W., & Jordan, P. (1926). Zur Quantenmechanik ii. *Zeitschrift für Physik, 35*, 557–615.

Born, M., & Jordan, P. (1925). Zur Quantenmechanik. *Zeitschrift für Physik, 34*, 858–888.

Boston Women's Health Collective. (1973). *Our bodies, ourselves*. New York: Simon and Schuster.

Boumans, M. (1999). Built-in justification. In M. Morgan & M. Morrison (eds.), *Models as mediators: Perspectives on natural and social science* (pp. 66–96). Cambridge: Cambridge University Press.

Bovens, L., & Hartmann, S. (2003). *Bayesian epistemology*. New York: Oxford University Press.
Boyd, N. (2017). Evidence enriched. *Philosophy of Science, 85*, 403–421.
Boyd, R. (1989). What realism implies and what it does not. *Dialectica, 43*, 5–29.
Boyle, R. ([1660] 1965). New experiments physico-mechanical. In T. Birch (ed.), *The works of the Honorable Robert Boyle* (Third edition) (vol. 1, pp. 1–117). Hildesheim: Georg Olms Verlagsbuchhandlung.
([1662] 1965). A defence of the doctrine touching the spring and weight of the air. In T. Birch (ed.), *The works of the Honorable Robert Boyle* (Third edition) (vol. 1, pp. 118–185). Hildesheim: Georg Olms Verlagsbuchhandlung.
Braithwaite, R. B. (1953). *Scientific explanation: A study of the function of theory, probability and law in science*. Cambridge: Cambridge University Press.
Bromberger, S. (1966). Why-questions. In B. Brody (ed.), *Readings in the philosophy of science* (pp. 66–84). Englewood Cliffs, NJ: Prentice Hall.
Brulle, R. J. (2014). Institutionalizing delay: Foundation funding and the creation of U.S. climate change counter-movement organizations. *Climatic Change, 122*, 681–694.
Burke, A., Okrent, A., & Hale, K. (2022). *Science and engineering indicators*. National Science Board. https://ncses.nsf.gov/pubs/nsb20221. Accessed September 17, 2023.
Burks, A. (1963). On the significance of Carnap's system of inductive logic for the philosophy of induction. In P. A. Schilpp (ed.), *The philosophy of Rudolf Carnap* (pp. 739–759). La Salle, IL: Open Court; London: Cambridge University Press.
Campbell, R. (1998). *Illusions of paradox: A feminist epistemology naturalized*. New York: Rowman and Littlefield.
Carnap, R. (1936). Testability and meaning. *Philosophy of Science*, 3(4), 419–471.
(1937). Testability and meaning-continued. *Philosophy of Science*, 4(1), 1–40.
(1950). Empiricism, semantics, and ontology. *Revue internationale de philosophie*, 4(11), 20–40.
(1952). *The continuum of inductive methods*. Chicago: University of Chicago Press.
(1955). Foundations of logic and mathematics. In O. Neurath, R. Carnap, & C. W. Morris (eds.), *International encyclopedia of unified science* (vol. 1, pp. 139–213). Chicago: University of Chicago Press.
(1962). *Logical foundations of probability*. Chicago: University of Chicago Press.
(1963). Replies and systematic expositions. In P. A. Schilpp (ed.), *The philosophy of Rudolf Carnap* (pp. 859–1013). LaSalle, IL: Open Court.
(1968). Inductive logic and inductive intuition. In I. Lakatos (ed.), *The problem of inductive logic* (pp. 258–267). Amsterdam: North Holland.

Carrión-González, O., Moreno, R., Lellouch, E., Cavalié, T., Guerlet, S., Milcareck, G., Spiga, A., Clément, N., & Leconte, J. (2023). Doppler wind measurements in Neptune's stratosphere with ALMA. *Astronomy & Astrophysics, 674*, Letters to the editor, L3.

Cartwright, N. (1983). *How the laws of physics lie.* New York: Oxford University Press.

(1999a). *The dappled world: A study of the boundaries of science.* Cambridge: Cambridge University Press.

(1999b). Models and the limits of theory: Quantum Hamiltonians and the BCS models of superconductivity. In M. Morgan & M. Morrison (eds.), *Models as mediators: Perspectives on natural and social science* (pp. 241–281). Cambridge: Cambridge University Press.

(2012). Will this policy work for you? Predicting effectiveness better: How philosophy helps. *Philosophy of Science, 79*(5), 973–989.

Cavendish, M. (2019). *Margaret Cavendish: Essential writings* (D. Cunning, ed.). New York: Oxford University Press.

Chakravartty, A. (2010). Perspectivism, inconsistent models, and contrastive explanation. *Studies in History and Philosophy of Science, 41*, 405–412.

Chaloner, C. (1997). The most wonderful experiment in the world: A history of the cloud chamber. *British Journal for the History of Science, 30*(3), 357–374.

Chang, H. (2017). Operationalism: Old lessons and new challenges. In N. Mößner & A. Nordmann (eds.), *Reasoning in measurement* (pp. 25–38). New York: Routledge.

(2022). *Realism for realistic people: A new pragmatist philosophy of science*. Cambridge: Cambridge University Press.

Chatrchyan, S., Barass, T. A., Bostock, F. J. D., et al. (2012). Observation of a new boson at a mass of 125 GeV with the CMS experiment at the LHC. *Physics Letters, B716*, 30–61.

Churchman, C. W. (1948). Statistics, pragmatics, induction. *Philosophy of Science, 15*(3), 249–268.

Clark, B. J. (2019). Collaborative archaeology as heritage process. *Archaeologies: Journal of the World Archaeological Congress, 15*(3), 466–480.

Cobb, A. (2009). Michael Faraday's "historical sketch of electro-magnetism" and the theory-dependence of experimentation. *Philosophy of Science, 76*, 624–636.

Collins, H. M. (1992). *Changing order: Replication and induction in scientific practice.* Chicago: University of Chicago Press.

(1994). A strong confirmation of the experimenters' regress. *Studies in the History and Philosophy of Science, 25*(3), 493–503.

Committee on the Use of Chimpanzees in Biomedical and Behavioral Research. (2011). *Chimpanzees in biomedical and behavioral research: Assessing the necessity* (B. M. Altevogt, D. E. Pankevich, M. K. Shelton-Davenport, & J. P. Kahn, eds.). Washington, DC: The National Academies Press.

Conant, J. B. (1957). *Harvard case studies in experimental science*, vol. 1. Cambridge, MA: Harvard University Press.

Cousins, R. D. (2017). The Jeffreys–Lindley paradox and discovery criteria in high energy physics. *Synthese, 194*, 395–432.

Cox, D. R. (2006). *Principles of statistical inference*. New York: Cambridge University Press.

Cox, D. R., & Hinkley, D. V. (1974). *Theoretical statistics*. London: Chapman and Hall.

Cox, D. R., & Mayo, D. G. (2010). Objectivity and conditionality in frequentist inference. In D. G. Mayo & A. Spanos (eds.), *Error and inference: Recent exchanges on experimental reasoning, reliability, objectivity, and rationality* (pp. 276–304). New York: Cambridge University Press.

Crasnow, S. (2023). Feminist perspectives on science. *The Stanford encyclopedia of philosophy* (E. N. Zalta & U. Nodelman, eds.). https://plato.stanford.edu/archives/fall2023/entries/feminist-science/.

Crasnow, S., & Intemann, K. (eds.). (2021). *The Routledge handbook of feminist philosophy of science*. New York: Routledge.

Craver, C. F. (2007). *Explaining the brain: Mechanisms and the mosaic unity of neuroscience*. New York: Oxford University Press.

(2008). Physical law and mechanistic explanation in the Hodgkin and Huxley model of the action potential. *Philosophy of Science, 75*(5), 1022–1033.

Craver, C. F., & Darden, L. (2013). *In search of mechanisms: Discoveries across the life sciences*. Chicago: University of Chicago Press.

Dahlberg, F. (ed.). (1981). *Woman the gatherer*. New Haven, CT: Yale University Press.

Darwin, C. (1859). *On the origin of species by means of natural selection or the preservation of favoured races in the struggle for life*. London: John Murray.

Davidson, J. (2013, January 17). Federal scientists can again research gun violence. *Washington Post*.

Dawid, R. (2017). Bayesian perspectives on the discovery of the Higgs particle. *Synthese, 194*, 377–394.

De Courtenay, N., & Grégis, F. (2017). The evaluation of measurement uncertainties and its epistemological ramifications. *Studies in History and Philosophy of Science, 65–66*, 21–32.

De Finetti, B. ([1937] 1964). Foresight: Its logical laws, its subjective sources (H. E. Kyburg, trans.). In H. E. Kyburg & H. E. Smokler (eds.), *Studies in subjective probability* (pp. 93–158). New York: John Wiley and Sons.

(1972). *Probability, induction, and statistics: The art of guessing.* New York: John Wiley.

De Melo-Martin, I., & Intemann, K. (2018). *The fight against doubt: How to bridge the gap between scientists and the public.* New York: Oxford University Press.

de Regt, H. W. (2004). Review of James Woodward, *Making things happen. Notre Dame Philosophical Reviews.* http://ndpr.nd.edu/news/23818-making-things-happen-a-theory-of-causal-explanation.

(2017). *Understanding scientific understanding.* New York: Oxford University Press.

Dennett, D. (1991) Real patterns. *Journal of Philosophy, 88*(1), 27–51.

Derrida, J. (1977). *Of grammatology.* Baltimore, MD: The Johns Hopkins University Press.

Detlefsen, K. (2021). Émilie du Châtelet: Feminism, epistemology, and natural philosophy. In S. Crasnow & K. Intemann (eds.), *The Routledge handbook of feminist philosophy of science* (pp. 41–52). New York: Routledge.

Detlefsen, K., & Janiak, A. (2018). Émilie du Châtelet. *The Stanford encyclopedia of philosophy* (E. N. Zalta, ed.). https://plato.stanford.edu/archives/win2018/entries/emilie-du-chatelet/.

Dirac, P. A. M. (1928). The quantum theory of the electron. *Proceedings of the Royal Society of London. Series A, Containing Papers of a Mathematical and Physical Character, 117*(778), 610–624.

(1931). Quantised singularities in the electromagnetic field. *Proceedings of the Royal Society of London, 133A*, 60–72.

Doob, J. L. (1971). What is a martingale? *American Mathematical Monthly, 78*, 451–462.

Dorling, J. (1979). Bayesian personalism, the methodology of scientific research programmes, and Duhem's problem. *Studies in History and Philosophy of Science, 10*, 177–187.

Douglas, H. (2009). *Science, policy, and the value-free ideal.* Pittsburgh, PA: University of Pittsburgh Press.

(2021). *The rightful place of science: Science, values, and democracy: The 2016 Descartes lectures.* Tempe, AZ: Consortium for Science, Policy & Outcomes.

Dowe, P. (1992). Wesley Salmon's process theory of causality and the conserved quantity theory. *Philosophy of Science, 59*(2), 195–216.

(1995). Causality and conserved quantities: A reply to Salmon. *Philosophy of Science, 62*(2), 321–333.

Dring, K. S., Silliman, S. W., Gambrell, N., Sebastian, S., & Sidberry, R. S. (2019). Authoring and authority in Eastern Pequot community heritage and archaeology. *Archaeologies: Journal of the World Archaeological Congress, 15*(3), 352–370.

Drouet, L., Bosetti, V., & Tavoni, M. (2015). Selection of climate policies under the uncertainties in the Fifth Assessment Report of the IPCC. *Nature Climate Change, 5*(October), 937–943.

Duhem, P. ([1906] 1991). *The aim and structure of physical theory*. Princeton, NJ: Princeton University Press.

Earman, J. (1986). *A primer on determinism*. Dordrecht: Reidel.

(1992). *Bayes or bust? A critical examination of Bayesian confirmation theory*. Cambridge, MA: MIT Press.

Ehrenreich, B., & English, D. (2010). *Witches, midwives, and nurses* (Second edition). New York: Feminist Press.

Einstein, A. (1918). Prinzipielles zur allgemeinen Relativitatstheorie. *Annalen der Physik, 55*, 241–244.

Elgin, C. (2017). *True enough*. Cambridge, MA: MIT Press.

Elliott, K. C. (2011). Direct and indirect roles for values in science. *Philosophy of Science, 78*, 303–324.

(2017). *A tapestry of values: An introduction to values in science*. New York: Oxford University Press.

Enz, C. P. (2002). *No time to be brief: A scientific biography of Wolfgang Pauli*. New York: Oxford University Press.

Epstein, S. S. (1998). *The politics of cancer revisited*. Fremont Center, NY: East Ridge.

Fausto-Sterling, A. (2000). *Sexing the body: Gender politics and the construction of sexuality*. New York: Basic Books.

Ferber, M. A., & Nelson, J. A. (2003). *Feminist economics today: Beyond economic man*. Chicago: University of Chicago Press.

Fernandez Pinto, M. (2020a). Ignorance, science, and feminism. In S. Crasnow & K. Intemann (eds.), *The Routledge handbook of feminist philosophy of science* (pp. 225–235). New York: Routledge.

(2020b). Commercial interests and the erosion of trust in science. *Philosophy of Science, 87*, 1003–1013.

Feyerabend, P. (1970). Against method: Outline of an anarchistic theory of knowledge. In *Analyses of theories and methods of physics and psychology* (vol. 4, pp. 17–130). Minneapolis, MN: University of Minnesota Press.

(1978). *Science in a free society*. London: Verso.

(1981). Two models of epistemic change: Mill and Hegel. In *Problems of empiricism* (pp. 65–79). Cambridge: Cambridge University Press.

(1988). *Against method* (Revised edition). London: Verso.

(1991). *Three dialogues on knowledge*. Oxford: Basil Blackwell.

(1995). *Killing time: The autobiography of Paul Feyerabend*. Chicago: University of Chicago Press.

Fine, A. (1984). The natural ontological attitude. In J. Leplin (ed.), *Scientific realism* (pp. 83–107). Berkeley: University of California Press.

(1986). Unnatural attitudes: Realist and instrumentalist attachments to science. *Mind, 95*, 149–179.

(1996). Science made up: Constructivist sociology of scientific knowledge. In P. Galison & D. Stump (eds.), *The disunity of science: Boundaries, contexts, and power* (pp. 231–254). Stanford: Stanford University Press.

Finlayson, L. (2016). *An introduction to feminism*. Cambridge: Cambridge University Press.

Finocchiaro, M. (1989). *The Galileo affair: A documentary history.* Berkeley: University of California Press.

Fiorin, G., Carnevale, V., & DeGrado, W. F. (2010). The flu's proton escort. *Science, 330*(6003), 456–458.

Fisher, R. A. (1949). *The design of experiments* (Fifth edition). New York: Hafner.

Franklin, A. (1986). *The neglect of experiment.* New York: Cambridge University Press.

(1994). How to avoid the experimenters' regress. *Studies in the History and Philosophy of Modern Physics*, 25, 463–491.

(1997). Calibration. *Perspectives on Science, 5*, 31–80.

(2005). *No easy answers: Science and the pursuit of knowledge.* Pittsburgh, PA: University of Pittsburgh Press.

(2012). Experiment in physics. In E. N. Zalta (ed.), *The Stanford encyclopedia of philosophy* (Winter 2012 edition). http://plato.stanford.edu/archives/win2012/entries/physics-experiment/.

(2017). The missing piece of the puzzle: The discovery of the Higgs boson. *Synthese, 194*, 259–274.

Franklin, A., & Collins, H. (2016). Two kinds of case study and a new agreement. In T. Sauer & R. Scholl (eds.), *The philosophy of historical case studies.* Boston Studies in the Philosophy and History of Science 319 (pp. 95–121). Cham: Springer.

Fricker, M. (2007). *Epistemic injustice: Power and the ethics of knowing*. Oxford: Oxford University Press.

Friedman, M. (1974). Explanation and scientific understanding. *Journal of Philosophy, 71*(1), 5–19.

(1999). *Reconsidering logical positivism.* Cambridge: Cambridge University Press.

Frigg, R., & Votsos, I. (2011). Everything you always wanted to know about structural realism but were afraid to ask. *European Journal for Philosophy of Science, 1*, 227–276.

Gaifman, H., & Snir, M. (1982). Probabilities over rich languages. *Journal of Symbolic Logic, 47*, 495–548.

Galilei, G. ([1632] 1967). *Dialogue concerning the two chief world systems* (Revised edition; S. Drake, trans.). Berkeley: University of California Press.

Galison, P. (1987). *How experiments end.* Chicago: University of Chicago Press.

Galison, P., & Assmus, A. (1989). Artificial clouds, real particles. In D. Gooding, T. Pinch, & S. Schaffer (eds.), *The uses of experiment: Studies in the natural sciences* (pp. 225–273). New York: Cambridge University Press.

Gardner, M. (1970). The fantastic combinations of John Conway's new solitaire game "life." *Scientific American, 223*(4), 120–123.

Gelfert, A. (2016). *How to do science with models: A philosophical primer.* Cham: Springer.

Giere, R. N. (1988). *Explaining science: A cognitive approach.* Chicago: University of Chicago Press.

(2004). How models are used to represent reality. *Philosophy of Science, 71*, 742–752.

(2006). *Scientific perspectivism.* Chicago: University of Chicago Press.

Gilbert, N., & Troitzsch, K. (2005). *Simulation for the social scientist.* Berkshire: Open University Press.

Gillies, D. (1993). The Duhem thesis and the Quine thesis. In M. Curd & J. A. Cover (eds.), *Philosophy of science: The central issues* (pp. 302–319). New York: W.W. Norton.

Glennan, S. S. (1996). Mechanisms and the nature of causation. *Erkenntnis, 44*(1), 49–71.

Goldin, C. (1990). *Understanding the gender gap: An economic history of American women.* Oxford: Oxford University Press.

Gooding, D. (1990). *Experiment and the making of meaning: Human agency in scientific observation and experiment.* Dordrecht: Kluwer.

Goodman, N. (1955). *Fact, fiction, and forecast.* Cambridge, MA: Harvard University Press.

Gorman, J. (2011, December 15). U.S. will not finance new research on chimps. *New York Times.*

(2013, January 22). Agency moves to retire most research chimps. *New York Times.*

Grégis, F. (2019). On the meaning of measurement uncertainty. *Measurement, 133*, 41–46.

Grene, M. (1966). *The knower and the known.* New York: Basic Books.

(1985). Perception, interpretation, and the sciences. In D. Depew & B. Weber (eds.), *Evolution at a crossroads* (pp. 1–20). Cambridge, MA: MIT Press.

Grimm, S. (2014). Understanding as knowledge of causes. In A. Fairweather (ed.), *Virtue epistemology naturalized: Bridges between virtue epistemology and philosophy of science* (pp. 329–345). Cham: Springer.

(ed.). (2019). *Varieties of understanding: New perspectives from philosophy, psychology, and theology*. New York: Oxford University Press.

(2021). Understanding. *The Stanford encyclopedia of philosophy* (E. N. Zalta, ed.). https://plato.stanford.edu/archives/sum2021/entries/understanding/.

Guerlac, H. (1983). Can we date Newton's early optical experiments? *Isis, 74*, 74–80.

Haack, S. (2014). Do not block the way of inquiry. *Transactions of the Charles S. Peirce Society, 50*(3), 319–339.

Hacking, I. (1975). *The emergence of probability*. Cambridge: Cambridge University Press.

(1981). Do we see through a microscope? *Pacific Philosophical Quarterly, 62*, 305–322.

(1982). Experimentation and scientific realism. *Philosophical Topics, 13*, 71–87.

(1983). *Representing and intervening*. Cambridge: Cambridge University Press.

(1990). *The taming of chance*. Cambridge: Cambridge University Press.

(1999). *The social construction of what?* Cambridge, MA: Harvard University Press.

(2001). *An introduction to probability and inductive logic*. Cambridge: Cambridge University Press.

Hajek, A. (2008). Arguments for – or against – probabilism? *British Journal for the Philosophy of Science, 59*, 793–819.

Hanson, N. R. (1958). *Patterns of discovery*. Cambridge: Cambridge University Press.

(1962). Leverrier: The zenith and nadir of Newtonian mechanics. *Isis, 53*(3), 359–378.

Haraway, D. (1988). The science question in feminism and the privilege of partial perspective. *Feminist Studies, 14*(3), 575–599.

Harding, S. (1986). *The science question in feminism*. Ithaca, NY: Cornell University Press.

(1991). *Whose science? Whose knowledge? Thinking from women's lives*. Ithaca, NY: Cornell University Press.

(ed.). (2004). *The feminist standpoint theory reader*. New York: Routledge.

(2015). *Objectivity and diversity: Another logic of scientific research*. Chicago: University of Chicago Press.

Harman, G. (1965). The inference to the best explanation. *Philosophical Review, 74*, 88–95.

Harper, W. (2002). Newton's argument for universal gravitation. In I. B. Cohen & G. E. Smith (eds.), *The Cambridge companion to Newton* (pp. 174–201). New York: Cambridge University Press.

(2007). Newton's methodology and Mercury's perihelion before and after Einstein. *Philosophy of Science, 74*(5), 932–942.

(2011). *Isaac Newton's scientific method: Turning data into evidence about gravity and cosmology.* New York: Oxford University Press.

Harris, T. (2003). Data models and the acquisition and manipulation of data. *Philosophy of Science, 70*(5), 1508–1517.

Hartmann, S. (1996). The world as a process: Simulation in the natural and social sciences. In R. Hegelsmann, U. Muller, & K. Troitzsch (eds.), *Modelling and simulation in the social sciences from the philosophy of science point of view* (pp. 77–100). Dordrecht: Kluwer Academic.

Hartsock, N. C. M. (1983). The feminist standpoint: Developing the ground for a specifically feminist historical materialism. In M. B. Hintikka & S. Harding (eds.), *Discovering reality* (pp. 283–310). Boston, MA: D. Reidel. (Rpt. in Harding 2004.)

Haslanger, S. (2017). Jane Addams's "Women and public housekeeping." In E. Schliesser (ed.), *Ten neglected classics of philosophy.* New York: Oxford University Press.

Heisenberg, W. (1925). Über quantentheoretische Umdeutung kinematischer und mechanischer Beziehungen. *Zeitschrift für Physik, 33*, 261–276.

Hempel, C. G. (1945). Studies in the logic of confirmation. *Mind, 54*, 1–26, 97–121.

(1965). Empiricist criteria of cognitive significance: Problems and changes. In *Aspects of scientific explanation and other essays in the philosophy of science* (pp. 101–119). New York: Free Press.

(1966). *Philosophy of natural science.* Englewood Cliffs, NJ: Prentice Hall.

Hempel, C. G., & Oppenheim, P. (1948). Studies in the logic of explanation. *Philosophy of Science, 15*, 135–175.

Hesse, M. (1980). *Revolutions and reconstructions in the philosophy of science.* Bloomington, IN: Indiana University Press.

(1986). Changing concepts and stable order. *Social Studies of Science, 16*(4), 714–726.

Hitchcock, C. R. (1995). Salmon on explanatory relevance. *Philosophy of Science, 62* (2), 304–320.

Hodgkin, A., & Huxley, A. (1952). A quantitative description of membrane current and its application to conduction and excitation in nerve. *Journal of Physiology, 117*, 500–544.

Hon, G. (1987). H. Hertz: "The electrostatic and electromagnetic properties of the cathode rays are either nil or very feeble." (1883) A case-study of an experimental error. *Studies in History and Philosophy of Science Part A, 18*(3), 367–382.

(1989). Towards a typology of experimental errors: An epistemological view. *Studies in the History and Philosophy of Science, 20*, 469–504.

hooks, b. (2000). Feminism: A movement to end sexist oppression. In *Feminist theory: From margin to center* (Second edition). Cambridge, MA: South End Press.

Hooper, W. (1998). Inertial problems in Galileo's preinertial framework. In P. Machamer (ed.), *The Cambridge companion to Galileo* (pp. 146–174). New York: Cambridge University Press.

Howson, C. (1997). Error probabilities in error. *Philosophy of Science, 64*, S185–S194.

Howson, C., & Urbach, P. (2000). Paul K. Feyerabend: An obituary. In J. Preston, G. Munévar, & D. Lamb (eds.), *The worst enemy of science? Essays in memory of Paul Feyerabend* (pp. 3–15). New York: Oxford University Press.

(2005). *Scientific reasoning: The Bayesian approach* (Third edition). Chicago: Open Court.

Hoyningen-Huene, P. (2000). Paul K. Feyerabend: An obituary. In J. Preston, G. Munévar, & D. Lamb (eds.), *The worst enemy of science? Essays in memory of Paul Feyerabend* (pp. 3–15). New York: Oxford University Press.

Hughes, R. I. G. (1999). The Ising model, computer simulation, and universal physics. In M. Morgan & M. Morrison (eds.), *Models as mediators: Perspectives on natural and social science* (pp. 97–145). Cambridge: Cambridge University Press.

Hume, D. ([1748] 1993). *An enquiry concerning human understanding* (Second edition; E. Steinberg, ed.). Indianapolis, IN: Hackett.

Humphreys, P. (2000). Analytic versus synthetic understanding, in J. Fetzer (ed.), *Science, explanation, and rationality: The philosophy of Carl G. Hempel* (pp. 267–286). Oxford: Oxford University Press.

(2004). *Extending ourselves: Computational science, empiricism, and scientific method.* New York: Oxford University Press.

(2008). *Notes on the origin of the term "Dutch book."* http://people.virginia.edu/~pwh2a/dutch%2520book%2520origins.doc.

Intemann, K. (2005). Feminism, underdetermination, and values in science. *Philosophy of Science, 72*(5), 1001–1012.

(2010). 25 years of feminist empiricism and standpoint theory: Where are we now? *Hypatia, 25*(4), 778–796.

Intergovernmental Panel on Climate Change (IPCC). (2005). Guidance notes for lead authors of the IPCC fourth assessment report on addressing uncertainties. https://www.ipcc.ch/report/ar4/wg1.

(2023). Summary for policymakers. In Core Writing Team, H. Lee, & J. Romero (eds.), *Climate change 2023: Synthesis report. contribution of working groups I, II and III to the sixth assessment report of the Intergovernmental Panel on Climate Change* (pp. 1–34). Geneva: IPCC.

Jammer, M. (2000). *Concepts of mass in contemporary physics and philosophy.* Princeton, NJ: Princeton University Press.

Jasanoff, S. S. (1987). Contested boundaries in policy-relevant science. *Social Studies of Science, 17*(2), 195–230.

Jaynes, E. T. (2003). *Probability theory: The logic of science.* New York: Cambridge University Press.

Jeffrey, R. C. (1956). Valuation and acceptance of scientific hypotheses. *Philosophy of Science, 23*(3), 237–246.

Kadvany, J. (2001). *Imre Lakatos and the guises of reason.* Durham, NC: Duke University Press.

Karaca, K. (2023). A network account of models in high energy physics. *Philosophy of Science, 90*(4), 777–796.

Keller, E. F. (1977). The anomaly of a woman in physics. In S. Ruddick & P. Daniels (eds.), *Working it out: Twenty-three women writers, artists, scientists, and scholars talk about their lives and work* (pp. 78–91). New York: Pantheon.

(1985). *Reflections on gender and science.* New Haven, CT: Yale University Press.

(1993). *A feeling for the organism: The life and work of Barbara McClintock.* New York: W. H. Freeman.

(2003). Models, simulation, and "computer experiments." In H. Radder (ed.), *The philosophy of scientific experimentation* (pp. 198–215). Pittsburgh, PA: University of Pittsburgh Press.

Keller, E. F., & Longino, H. E. (eds.). (1996). *Feminism and science.* New York: Oxford University Press.

Kellermann, A. L., & Rivara, F. P. (2013). Silencing the science on gun research. *JAMA, 309*(6), 549–550.

Kellert, S., Longino, H., & Waters, C. K. (2006). *Scientific pluralism.* Minneapolis: University of Minnesota Press.

Kerns, W. D., Pavkov, K. L., Donofrio, D. J., Gralla, E. J., & Swenberg, J. A. (1983). Carcinogenicity of formaldehyde in rats and mice after long-term inhalation exposure. *Cancer Research, 43*(9), 4382–4392.

Khalifa, K. (2017). *Understanding, explanation, and scientific knowledge.* New York: Cambridge University Press.

Kitcher, P. (1976). Explanation, conjunction, and unification. *Journal of Philosophy, 73*(8), 207–212.

(1981). Explanatory unification. *Philosophy of Science, 48*(4), 507–531.

(1989). Explanatory unification and the causal structure of the world. In P. Kitcher & W. Salmon (eds.), *Scientific explanation* (vol. 13, pp. 410–506). Minneapolis: University of Minnesota Press.

(2001). *Science, truth, and democracy.* New York: Oxford University Press.

Kleingeld, P., & Brown, E. (2019). Cosmopolitanism. *The Stanford encyclopedia of philosophy* (E. N. Zalta, ed.). https://plato.stanford.edu/archives/win2019/entries/cosmopolitanism.

Koertge, N. (2004). How might we put gender politics into science? *Philosophy of Science, 71*(5), 868–879.

Kolmogorov, A. N. (1950). *Foundations of the theory of probability* (N. Morrison, trans.). New York: Chelsea.

Kourany, J. (2010). *Philosophy of science after feminism.* New York: Oxford University Press.

Kronig, R. (1960). The turning point. In M. Fierz & V. F. Weisskopf (eds.), *Theoretical physics in the twentieth century: A memorial volume to Wolfgang Pauli* (pp. 5–39). New York: Interscience Publishers.

Kuhn, T. (1957). *The Copernican revolution: Planetary astronomy in the development of western thought.* Cambridge, MA: Harvard University Press.

(1962). *The structure of scientific revolutions.* Chicago: University of Chicago Press.

(1977). Objectivity, value judgment, and theory choice. In *The essential tension* (pp. 320–339). Chicago: University of Chicago Press.

(1996). *The structure of scientific revolutions* (Third edition). Chicago: University of Chicago Press.

Kusch, M. (2020). Introduction: A primer on relativism. In M. Kusch (ed.), *The Routledge handbook of philosophy of relativism* (pp. 1–7). New York: Routledge.

Kyburg, H. E. (1965). Salmon's paper. *Philosophy of Science, 32*(2), 147–151.

(1978). Subjective probability: Criticisms, reflections, and problems. *Journal of Philosophical Logic, 7*(1), 157–180.

(1993). Peirce and statistics. In E. C. Moore (ed.), *Charles S. Peirce and the philosophy of science: Papers from the Harvard sesquicentennial congress* (pp. 130–138). Tuscaloosa, AL: University of Alabama Press.

Lacey, H. (1999). *Is science value free? Values and scientific understanding.* London: Routledge.

(2005). *Values and objectivity in science: The current controversy about transgenic crops.* Oxford: Lexington Books.

Ladyman, J. (2023). Structural realism. *The Stanford encyclopedia of philosophy* (Summer 2023 edition; E. N. Zalta & U. Nodelman, eds.). https://plato.stanford.edu/archives/sum2023/entries/structural-realism.

Ladyman, J., Ross, D., Spurrett, D., & Collier, J. (2007). *Every thing must go: Metaphysics naturalized.* New York: Oxford University Press.

Lakatos, I. (1970). Falsification and the methodology of scientific research programmes. In I. Lakatos & A. Musgrave (eds.), *Criticism and the growth of knowledge* (pp. 91–196). Cambridge: Cambridge University Press.

(1978). *The methodology of scientific research programmes: Philosophical papers* (vol. 1; J. Worrall & G. Currie, eds.). Cambridge: Cambridge University Press.

Lakatos, I., & Musgrave, A. (eds.). (1970). *Criticism and the growth of knowledge.* Cambridge: Cambridge University Press.

Landau, I. (1998). Feminist criticisms of metaphors in Bacon's philosophy of science. *Philosophy, 73*(283), 47–61.

Lange, M. (2009). *Laws and lawmakers: Science, metaphysics, and the laws of nature.* New York: Oxford University Press.

Larvor, B. (1998). *Lakatos: An introduction.* New York: Routledge.

Lascano, M. P. (2021). Margaret Cavendish and the new science. In S. Crasnow & K. Intemann (eds.), *The Routledge handbook of feminist philosophy of science* (pp. 28–40). New York: Routledge.

Latour, B., & Woolgar, S. (1979). *Laboratory life: The social construction of scientific facts.* Beverly Hills, CA: Sage.

(1986). *Laboratory life: The construction of scientific facts* (Second edition). Princeton, NJ: Princeton University Press.

Laudan, L. (1981). A confutation of convergent realism. *Philosophy of Science, 48*, 19–49.

(1996a). Demystifying underdetermination. In *Beyond positivism and relativism* (pp. 29–54). Boulder, CO: Westview Press.

(1996b). *Beyond positivism and relativism.* Boulder, CO: Westview Press.

Lemmens, T., & Waring, D. R. (eds.). (2006). *Law and ethics in biomedical research: Regulation, conflict of interest, and liability.* Toronto: University of Toronto Press.

Lenhard, J. (2007). Computer simulation: The cooperation between experimenting and modeling. *Philosophy of Science, 74*(2), 176–194.

Leonelli, S. (2016). *Data-centric biology: A philosophical study.* Chicago: University of Chicago Press.

(2019). What distinguishes data from models? *European Journal for Philosophy of Science, 9*, 22.

Leplin, J. (1997). *A novel defense of scientific realism.* New York: Oxford University Press.

Levi, I. (1962). On the seriousness of mistakes. *Philosophy of Science,* 29(1), 47–65.

(1967). *Gambling with truth: An essay on induction and the aims of science.* New York: Alfred A. Knopf.

Lewis, D. (1973). *Counterfactuals.* Oxford: Blackwell.

(1986). *Philosophical papers: Volume II.* New York: Oxford University Press.

Lewis, P. J. (2001). Why the pessimistic induction is a fallacy. *Synthese, 129*, 371–380.

Lilford, R. J., & Jackson, J. (1995). Equipoise and the ethics of randomization. *Journal of the Royal Society of Medicine, 88*, 552–559.

Lin, H. (2022). Bayesian epistemology. *The Stanford encyclopedia of philosophy* (Fall 2022 edition; Edward N. Zalta & Uri Nodelman, eds.). https://plato.stanford.edu/archives/fall2022/entries/epistemology-bayesian.

Lipton, P. (2004). *Inference to the best explanation* (Second edition). London: Routledge.

Lloyd, E. (1994). *The structure and confirmation of evolutionary theory*. Princeton, NJ: Princeton University Press.

(1997). Feyerabend, Mill, and pluralism. *Philosophy of Science, 64*, S396–S407.

(2005). *The case of the female orgasm: Bias in the science of evolution*. Cambridge, MA: Harvard University Press.

(2010). Confirmation and robustness of climate models. *Philosophy of Science, 77*, 971–984.

Longino, H. (1990). *Science as social knowledge*. Princeton, NJ: Princeton University Press.

(1993). Subjects, power, and knowledge: Description and prescription in feminist philosophies of science. In L. Alcoff & E. Potter (eds.), *Feminist epistemologies* (pp. 101–120). New York: Routledge. Reprinted in Keller and Longino, 1996, pp. 264–279.

(1994). The fate of knowledge in social theories of science. In F. Schmitt (ed.), *Socializing epistemology: The social dimensions of knowledge* (pp. 135–157). Lanham, MD: Rowman and Littlefield.

(2002). *The fate of knowledge*. Princeton, NJ: Princeton University Press.

Lundh, A., Lexchin, J., Mintzes, B., Schroll, J. B., & Bero, L. (2017). Industry sponsorship and research outcome. *Cochrane Database of Systematic Reviews, No. 2*. https://doi.org//10.1002/14651858.MR000033.pub3.

Machamer, P., Darden, L., & Craver, C. F. (2000). Thinking about mechanisms. *Philosophy of Science, 67*(1), 1–25.

MacIntyre, A. (1981). *After virtue*. Notre Dame, IN: University of Notre Dame Press.

Magnus, P. D., & Callender, C. (2004). Realist ennui and the base rate fallacy. *Philosophy of Science, 71*(3), 320–338.

Maley, C. J. (2023). Analogue computation and representation. *British Journal for the Philosophy of Science*, 74(3), 739–769.

Massimi, M. (2004). Non-defensible middle ground for experimental realism: Why we are justified to believe in colored quarks. *Philosophy of Science, 71*, 36–60.

(2022). *Perspectival realism*. New York: Oxford University Press.

Massimi, M., & Bhimji, W. (2015). Computer simulations and experiments: The case of the Higgs boson. *Studies in History and Philosophy of Modern Physics, 51*, 71–81.

Masterman, M. (1970). The nature of a paradigm. In I. Lakatos & A. Musgrave (eds.), *Criticism and the growth of knowledge* (pp. 59–89). Cambridge: Cambridge University Press.

Matthews, M. R. (1989). *The scientific background to modern philosophy*. Indianapolis, IN: Hackett.

Maxwell, G. (1962). The ontological status of theoretical entities. In H. Feigl & G. Maxwell (eds.), *Minnesota studies in the philosophy of science* (vol. 3: Scientific Explanation, Space, and Time, pp. 3–27). Minneapolis: University of Minnesota Press.

Maxwell, J. C. (1861). On physical lines of force. Parts I & II. *Philosophical Magazine, 21*, 161–175, 281–291, 338–348.

Mayo, D. G. (1985). Behavioristic, evidentialist, and learning models of statistical testing. *Philosophy of Science, 52*, 493–516.

(1991). Sociological versus metascientific views of risk assessment. In D. G. Mayo & R. Hollander (eds.), *Acceptable evidence: Science and values in risk management* (pp. 249–279). New York: Oxford University Press.

(1993). The test of experiment: C. S. Peirce and E. S. Pearson. In E. C. Moore (ed.), *Charles S. Peirce and the philosophy of science: Papers from the Harvard sesquicentennial congress* (pp. 161–174). Tuscaloosa: University of Alabama Press.

(1996). *Error and the growth of experimental knowledge*. Chicago: University of Chicago Press.

(1997a). Duhem's problem, the Bayesian way, and error statistics, or "what's belief got to do with it?" *Philosophy of Science, 64*, 222–244.

(1997b). Response to Howson and Laudan. *Philosophy of Science, 64*(2), 323–333.

(2002). Theory testing, statistical methodology, and the growth of experimental knowledge. In P. Gardenfors, J. Wolinski, & K. Kijania-Placek (eds.), *In the scope of logic, methodology, and philosophy of science* (pp. 171–190). Dordrecht: Kluwer.

(2018). *Statistical inference as severe testing: How to get beyond the statistics wars*. Cambridge: Cambridge University Press.

(2022). The statistics wars and intellectual conflicts of interest. *Conservation Biology, 36*, e13861.

Mayo, D. G., & Spanos, A. (2004). Methodology in practice: Statistical misspecification testing. *Philosophy of Science, 71*, 1007–1025.

(2006). Severe testing as a basic concept in a Neyman–Pearson philosophy of induction. *British Journal for the Philosophy of Science, 57*(2), 323–357.

Mayo, D. G., & Spanos, A. (eds.). (2009). *Error and inference: Recent exchanges on experimental reasoning, reliability, objectivity, and rationality*. New York: Cambridge University Press.

McAllister, J. A. (1999). *Beauty and revolution in science*. Ithaca, NY: Cornell University Press.

McCright, A. M., & Dunlap, R. E. (2010). Anti-reflexivity: The American conservative movement's success in undermining climate science and policy. *Theory, Culture, & Society, 27*(2–3), 100–133.

McLaughlin, A. (2009). Peircean polymorphism: Between realism and anti-realism. *Transactions of the Charles S. Peirce Society, 45*(3), 402–421.

(2011). In pursuit of resistance: Pragmatic recommendations for doing science within one's means. *European Journal for Philosophy of Science, 1*(3), 353–371.

McMullin, E. (1982). Values in science. In P. D. Asquith & T. Nickles (eds.), *PSA 1982: Proceedings of the biennial meeting of the Philosophy of Science Association* (vol. 2: Symposia and Invited Papers, pp. 3–28). Chicago: University of Chicago Press.

(1985). Galilean idealization. *Studies in History and Philosophy of Science, 16*(3), 247–273.

(1993). Rationality and paradigm change in science. In P. Horwich (ed.), *World changes* (pp. 55–78). Cambridge, MA: MIT Press.

Menon, T., & Stegenga, J. (2023). Sisyphean science: Why value freedom is worth pursuing. *European Journal for Philosophy of Science, 13*, 48.

Merchant, C. (1980). *The death of nature: Women, ecology, and the scientific revolution*. San Francisco: Harper.

(2008). Secrets of nature: The Bacon debates revisited. *Journal of the History of Ideas, 69*(1), 147–162.

Merton, R. K. (1938). Science, technology and society in seventeenth century England. *Osiris, 4*, 360–632.

(1973). *The sociology of science* (N. W. Storer, ed.). Chicago: University of Chicago Press.

Metropolis, N. (1987). The beginning of the Monte Carlo method. *Los Alamos Science, 15*, 125–130.

Metropolis, N., & Ulam, S. (1949). The Monte Carlo method. *Journal of the American Statistical Association, 44*, 335–341.

Michelson, A. A. (1879). Experimental determination of the velocity of light. *Nature, 21*, 94–96, 120–122, 226.

Mill, J. S. ([1859] 1963). On liberty. In J. M. Robson (ed.), *Collected works: Essays on politics and society* (vol. 18, pp. 213–310). Toronto: University of Toronto Press.

Miller, D. E. (2022). Harriet Taylor Mill. *The Stanford encyclopedia of philosophy* (E. N. Zalta & U. Nodelman, eds.). https://plato.stanford.edu/archives/fall2022/entries/harriet-mill.

Morgan, M. (2003). Experiments without material intervention: Model experiments, virtual experiments, and virtually experiments. In H. Radder (ed.),

The philosophy of scientific experimentation (pp. 216–235). Pittsburgh, PA: University of Pittsburgh Press.

Morgan, M., & Boumans, M. (2004). Secrets hidden by two-dimensionality: The economy as a hydraulic machine. In S. de Chadarevian & N. Hopwood (eds.), *Models: The third dimension of science* (pp. 369–401). Stanford: Stanford University Press.

Morgan, M., & Morrison, M. (eds.). (1999). *Models as mediators: Perspectives on natural and social science*. Cambridge: Cambridge University Press.

Morrison, D. E., & Henkel, R. E. (eds.). (1970). *The significance test controversy: A reader*. Chicago, IL: Aldine.

Morrison, M. (2000). *Unifying scientific theories: Physical concepts and mathematical structures*. New York: Cambridge University Press.

(2009). Models, measurement and computer simulation: The changing face of experimentation. *Philosophical Studies, 143*, 33–57.

(2011). One phenomenon, many models: Inconsistency and complementarity. *Studies in History and Philosophy of Science, 42*, 342–351.

(2015). *Reconstructing reality: Models, mathematics, and simulations*. Oxford: Oxford University Press.

Morrison, M., & Morgan, M. (1999). Models as mediating instruments. In M. Morgan & M. Morrison (eds.), *Models as mediators: Perspectives on natural and social science*. Cambridge: Cambridge University Press.

Musgrave, A. (1974). Logical versus historical theories of confirmation. *British Journal for the Philosophy of Science, 25*(1), 1–23.

(1982). Constructive empiricism versus scientific realism. *Philosophical Quarterly, 128*, 262–271.

(2010). Critical rationalism, explanation, and severe tests. In D. G. Mayo & A. Spanos (eds.), *Error and inference: Recent exchanges on experimental reasoning, reliability, objectivity, and rationality* (pp. 88–112). New York: Cambridge University Press.

Nelson, L. H. (1990). *Who knows? From Quine to a feminist empiricism*. Philadelphia, PA: Temple University Press.

Neurath, O. ([1938] 1955). Unified science as encyclopedic integration. In O. Neurath, R. Carnap, & C. W. Morris (eds.), *International encyclopedia of unified science* (vol. 1, pp. 1–27). Chicago: University of Chicago Press.

Newton, I. ([1687] 1999). *The* principia: *Mathematical principles of natural philosophy* (I. B. Cohen & A. Whitman, eds. & trans.). Berkeley: University of California Press.

([1730] 1979). *Opticks* (Fourth edition). New York: Dover.

(1978). *Isaac Newton's papers and letters on natural philosophy* (Second edition; I. B. Cohen, ed.). Cambridge, MA: Harvard University Press.

(1984). *The optical papers of Isaac Newton* (vol. 1; A. E. Shapiro, ed.). Cambridge: Cambridge University Press.

Neyman, J. (1950). *First course in probability and statistics.* New York: Henry Holt.

(2014). *Philosophical writings* (A. Janiak, ed.). Cambridge: Cambridge University Press.

Neyman, J., & Pearson, E. S. (1928). On the use and interpretation of certain test criteria for purposes of statistical inference: Part I. *Biometrika, 20A*(1/2), 175–240.

(1933). On the problem of the most efficient tests of statistical hypotheses. *Philosophical Transactions of the Royal Society of London. Series A, Containing Papers of a Mathematical or Physical Character, 231*, 289–337.

Norton, J. D. (1993). General covariance and the foundations of general relativity: Eight decades of dispute. *Reports of Progress in Physics, 56*, 791–861.

(2010). Cosmic confusions: Not supporting versus supporting not. *Philosophy of Science, 77*, 501–523.

(2012). Approximation and idealization: Why the difference matters. *Philosophy of Science, 79*, 207–232.

Okruhlik, K. (1994). Gender and the biological sciences. *Canadian Journal of Philosophy, 20* (suppl.), 21–42.

Okun, L. B. (1989). The concept of mass. *Physics Today, 42*, 6, 31–36.

Oreskes, N. (2021). *Why trust science?* Princeton, NJ: Princeton University Press.

Oreskes, N., & Conway, E. (2010). *Merchants of doubt: How a handful of scientists obscured the truth on issues from tobacco smoke to global warming*. New York: Bloomsbury.

Oreskes, N., Sainforth, D. A., & Smith, L. A. (2010). Adaptation to global warming: Do climate models tell us what we need to know? *Philosophy of Science, 77*, 1012–1028.

Park, K. (2008). Response to Brian Vickers, "Francis Bacon, feminist historiography, and the dominion of nature." *Journal of the History of Ideas, 69*(1), 143–146.

Parker, W. (2008). Computer simulation through an error-statistical lens. *Synthese, 163*(3), 371–384.

(2009). Does matter really matter? Computer simulations, experiments, and materiality. *Synthese, 169*, 483–496.

(2010). Whose probabilities? Predicting climate change with ensembles of models. *Philosophy of Science, 77*, 985–997.

(2011). When climate models agree: The significance of robust model predictions. *Philosophy of Science, 78*(4), 579–600.

(2017). Computer simulation, measurement, and data assimilation. *British Journal for the Philosophy of Science, 68*, 273–304.

Pashby, T. (2012). Dirac's prediction of the positron: A case study for the current realism debate. *Perspectives on Science, 20*(4), 440–475.

Paul. L. A. (2006). In defense of essentialism, *Philosophical Perspectives, 20*, 333–372.

Peirce, C. S. (1883). A theory of probable inference. In C. S. Peirce (ed.), *Studies in logic: By members of the Johns Hopkins University* (pp. 126–181). Boston, MA: Little, Brown.

(1931–1958). *Collected papers* (vols. 1–8; C. Hartshorne & P. Weiss, eds.). Cambridge, MA: Harvard University Press.

(1992). *Reasoning and the logic of things* (H. Putnam & K. L. Ketner, eds.). Cambridge, MA: Harvard University Press.

Perrin, J. (1916). *Atoms* (D. L. Hammick, trans.). New York: D. Van Nostrand.

Peschard, I. (2010). Modeling and experimenting. In P. Humphreys & C. Imbert (eds.), *Models, simulations, and representation* (pp. 42–61). London: Routledge.

Pielke, R. A., Jr. (2007). *The honest broker: Making sense of science in policy and politics*. New York: Cambridge University Press.

Pinnick, C. L. (1994). Feminist epistemology: Implications for philosophy of science. *Philosophy of Science, 61*(4), 646–657.

(1998). "What is wrong with the strong programme's case study of the 'Hobbes-Boyle dispute'?" In N. Koertge (ed.), *A house built on sand: Exposing postmodernist myths about science* (pp. 227–239). New York: Oxford University Press.

Pinnick, C. L., Koertge, N., & Almeder, R. (eds.). (2003). *Scrutinizing feminist epistemology: An examination of gender in science.* New Brunswick, NJ: Rutgers University Press.

Plato. (1992). *Republic* (C. D. C. Reeve, trans., G. M. A. Grube, rev.). Indianapolis, IN: Hackett.

Poincaré, H. ([1905] 1952). *Science and hypothesis* (F. Maitland, trans.). Mineola, NY: Dover.

Popper, K. (1965). *Conjectures and refutations.* New York: Harper.

([1959] 1992). *The logic of scientific discovery.* New York: Routledge.

Porter, T. (1988). *The rise of statistical thinking: 1820–1900.* Princeton, NJ: Princeton University Press.

Potochnik, A. (2017). *Idealization and the aims of science.* Chicago: University of Chicago Press.

Potters, J., & Simons, M. (2023). We have never been "New Experimentalists:" On the rise and fall of the turn to experimentation in the 1980s. *HOPOS: The Journal of the International Society for the History of Philosophy of Science, 13*, 91–119.

Pratt, J. W. (1977). "Decisions" as statistical evidence and Birnbaum's "confidence concept." *Synthese, 36*(1), 59–69.

Preston, T. (1890). *The theory of light.* New York: Macmillan.
Psillos, S. (1999). *Scientific realism: How science tracks truth.* New York: Routledge.
Putnam, H. (1978). *Meaning and the moral sciences.* London: Routledge.
(1979). What is mathematical truth? In *Mathematics, matter, and method: Philosophical papers* (vol. 1, pp. 60–78). New York: Cambridge University Press.
Quine, W. v. O. (1951). Two dogmas of empiricism. *Philosophical Review, 60,* 20–43.
(1980). *From a logical point of view* (Second edition). Cambridge, MA: Harvard University Press.
(1991). Two dogmas in retrospect. *Canadian Journal of Philosophy, 21,* 265–274.
Ramsey, F. P. (1950). Truth and probability. In R. B. Braithwaite (ed.), *The foundations of mathematics and other logical essays* (pp. 156–198). London: Routledge and Kegan Paul.
Rehg, W., & Staley, K. (2017). "Agreement" in the IPCC confidence measure. *Studies in History and Philosophy of Modern Physics, 57,* 126–134.
Reichenbach, H. (1938). *Experience and prediction.* Chicago: University of Chicago Press.
Reisch, G. A. (1991). Did Kuhn kill logical empiricism? *Philosophy of Science,* 58(2), 264–277.
Resnik, D. B. (1994). Hacking's experimental realism. *Canadian Journal of Philosophy, 24,* 395–411.
Reynolds, J. C., & Tapper, S. C. (1995). The ecology of the red fox *Vulpes vulpes* in relation to small game in rural southern England. *Wildlife Biology, 1,* 105–119.
Rhodes, R. (1986). *The making of the atomic bomb.* New York: Simon and Schuster.
Richardson, A. W. (2002). Engineering philosophy of science: American pragmatism and logical empiricism in the 1930s. *Philosophy of Science, 69*(S3), S36–S47.
Richardson, A., & Uebel, T. (eds.). (2007). *The Cambridge companion to logical empiricism.* New York: Cambridge University Press.
Rinard, S. (2014). A new Bayesian solution to the problem of the ravens. *Philosophy of Science, 81,* 81–100.
Ritson, S., & Staley, K. (2021). How uncertainty can save measurement from circularity and holism. *Studies in History and Philosophy of Science, 85,* 155–165.
Roberts, J. T. (2008). *The law-governed universe.* New York: Oxford University Press.
Rolin, K. (2006). The bias paradox in feminist standpoint epistemology. *Episteme, 3* (1–2), 125–136.
(2009). Standpoint theory as a methodology for the study of power relations. *Hypatia, 24*(4), 218–226.
Roqué, X. (1997). The manufacture of the positron. *Studies in History and Philosophy of Modern Physics, 28,* 73–129.

Rosenstock, L., & Lee, L. J. (2002). Attacks on science: The risks to evidence-based policy. *American Journal of Public Health, 92*(1), 14–18.

Rowbottom, D. P. (2019). *The instrument of science: Scientific anti-realism revitalized.* New York: Routledge.

Rudge, D. W. (1998). A Bayesian analysis of strategies in evolutionary biology. *Perspectives on Science, 6*, 341–360.

(2001). Kettlewell from an error statistician's point of view. *Perspectives on Science, 9*, 59–77.

Rudner, R. (1953). The scientist qua scientist makes value judgments. *Philosophy of Science, 20*(1), 1–6.

Rudwick, M. (1985). *The meaning of fossils: Episodes in the history of paleontology* (Second edition). Chicago: University of Chicago Press.

Rueger, A. (2005). Perspectival models and theory unification. *British Journal for the Philosophy of Science, 56*, 579–594.

Russell, B. (1906–1907). On the nature of truth. *Proceedings of the Aristotelian Society, 7*, 28–49.

(1918). The philosophy of logical atomism. In R. C. Marsh (ed.), *Logic and knowledge: Essays 1901–1950* (pp. 177–281). London: George Allen and Unwin.

Salmon, W. C. (1981). Rational prediction. *British Journal for the Philosophy of Science, 32*, 115–125.

(1984). *Scientific explanation and the causal structure of the world.* Princeton, NJ: Princeton University Press.

(1990). Rationality and objectivity in science or Tom Kuhn meets Tom Bayes. In C. W. Savage (ed.), *Scientific theories* (pp. 175–204). Minneapolis: University of Minnesota Press.

(1994). Causality without counterfactuals. *Philosophy of Science, 61*(2), 297–312.

Sargent, R.-M. (1995). *The diffident naturalist: Robert Boyle and the philosophy of experiment*. Chicago: University of Chicago Press.

Savage, L. (ed.). (1962). *The foundations of statistical inference: A discussion.* London: Methuen.

(1972). *Foundations of statistics* (Second edition). New York: Dover.

Scheiber, N. (2023, October 1). The Harvard professor and the bloggers. *New York Times*.

Schiebinger, L. (1989). *The mind has no sex? Women in the origins of modern science.* Cambridge, MA: Harvard University Press.

Schweber, S. (1994). *QED and the men who made it.* Princeton, NJ: Princeton University Press.

Scriven, M. (1959). Explanation and prediction in evolutionary theory. *Science, 130* (3374), 477–482.

(1962). Explanations, predictions, and laws. In H. Feigl & G. Maxwell (eds.), *Minnesota studies in the philosophy of science* (vol. 3: Scientific Explanation, Space, and Time, pp. 170–230). Minneapolis: University of Minnesota Press.

Searle, J. (1969). *Speech acts: An essay in the philosophy of knowledge*. Cambridge: Cambridge University Press.

Searle, J., & Vanderken, D. (1985). *Foundations of illocutionary logic*. Cambridge: Cambridge University Press.

Senn, S. (2001). Two cheers for p-values? *Journal of Epidemiology and Biostatistics, 6*, 193–204.

(2011). You may believe you are a Bayesian but you are probably wrong. *Rationality, Markets, and Morals, 2*, 48–66.

Shapere, D. (1964). The structure of scientific revolutions. *Philosophical Review, 73* (3), 383–394.

(1966). Meaning and scientific change. In R. G. Colodny (ed.), *Mind and cosmos: Essays in contemporary science and philosophy* (pp. 41–85). Pittsburgh, PA: University of Pittsburgh Press.

(1993). Astronomy and anti-realism. *Philosophy of Science, 60*, 134–150.

Shapin, S., & Schaffer, S. (1985). *Leviathan and the air-pump: Hobbes, Boyle, and the experimental life*. Princeton, NJ: Princeton University Press.

Shapiro, A. (2002). Newton's optics and atomism. In *The Cambridge companion to Newton*. New York: Cambridge University Press.

Sheehan, W., Kollerstrom, N., & Waff, C. B. (2004). The case of the pilfered planet: Did the British steal Neptune? *Scientific American, 291*, 92–99.

Silver, N. (2012, July). Measuring the effects of voter identification laws. http://fivethirtyeight.blogs.nytimes.com/2012/07/15/measuring-the-effects-of-voter-identification-laws.

Simoncelli, T., & Park, S. (2015). Making the case against gene patents. *Perspectives on Science, 23*, 106–145.

Smith, L. A. (2007). *Chaos: A very short introduction*. New York: Oxford University Press.

Smith, L. A., & Peterson, A. C. (2014). Variations on reliability: Connecting climate predictions to climate policy. In M. Boumans, G. Hon, & A. Petersen (eds.), *Error and uncertainty in scientific practice* (pp. 137–156). London: Pickering and Chatto.

Sneed, J. (1971). *The logical structure of mathematical physics*. Dordrecht: Reidel.

Snyder, L. (2006). *Reforming philosophy: A Victorian debate on science and society*. Chicago: University of Chicago Press.

(2011). *The philosophical breakfast club: Four remarkable friends who transformed science and changed the world*. New York: Broadway Books.

Solomon, M. (2001). *Social empiricism*. Cambridge, MA: MIT Press.

Spanos, A. (1999). *Probability theory and statistical inference*. Cambridge: Cambridge University Press.

(2010). Is frequentist testing vulnerable to the base-rate fallacy? *Philosophy of Science, 77*(4), 565–583.

Stadler, F. (ed.). (2003). *The Vienna Circle and logical empiricism*. Dordrecht: Kluwer.

Staley, K. W. (1999a). Golden events and statistics: What's wrong with Galison's image/logic distinction? *Perspectives on Science, 7*(2), 196–230.

(1999b). Logic, liberty, and anarchy: Mill and Feyerabend on scientific method. *Social Science Journal, 36*, 603–614.

(2008). Error-statistical elimination of alternative hypotheses. *Synthese, 163*, 397–408.

(2012). Strategies for securing evidence through model criticism. *European Journal for Philosophy of Science, 2*, 21–43.

(2014). Experimental knowledge in the face of theoretical error. In M. Boumans, G. Hon, & A. Petersen (eds.), *Error and uncertainty in scientific practice* (pp. 39–55) London: Pickering and Chatto.

(2017a). Pragmatic warrant for frequentist statistical practice: The case of high energy physics. *Synthese, 194*, 355–376.

(2017b). Decisions, decisions: Inductive risk and the Higgs boson. In K. C. Elliott and T. Richards (eds.), *Exploring inductive risk: Case studies of values in science* (pp. 37–55). New York: Oxford University Press.

(2020). Securing the empirical value of measurement results. *British Journal for the Philosophy of Science, 71*, 87–113.

Staley, K. W., & Cobb, A. (2011). Internalist and externalist aspects of justification in scientific inquiry. *Synthese, 182*, 475–492.

Stalker, D. F. (ed.). (1994). *Grue! The new riddle of induction*. Chicago, IL: Open Court.

Standage, T. (2000). *The Neptune file: A story of mathematical cunning, astronomical rivalry, and the discovery of new worlds*. New York: Walker.

Stanford, K. (2006). *Exceeding our grasp: Science, history, and the problem of unconceived alternatives*. New York: Oxford University Press.

(2010). Getting real: The hypothesis of organic fossil origins. *Modern Schoolman, 87*, 219–241.

Steel, D. (2010). Epistemic values and the argument from inductive risk. *Philosophy of Science, 77*(1), 14–34.

Stegmuller, W. (1976). *The structure and dynamics of theories*. New York: Springer-Verlag.

Stein, H. (1989). Yes, but . . .: Some skeptical remarks on realism and anti-realism. *Dialectica, 43*, 47–65.

Steno, N. ([1667] 1958). *The earliest geological treatise* (A. Garboe, trans.). London: Macmillan.

Stigler, S. (1990). *The history of statistics: The measurement of uncertainty before 1900.* Cambridge, MA: Belknap Press of Harvard University Press.

Strevens, M. (2007). Review of Woodward, *Making things happen*. *Philosophy and Phenomenological Research, 74*(1), 233–249.

(2008). *Depth: An account of scientific explanation*. Cambridge, MA: Harvard University Press.

Suárez, M. (2004). An inferential conception of scientific representation. *Philosophy of Science, 71*(5), 767–779.

(2015). Deflationary representation, inference, and practice. *Studies in History and Philosophy of Science, 49*, 36–47.

Suppe, F. (1977). The search for philosophic understanding of scientific theories. In F. Suppe (ed.), *The structure of scientific theories* (Second edition, pp. 3–241). Urbana: University of Illinois Press.

Suppes, P. (1962). Models of data. In E. Nagel, P. Suppes, & A. Tarski (eds.), *Logic, methodology, and philosophy of science: Proceedings of the 1960 International Congress* (pp. 252–261). Stanford: Stanford University Press.

(2002). *Representation and invariance of scientific structures.* Stanford, CA: CSLI Publications.

Tal, E. (2017). A model-based epistemology of measurement. In N. Mößner and A. Nordmann (eds.), *Reasoning in measurement* (pp. 233–253). New York: Routledge.

Teller, P. (1973). Conditionalisation and observation. *Synthese, 26*, 218–258.

(2019). What is perspectivism, and does it count as realism? In M. Massimi & C. D. McCoy (eds.), *Understanding perspectivism: Scientific challenges and methodological prospects* (pp. 49–64). New York: Routledge.

Thagard, P. (1978). The best explanation: Criteria for theory choice. *Journal of Philosophy, 75*, 76–92.

Tobin, W. (2003). *The life and science of Léon Foucault: The man who proved the earth rotates*. Cambridge: Cambridge University Press.

Tukey, J. (1960). A survey of sampling from contaminated distributions. In I. Olkin (ed.), *Contributions to probability and statistics: Essays in honor of Harold Hotelling* (pp. 448–485). Stanford: Stanford University Press.

US Department of Health and Human Services Public Health Service. (2011). *Report on carcinogens* (Twelfth edition). Washington, DC: US Department of Health and Human Services.

(2021). *Report on carcinogens* (Fifteenth edition). Washington, DC: US Department of Health and Human Services. https://ntp.niehs.nih.gov/go/roc15.

US Environmental Protection Agency. (1990). *Formaldehyde; CASRN 50-00-0*. https://iris.epa.gov/ChemicalLanding/&substance_nmbr=419.

Valles, S. (2020). The predictable inequities of COVID-19 in the US: Fundamental causes and broken institutions. *Kennedy Institute of Ethics Journal, 30*(3–4), 191–214.

van Fraassen, B. (1972). A formal approach to the philosophy of science. In R. G. Colodny (ed.), *Paradigms and paradoxes*. Pittsburgh, PA: University of Pittsburgh Press.

(1980). *The scientific image*. New York: Oxford University Press.

(1989). *Laws and symmetry*. New York: Oxford University Press.

(2002). *The empirical stance*. New Haven, CT: Yale University Press.

Vickers, B. (2008). Francis Bacon, feminist historiography, and the dominion of nature. *Journal of the History of Ideas, 69*(1), 117–141.

Wasserstein, R. L., Schirm, A. L., & Lazar, N. A. (2019). Moving to a world beyond "$p < 0.05$." *American Statistician, 73*(suppl. 1), 1–19.

Weber, M. (2005). *Philosophy of experimental biology*. New York: Cambridge University Press.

Weinberg, S. (2012, July 13). Why the Higgs boson matters. *New York Times*.

Weisberg, J. (2021). Formal epistemology. *The Stanford encyclopedia of philosophy* (Spring 2021 edition; Edward N. Zalta, ed.). https://plato.stanford.edu/archives/spr2021/entries/formal-epistemology.

Weisberg, M. (2007). Three kinds of idealization. *Journal of Philosophy, 104*, 639–659.

(2013). *Simulation and similarity*. New York: Oxford University Press.

Whewell, W. (1847). *The philosophy of the inductive sciences: Founded upon their history*. London: John W. Parker.

Wilholt, T. (2009). Bias and values in scientific research. *Studies in History and Philosophy of Science, 40*, 92–101.

Will, C. M. (1993a). *Theory and experiment in gravitational physics* (Second edition). New York: Cambridge University Press.

(1993b). *Was Einstein right? Putting general relativity to the test* (Second edition). New York: Basic Books.

Williamson, J. (2010). *In defence of objective Bayesianism*. New York: Oxford University Press.

Willink, R. (2013). *Measurement uncertainty and probability*. Cambridge: Cambridge University Press.

Wimsatt, W. (2007). *Re-engineering philosophy for limited beings*. Cambridge, MA: Harvard University Press.

Winsberg, E. (2010). *Science in the age of computer simulation*. Chicago: University of Chicago Press.

(2022). Computer simulations in science. *The Stanford encyclopedia of Philosophy* (E. N. Zalta & U. Nodelman, eds.). https://plato.stanford.edu/archives/win2022/entries/simulations-science.

Wittgenstein, L. (1921). *Tractatus Logico-Philosophicus* (D. F. Pears & B. F. McGuinnnes, trans.). London: Routledge and Kegan Paul.

Woodward, J. (2003). *Making things happen: A theory of causal explanation*. New York: Oxford University Press.

Worrall, J. (1985). Scientific discovery and theory-confirmation. In J. Pitt (ed.), *Change and progress in modern science* (pp. 301–332). Dordrecht: Reidel.

(1989a). Fresnel, Poisson and the white spot: The role of successful predictions in the acceptance of scientific theories. In D. Gooding, T. Pinch, & S. Schaffer (eds.), *The uses of experiment: Studies in the natural sciences* (pp. 137–157). New York: Cambridge University Press.

(1989b). Structural realism: The best of both worlds? *Dialectica, 43*, 99–124.

(2008). Evidence and ethics in medicine. *Perspectives in Biology and Medicine, 51*, 418–431.

Wray, K. B. (1999). A defense of Longino's social epistemology. *Philosophy of Science, 66*, S538–S552.

(2011). *Kuhn's evolutionary social epistemology*. New York: Cambridge University Press.

Wüthrich, A. (2017). The Higgs discovery as a diagnostic causal inference. *Synthese, 194*, 461–476.

Wylie, A. (2003). Why standpoint matters. In R. Figueroa & S. Harding (eds.), *Science and other cultures: Issues in philosophies of science and technology* (pp. 26–48). New York: Routledge.

(2019). Crossing a threshold: Collaborative archaeology in global dialogue. *Archaeologies: Journal of the World Archaeological Congress, 15*(3), 570–587.

Wylie, A., & Hankinson Nelson, L. (2007). Coming to terms with the values of science: Insights from feminist science studies scholarship. In H. Kincaid, J. Dupré, & A. Wylie (eds.), *Value-free science? Ideals and illusions* (pp. 58–86). New York: Oxford University Press.

Zahar, E. (1973). Why did Einstein's programme supersede Lorentz's? (i). *British Journal for the Philosophy of Science, 24*(2), 95–123.

Ziliak, S. T., & McCloskey, D. N. (2008). *The cult of statistical significance: How the standard error costs us jobs, justice, and lives*. Ann Arbor, MI: University of Michigan Press.

Index

Printed in the United States
by Baker & Taylor Publisher Services